Statistical Topics and Stochastic Models for Dependent Data with Applications

Series Editor
Nikolaos Limnios

Statistical Topics and Stochastic Models for Dependent Data with Applications

Edited by

Vlad Stefan Barbu
Nicolas Vergne

ISTE

WILEY

First published 2020 in Great Britain and the United States by ISTE Ltd and John Wiley & Sons, Inc.

ISTE Ltd
27-37 St George's Road
London SW19 4EU
UK

www.iste.co.uk

John Wiley & Sons, Inc.
111 River Street
Hoboken, NJ 07030
USA

www.wiley.com

Library of Congress Control Number: 2020938718

British Library Cataloguing-in-Publication Data
A CIP record for this book is available from the British Library
ISBN 978-1-78630-603-6

MIX
Paper from responsible sources
FSC® C013604

Contents

Preface

This collective book stems from a workshop that took place in Rouen, October 3–5, 2018, within the framework of the project *Random Models and Statistical Tools, Informatics and Combinatorics (MOUSTIC – Modèles aléatoires et Outils Statistiques, Informatiques et Combinatoires)*, financed by the Region of Normandy and the European Regional Development Fund. The main idea was to bring together leading scientists working on probabilistic and statistical topics for dependent data, as well as on associated applications.

This is the way this book was written, with the intention to offer to the scientific community a part of the latest advances in the field of stochastic modeling for dependent data, authored by leading experts in the field. This is a crucial aspect in a time when we face an increasing need for more and more complex models capable of capturing the main features of increasingly complex applications. From a technical point of view, our book is important for gathering theoretical developments and applications related to Markov type models (semi-Markov processes, autoregressive processes, piecewise deterministic Markov processes, and variable length Markov chains) as well as probabilistic/statistical techniques issued from information theory, based on divergence measures and entropies.

This volume is divided into three parts: the first one examines Markov and semi-Markov processes, the second one deals with autoregressive processes and the last one presents divergence measures and entropies. Particular attention is given to applications of these methods in various fields such as finance, DNA analysis, quantum physics and survival analysis.

The workshop in Rouen and, consequently, the present book, would not have been possible without the support of the Laboratory of Mathematics Raphaël Salem, the Laboratory of Mathematics Nicolas Oresme, the University of Rouen Normandy, the Region of Normandy, the French Statistical Society-SFdS, the MAS group of the

French Society of Applied and Industrial Mathematics-SMAI, the Normandie-Mathématiques Research Federation and the Normastic Research Federation.

We would like to thank all the speakers of the workshop who contributed, although some of them indirectly, to the quality of the present volume. Our thanks go also to the anonymous reviewers for their valuable work: without their support, this volume could not have been successfully completed.

We would also like to thank Professor Nikolaos Limnios for proposing and encouraging us to elaborate this collective volume as well as the editorial staff of ISTE Ltd for their technical support.

Vlad Stefan BARBU
Nicolas VERGNE
Rouen, May 2020

PART 1

Markov and Semi-Markov Processes

1

Variable Length Markov Chains, Persistent Random Walks: A Close Encounter

We consider a walker on the line that at each step keeps the same direction with a probability that depends on the time already spent in the direction the walker is currently moving. These walks with memories of variable length can be seen as generalizations of directionally reinforced random walks (DRRWs) introduced in Mauldin *et al.* (1996). We give a complete and usable characterization of the recurrence or transience in terms of the probabilities to switch the direction. These conditions are related to some characterizations of existence and uniqueness of a stationary probability measure for a particular Markov chain: in this chapter, we define the general model for words produced by a variable length Markov chain (VLMC) and we introduce a key combinatorial structure on words. For a subclass of these VLMC, this provides necessary and sufficient conditions for existence of a stationary probability measure.

1.1. Introduction

This is the story of the encounter between two worlds: the world of random walks and the world of VLMCs. The meeting point turns around the semi-Markov property of underlying processes.

In a VLMC, unlike fixed-order Markov chains, the probability to predict the next symbol depends on a possibly unbounded part of the past, the length of which depends on the past itself. These relevant parts of pasts are called *contexts*. They are stored in a *context tree*. With each context, a probability distribution is associated, prescribing the conditional probability of the next symbol, given this context.

Chapter written by Peggy CÉNAC, Brigitte CHAUVIN, Frédéric PACCAUT and Nicolas POUYANNE.

VLMCs are now widely used as random models for character strings. They were introduced in Rissanen (1983) to perform data compression. When they have a finite memory, they provide a parsimonious alternative to fixed-order Markov chain models, in which the number of parameters to estimate grows exponentially fast with the order; they are also able to capture finer properties of character sequences. When they have infinite memory – this will be our case of study in this chapter – they are a tractable way to build non-Markov models and they may be considered as a subclass of "*chaînes à liaisons complètes*" (Doeblin and Fortet 1937) or "chains with infinite order" (Harris 1955).

VLMCs are used in bioinformatics, linguistics and coding theory to model how words grow or to classify words. In bioinformatics, both for protein families and DNA sequences, identifying patterns that have a biological meaning is a crucial issue. Using VLMC as a model enables one to quantify the influence of a meaning pattern by giving a transition probability on the following letter of the sequence. In this way, these patterns appear as contexts of a context tree. Note that their length may be unbounded (Bejerano and Yona 2001).

In addition, if the context tree is recognized to be a signature of a family (say, of proteins), this gives an efficient statistical method to test whether or not two samples belong to the same family (Busch *et al.* 2009).

Therefore, estimating a context tree is an issue of interest and many authors (statisticians or not, applied or not) stress the fact that the height of the context tree should not be supposed to be bounded. This is the case in Galves and Leonardi (2008) where the algorithm `CONTEXT` is used to estimate an unbounded context tree, or in Garivier and Leonardi (2011). Furthermore, as explained in Csiszár and Talata (2006), the height of the estimated context tree grows with the sample size, so that estimating a context tree by assuming *a priori* that its height is bounded is not realistic.

There is extensive literature on the construction of efficient estimators of context trees, as well for finite or infinite context trees. This chapter is not a review of statistics issues, which would already be relevant for finite memory VLMC. This is a study of the probabilistic properties of infinite memory VLMC as random processes, and more specifically of the main property of interest for such processes: existence and uniqueness of a stationary measure.

As has already been said, VLMC are a natural generalization to infinite memory of Markov chains. It is usual to index a sequence of random variables forming a Markov chain with positive integers and to make the process grow to the right. The main drawback of this habit for an infinite memory process is that the sequence of the process is read from left to right, whereas the (possibly infinite) sequence giving the past needed to predict the next symbol is read in the context tree from right to left, thus giving rise to confusion and lack of readability. For this reason, in this chapter, the VLMC grows to the left. In this way, both the process sequence and the memory in the context tree are read from left to right.

Classical random walks have *independent* and identically distributed increments. In the literature, *persistent* random walks (PRMs), also called *Goldstein-Kac random walks* or *correlated random walks*, refer to random walks having a Markov chain of finite order as an increment process. For such walks, the dynamics of trajectories has a short memory of given length and the random walk itself is not Markovian anymore. What happens whenever the increments depend on a *non-bounded* past memory?

Consider a walker on $\mathbb{Z}$, allowed to increment its trajectory by -1 or 1 at each step of time. Assume that the probability to keep the current direction ± 1 depends on the time already spent in the said direction – the distribution of increments thus acts as a reinforcement of the dependency from the past. More precisely, the process of increments of such a one-dimensional random walk is a Markov chain on the set of (right-)infinite words, with variable – and unbounded – length memory: a VLMC. The concerned VLMC is defined in section 1.3.1. It is based on a context tree called a *double comb*. Later, section 1.3.2 deals with a two-dimensional persistent random walk defined in an analogous manner on $\mathbb{Z}^2$ by a VLMC based on a context tree called a *quadruple comb*.

These random walks that have an unbounded past memory can be seen as a generalization of "directionally reinforced random walks (DRRW)" introduced by Mauldin *et al.* (1996), in the sense that the persistence times are anisotropic ones. For a one-dimensional random walk associated with a double comb, a complete characterization of recurrence and transience, in terms of changing (or not) direction probabilities, is given in section 1.3.1. More precisely, when one of the random times spent in a given direction (the so-called *persistence times*) is an integrable random variable, the recurrence property is equivalent to a classical drift-vanishing. In all other cases, the walk is transient unless the weight of the tail distributions of both persistent times are equal. In two-dimensional random walk, sufficient conditions of transience of recurrence are given in section 1.3.2.

Actually, because of the very specific form of the underlying driving VLMC, these PRWs turn out to be in one-to-one correspondence with so-called *Markov additive processes*. Section 1.5 examines the close links between PRWs, Markov additive processes, semi-Markov chains and VLMC.

In section 1.2, the definition of a general VLMC and a couple of examples are given. In section 1.3, the PRWs are defined and known results on their recurrence properties are collected. In view of section 1.5 where we show how PRW and VLMC meet through the world of semi-Markov chains, section 1.4 is devoted to results – together with a heuristic approach – on the existence and unicity of stationary measures for a VLMC.

1.2. VLMCs: definition of the model

Let $\mathcal{A}$ be a finite set, called the *alphabet*. Here $\mathcal{A}$ will most often be the standard alphabet $\mathcal{A} = \{0, 1\}$, but also $\mathcal{A} = \{d, u\}$ (for *down* and *up*) or $\mathcal{A} = \{\mathtt{n}, \mathtt{e}, \mathtt{w}, \mathtt{s}\}$ (for the cardinal directions). Let

$$\mathcal{R} = \{\alpha\beta\gamma\cdots : \alpha, \beta, \gamma, \cdots \in \mathcal{A}\}$$

be the set of *right-infinite* words over $\mathcal{A}$, written by simple concatenation. A VLMC on $\mathcal{A}$, defined below and most often denoted by $(U_n)_{n\in\mathbb{N}}$, is a particular type of $\mathcal{R}$-valued discrete time Markov chain where:

– the process evolves between time n and time $n+1$ by adding one letter *on the left* of U_n;

– the transition probabilities between time n and time $n+1$ depend on a finite – but not bounded – prefix[1] of the current word U_n.

Giving a formal frame of such a process leads to the following definitions. For a complete presentation of VLMC, one can also refer to Cénac *et al.* (2012).

As usual, a *tree on* $\mathcal{A}$ is a set $\mathcal{T}$ of finite words – namely a subset of $\cup_{n\in\mathbb{N}}\mathcal{A}^n$ – which contains the empty word $\emptyset$ (the *root* of $\mathcal{T}$) and which is prefix-stable: for all finite words u, v, $uv \in \mathcal{T} \Longrightarrow u \in \mathcal{T}$. A tree is made of *internal nodes* ($u \in \mathcal{T}$ is internal when $\exists\alpha \in \mathcal{A}$, $u\alpha \in \mathcal{T}$) and of *leaves* ($u \in \mathcal{T}$ is a leaf when it has no child: $\forall\alpha \in \mathcal{A}$, $u\alpha \notin \mathcal{T}$).

DEFINITION 1.1 (Context tree).– *A* context tree *on* $\mathcal{A}$ *is a saturated tree on* $\mathcal{A}$ *having an at most countable set of infinite branches.*

The tree $\mathcal{T}$ is *saturated* whenever any internal node has $\#(\mathcal{A})$ children: for any finite word u and for any $\alpha \in \mathcal{A}$, $u\alpha \in \mathcal{T} \Longrightarrow (\forall\beta \in \mathcal{A},\ u\beta \in \mathcal{T})$. A right-infinite word on $\mathcal{A}$ is an *infinite branch* of $\mathcal{T}$ when all its finite prefixes belong to $\mathcal{T}$.

Following the vocabulary introduced by Rissanen, a *context* of the tree is a leaf or an infinite branch. A finite or right-infinite word on $\mathcal{A}$ is an *external node* when it is neither internal nor a context. See Figure 1.1 which illustrates these definitions, as well as the pref function defined hereunder.

DEFINITION 1.2 (pref function).– *Let* $\mathcal{T}$ *be a context tree. If* w *is any external node or any context, the symbol* $\operatorname{pref} w$ *denotes the longest (finite or infinite) prefix of* w *that belongs to* $\mathcal{T}$.

1 In fact, an infinite prefix might be needed in a denumerable number of cases.

In other words, pref w is the only context c for which $w = c\cdots$ For a more visual presentation, hang w by its head (its left-most letter) and insert it into the tree; the only context through which the word goes out of the tree is its pref.

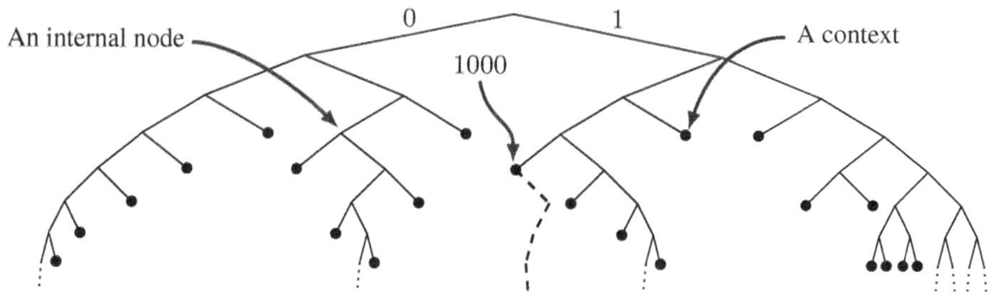

Figure 1.1. *A context tree on the alphabet $\mathcal{A} = \{0, 1\}$. The dotted lines are possibly the beginning of infinite branches. Any word that writes $1000\cdots$, like the one represented by the dashed line, admits 1000 as a pref. For a color version of this figure, see www.iste.co.uk/barbu/data.zip*

With these definitions, it is now possible to define a VLMC.

DEFINITION 1.3 (VLMC).– *Let $\mathcal{T}$ be a context tree. For every context c of $\mathcal{T}$, let q_c be a probability measure on $\mathcal{A}$. The* VLMC *defined by $\mathcal{T}$ and by the $(q_c)_c$ is the $\mathcal{R}$-valued discrete-time Markov chain $(U_n)_{n\in\mathbb{N}}$ defined by the following transition probabilities:* $\forall n \in \mathbb{N}$, $\forall \alpha \in \mathcal{A}$,

$$\mathbb{P}\left(U_{n+1} = \alpha U_n | U_n\right) = q_{\text{pref}(U_n)}\left(\alpha\right). \qquad [1.1]$$

To get a realization of a VLMC as a process on $\mathcal{R}$, take a (random) right infinite word

$$U_0 = X_0 X_{-1} X_{-2} X_{-3} \cdots$$

At each step of time $n \geq 0$, one gets U_{n+1} by adding a random letter X_{n+1} on the left of U_n:

$$\begin{aligned} U_{n+1} &= X_{n+1} U_n \\ &= X_{n+1} X_n \cdots X_1 X_0 X_{-1} X_{-2} \cdots \end{aligned}$$

under the conditional distribution [1.1].

REMARK 1.1.– *Probabilizing* a context tree consists, as in definition 1.3, of endowing it with a family of probability measures on the alphabet, indexed by the set of contexts. This vocabulary is used below.

REMARK 1.2.– Assume that the context tree is finite and denote its height by h; in this condition, the VLMC is just a Markov chain of order h on $\mathcal{A}$. On the contrary, when the context tree is infinite, and this is mainly our case of interest, the VLMC is generally *not* a Markov process on $\mathcal{A}$.

EXAMPLE 1.1.– Take $\mathcal{A} = \{\mathtt{n}, \mathtt{e}, \mathtt{w}, \mathtt{s}\}$ as an (ordered) alphabet, so that the daughters of an internal node are represented, as shown on the left side of Figure 1.2. Making the transition probabilities $\mathbb{P}(U_{n+1} = \alpha U_n | U_n)$ depend only on the length of the largest prefix of the form $\mathtt{n}^k$ ($k \geq 0$) of U_n amounts to taking a comb as a context tree, as shown on the right side of Figure 1.2. Its finite contexts are the $\mathtt{n}^k\alpha$ where $k \geq 0$ and $\alpha \in \mathcal{A} \setminus \{\mathtt{n}\}$.

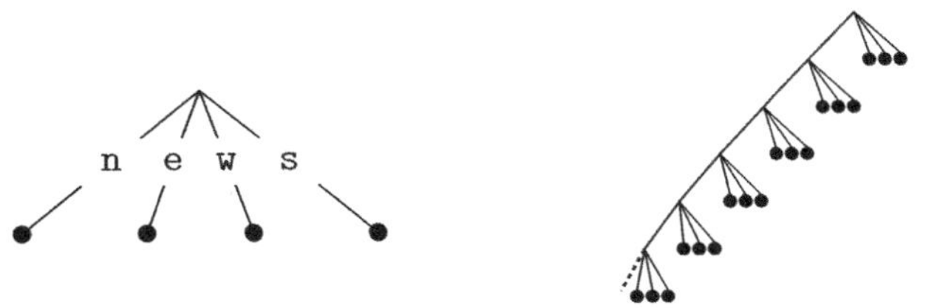

Figure 1.2. *On the left: how one can represent trees on* $\mathcal{A} = \{\mathtt{n}, \mathtt{e}, \mathtt{w}, \mathtt{s}\}$. *On the right, the so-called* left comb *on* $\mathcal{A} = \{\mathtt{n}, \mathtt{e}, \mathtt{w}, \mathtt{s}\}$

EXAMPLE 1.2.– Take again $\mathcal{A} = \{\mathtt{n}, \mathtt{e}, \mathtt{w}, \mathtt{s}\}$ as an alphabet. Making the transition probabilities $\mathbb{P}(U_{n+1} = \alpha U_n | U_n)$ depend only on the length of the largest prefix of the form α^k ($k \geq 1$) of U_n, where α is *any* letter, amounts to taking a *quadruple comb* as a context tree, as shown on the right side of Figure 1.3. In the same vein, if one takes $\mathcal{A} = \{u, d\}$, the *double comb* is the context tree, as shown on the left side of Figure 1.3. In the corresponding VLMC, the transitions depend only on the length of the last current run u^k or d^k, $k \geq 1$. The double comb and the quadruple comb are used below to define PRWs.

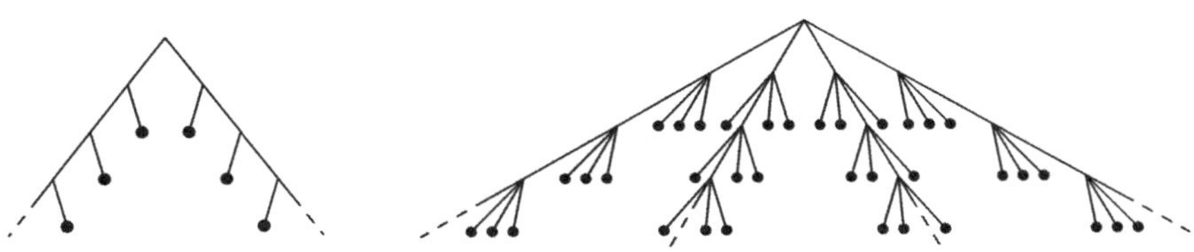

Figure 1.3. *The double comb and the quadruple comb*

EXAMPLE 1.3.– Take $\mathcal{A} = \{0, 1\}$ (naturally ordered for the drawings). The left comb of right combs, shown on the left side of Figure 1.4, is the context tree of a VLMC that makes its transition probabilities depend on the largest prefix of U_n of the form 0^p1^q.

If one has to take into consideration the largest prefix of the form 0^p1^q or 1^p0^q, one has to use the double comb of opposite combs, as shown on the right side of Figure 1.4.

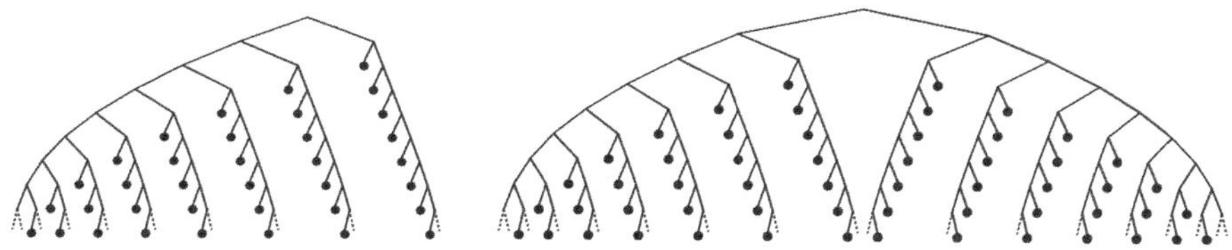

Figure 1.4. *Context trees on $\mathcal{A} = \{0, 1\}$: the left comb of right combs (on the left) and a double comb of opposite combs (on the right)*

DEFINITION 1.4 (Non-nullness).– *A VLMC is called* non-null *when no transition probability vanishes, i.e. when $q_c(\alpha) > 0$ for every context c and for every $\alpha \in \mathcal{A}$.*

Non-nullness appears below as an irreducibility-like assumption made on the driving VLMC of PRWs and for existence and unicity of an invariant probability measure for a general VLMC as well.

1.3. Definition and behavior of PRWs

In this section, the so-called *PRWs* are defined. A PRW is a random walk driven by some VLMC. In dimensions one and two, results on transience and the recurrence of PRW are given. These results are detailed and proven in Cénac *et al.* (2018b, 2013) in dimension one and in Cénac *et al.* (2020) in dimension two.

1.3.1. *PRWs in dimension one*

In this section, we deal with one-dimensional PRWs. Note that, contrary to the classical random walk, a PRW is generally not Markovian. Let $\mathcal{A} := \{d, u\} = \{-1, 1\}$ (d for down and u for up) and consider the *double comb* on this alphabet as a context tree, probabilize it and denote by $(U_n)_n$ a realization of the associated VLMC. The nth increment X_n of the PRW is given as the first letter of U_n: define the persistent random walk $S = (S_n)_{n \geq 0}$ by $S_0 = 0$ and, for $n \geq 1$,

$$S_n := \sum_{\ell=1}^{n} X_\ell, \qquad [1.2]$$

so that for any $n \geq 1$, $m \geq 0$,

$$\mathbb{P}(S_{m+1} = S_m + 1 | U_m = d^n u \ldots) = q_{d^n u}(u)$$
$$\mathbb{P}(S_{m+1} = S_m - 1 | U_m = u^n d \ldots) = q_{u^n d}(d).$$

Furthermore, for the sake of simplicity and without loss of generality, we condition the walk to start almost surely (a.s.) from $\{X_{-1} = u, X_0 = d\}$ – this amounts to changing the origin of time. In this model, a walker on a line keeps the same direction with a probability that depends on the discrete time already spent in the direction the walker is currently moving (see Figure 1.5). This model can be seen as a generalization of DRRWs introduced in Mauldin *et al.* (1996).

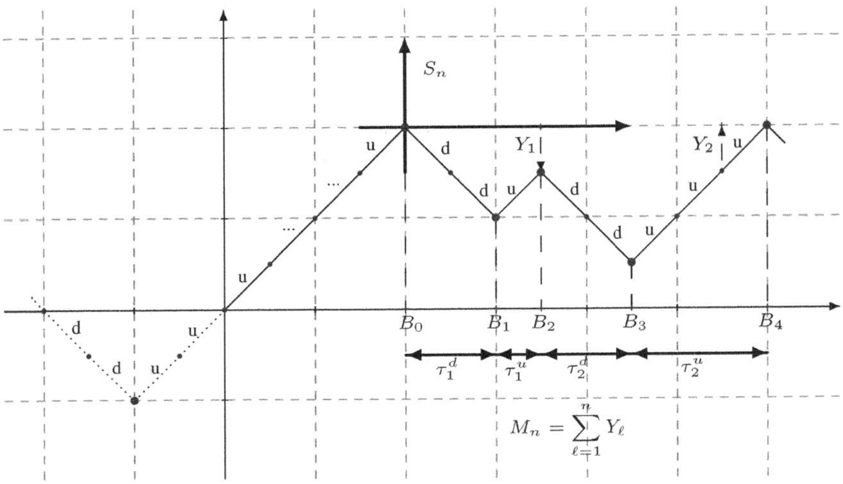

Figure 1.5. *A one-dimensional PRW. For a color version of this figure, see www.iste.co.uk/barbu/data.zip*

Taking different probabilized context trees would lead to different probabilistic impacts on the asymptotic behavior of resulting PRWs. Moreover, the characterization of the recurrent versus transient behavior is difficult in general. We state here exhaustive recurrence criteria for PRWs defined from a double comb.

In order to avoid trivial cases, we assume that S cannot be frozen in one of the two directions with a positive probability. Therefore, we make the following assumption.

ASSUMPTION 1.1 (Finiteness of the length of runs).– For any $\alpha, \beta \in \{u, d\}$, $\alpha \neq \beta$,

$$\lim_{n \to +\infty} \left(\prod_{k=1}^{n} q_{\alpha^k \beta}(\alpha) \right) = 0. \qquad [1.3]$$

Let τ_n^u and τ_n^d be, respectively, the length of the nth rise and of the nth descent.

Then, by a renewal-type property (see Cénac *et al.* 2013, proposition 2.3), $(\tau_n^d)_{n\geq 1}$ and $(\tau_n^u)_{n\geq 1}$ are independent sequences of i.d.d. random variables. Their distribution tails are straightforwardly given by: for any $\alpha, \beta \in \{u, d\}$, $\alpha \neq \beta$ and $n \geq 1$,

$$\mathbb{P}(\tau_1^\alpha \geq n) = \prod_{k=1}^{n-1} q_{\alpha^k \beta}(\alpha). \quad [1.4]$$

Note that assumption 1.1 amounts to supposing that the *persistence times* τ_n^d and τ_n^u are a.s. finite. The *jump times* (or breaking times) are: $B_0 = 0$ and, for $n \geq 1$,

$$B_{2n} := \sum_{k=1}^{n} \left(\tau_k^d + \tau_k^u\right) \text{ and } B_{2n+1} := B_{2n} + \tau_{n+1}^d. \quad [1.5]$$

In order to deal with a more tractable random walk built with the possibly unbounded but i.d.d. increments $Y_n := \tau_n^u - \tau_n^d$, we introduce the underlying *skeleton* random walk $(M_n)_{n\geq 1}$, which is the original walk observed at the random times of up-to-down turns:

$$M_n := \sum_{k=1}^{n} Y_k = S_{B_{2n}}. \quad [1.6]$$

Two main quantities play a key role in the asymptotic behavior, namely the expectations of the lengths of runs: with formula [1.4], let

$$\Theta_d := \mathbf{E}[\tau_1^d] = \sum_{n\geq 1} \prod_{k=1}^{n-1} q_{d^k u}(d) \text{ and } \Theta_u := \mathbf{E}[\tau_1^u] = \sum_{n\geq 1} \prod_{k=1}^{n-1} q_{u^k d}(u). \quad [1.7]$$

Actually, Θ_d and Θ_u are already discussed in Cénac *et al.* (2013, proposition B1), where it is shown that the driving VLMC of a one-dimensional PRW admits a unique invariant probability measure if and only if $\Theta_d < \infty$ and $\Theta_u < \infty$.

Note that the expectation of Y_1 is well defined in $[-\infty, +\infty]$ whenever at least one of the persistence times τ_1^u or τ_1^d is integrable. Thus, as soon as $\Theta_d < \infty$ or $\Theta_u < \infty$, let

$$\mathbf{d}_M := \mathbf{E}[Y_1] = \underbrace{\Theta_u - \Theta_d}_{\in[-\infty,+\infty]} \quad [1.8]$$

and

$$\mathbf{d}_S := \frac{\mathbf{E}[\tau_1^u] - \mathbf{E}[\tau_1^d]}{\mathbf{E}[\tau_1^u] + \mathbf{E}[\tau_1^d]} = \frac{\Theta_u - \Theta_d}{\Theta_u + \Theta_d} \in [-1, 1]. \quad [1.9]$$

An elementary computation shows that $\mathbb{E}(M_n) = n\mathbf{d}_M$ and $\mathbb{E}(S_n) \sim n\mathbf{d}_S$ when n tends to infinity. Thus, $\mathbf{d}_M$ and $\mathbf{d}_S$ appear as asymptotic drifts when the walks $(M_n)_n$ and $(S_n)_n$ respectively, turn out to be transient (see Table 1.1). The behavior of the walk also depends on quantities $J_{\alpha|\beta}$, defined for α and $\beta \in \mathcal{A}, \alpha \neq \beta$ by:

$$J_{\alpha|\beta} := \sum_{n=1}^{\infty} \frac{n\mathbb{P}(\tau_1^\alpha = n)}{\sum_{k=1}^{n} \mathbb{P}(\tau_1^\beta \geq k)}.$$

A complete and usable characterization of the recurrence and the transience of the PRW in terms of the probabilities to persist in the same direction or to switch is given in proposition 1.1. Its proof relies on a criterion of Erickson (1973), applied to the skeleton walk $(M_n)_n$, which is simpler to deal with because its increments are independent.

PROPOSITION 1.1.– Under the non-nullness assumption and assumption 1.1, the random walk $(S_n)_n$ is recurrent or transient as described in Table 1.1.

<table>
<tr><td></td><td colspan="2">$\Theta_u < \infty$</td><td colspan="2">$\Theta_u = \infty$</td></tr>
<tr><td rowspan="2">$\Theta_d < \infty$</td><td rowspan="2">Recurrent
$\mathbf{d}_S = 0$</td><td>Drifting $+\infty$
$\mathbf{d}_S > 0$</td><td colspan="2" rowspan="2">Drifting $+\infty$</td></tr>
<tr><td>Drifting $-\infty$
$\mathbf{d}_S < 0$</td></tr>
<tr><td rowspan="2">$\Theta_d = \infty$</td><td colspan="2" rowspan="2">Drifting $-\infty$</td><td rowspan="2">Recurrent
$J_{u|d} = J_{d|u} = \infty$</td><td>Drifting $+\infty$
$\infty = J_{u|d} > J_{d|u}$</td></tr>
<tr><td>Drifting $-\infty$
$\infty = J_{d|u} > J_{u|d}$</td></tr>
</table>

Table 1.1. *Recurrence versus transience (drifting) for $(S_n)_n$ in dimension one*

The most fruitful situation emerges when both running times τ_1^u and τ_1^d have infinite means. In that case, the recurrence properties of $(S_n)_n$ are related to the behavior of the skeleton random walk $(M_n)_n$ defined in [1.6], the drift of which, $\mathbf{d}_M$, is not defined. Thus, the behavior of $(S_n)_n$ depends on the comparison between the distribution tails of τ_1^u and τ_1^d defined in [1.4], expressed by the quantities $J_{\alpha|\beta}$. Note that the case when both $J_{u|d}$ and $J_{d|u}$ are finite does not appear in the table since it would imply that $\Theta_u < \infty$ and $\Theta_d < \infty$ (see Erickson 1973).

In all three other cases, the drift $\mathbf{d}_S$ is well defined and the PRW is recurrent if and only if $\mathbf{d}_S = 0$. In that case, $\lim\limits_{n\to\infty} \dfrac{S_n}{n} = \mathbf{d}_S = 0$. Note that modifying one transition q_c transforms a recurrent PRW into a transient one, since $\mathbf{d}_S$ becomes non-zero.

1.3.2. *PRWs in dimension two*

Take the alphabet $\mathcal{A} := \{\mathtt{n}, \mathtt{e}, \mathtt{w}, \mathtt{s}\}$. Here, $(\mathtt{e}, \mathtt{n})$ stands for the canonical basis of $\mathbb{Z}^2$, $\mathtt{w} = -\mathtt{e}$ and $\mathtt{s} = -\mathtt{n}$. Hence, the letters $\mathtt{e}$, $\mathtt{n}$, $\mathtt{w}$ and $\mathtt{s}$ stand for moves to the east, north, west and south, respectively. Having in mind a random walk with increments in $\mathcal{A}$, any word of the form $\alpha\beta$, $\alpha, \beta \in \mathcal{A}, \alpha \neq \beta$ is called a *bend*. For the sake of simplicity, we condition the walk to start a.s. with an $\mathtt{ne}$ bend: $\{X_{-1} = \mathtt{n}, X_0 = \mathtt{e}\}$.

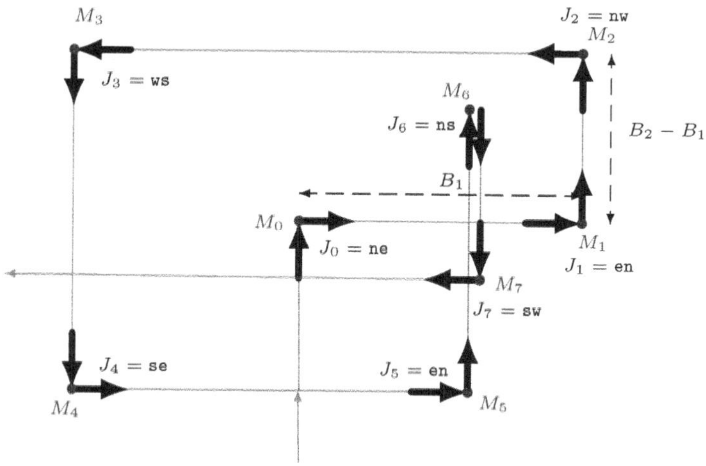

Figure 1.6. *A walk in dimension two. For a color version of this figure, see www.iste.co.uk/barbu/data.zip*

Take a non-null VLMC associated with a quadruple comb on $\mathcal{A}$, as shown in Figure 1.3: the contexts are $\alpha^n\beta$ for $\alpha, \beta \in \mathcal{A}, \alpha \neq \beta$, $n \geq 1$ and the attached probability distributions are denoted by $q_{\alpha^n\beta}$. The two-dimensional PRW $(S_n)_n$ is defined, using this VLMC, as in formula [1.2].

Contrary to the one-dimensional PRWs, as detailed below, the probability to change direction depends on the time spent in the current direction but also on the previous direction. As in dimension one, we intend to avoid that S remains frozen in one of the four directions with a positive probability. Therefore, we make the following assumption, analogous to assumption 1.1 in dimension two.

ASSUMPTION 1.2 (Finiteness of the length of runs).– For any $\alpha, \beta \in \{\mathtt{n}, \mathtt{e}, \mathtt{w}, \mathtt{s}\}$, $\alpha \neq \beta$,

$$\lim_{n \to +\infty} \left(\prod_{k=1}^{n} q_{\alpha^k\beta}(\alpha) \right) = 0. \qquad [1.10]$$

Let $(B_n)_{n\geq 0}$ be the *breaking times* defined inductively by

$$B_0 = 0 \quad \text{and} \quad B_{n+1} = \inf\left\{k > B_n : X_k \neq X_{k-1}\right\}. \qquad [1.11]$$

As in dimension one, assumption 1.2 implies that the breaking times B_n are a.s. finite.

Define the so-called *internal chain* $(J_n)_{n\geq 0}$ by $J_0 = \mathtt{ne}$ and, for all $n \geq 1$,

$$J_n := X_{B_{n-1}} X_{B_n}. \qquad [1.12]$$

Let us illustrate these random variables with a small example, in which: $B_1 = 4$, $B_2 = 7$, $J_0 = X_{-1}X_0$, $J_1 = X_{B_0}X_{B_1} = X_0X_4$, $J_2 = X_{B_1}X_{B_2} = X_4X_7$.

n:	-1	0	1	2	3	4	5	6	7
X_n:	n	e	e	e	e	n	n	n	w
		$B_0 = 0$				$B_1 = 4$			$B_2 = 7$
		$J_0 = \mathtt{ne}$				$J_1 = \mathtt{en}$			$J_2 = \mathtt{nw}$

The process $(J_n)_{n\geq 0}$ is an irreducible Markov chain on the set of bends $\mathcal{S} := \{\alpha\beta | \alpha \in \mathcal{A}, \beta \in \mathcal{A}, \alpha \neq \beta\}$. Its Markov kernel is defined by: for every $\beta, \alpha, \gamma \in \mathcal{A}$ with $\beta \neq \alpha$ and $\alpha \neq \gamma$,

$$P(\beta\alpha; \alpha\gamma) := \sum_{n=1}^{\infty} \left(\prod_{k=1}^{n-1} q_{\alpha^k\beta}(\alpha) \right) q_{\alpha^n\beta}(\gamma), \qquad [1.13]$$

the numbers $P(\alpha\beta, \gamma\delta)$ being 0 for every couple of bends not of the previous form. Remark that the non-nullness assumption (see definition 1.4) implies the irreducibility of $(J_n)_n$ and its aperiodicity. The state space $\mathcal{S}$ is finite so that $(J_n)_n$ is positive recurrent: it admits a unique invariant probability measure π_J.

Denote $T_0 = 0$ and $T_{n+1} := B_{n+1} - B_n$ for every $n \geq 0$. These waiting times (also called *persistence times*) are not independent, contrary to the one-dimensional case. The *skeleton random walk* $(M_n)_{n\geq 0}$ on $\mathbb{Z}^2$ – which is the PRW observed at the breaking times – is then defined as

$$M_n := S_{B_n} = \sum_{i=1}^{n} \left(\sum_{k=B_{i-1}+1}^{B_i} X_k \right) = \sum_{i=1}^{n} (B_i - B_{i-1}) X_{B_i}. \qquad [1.14]$$

Note that $(M_n)_n$ is generally not a classical RW with i.d.d. increments. Nevertheless, taking into account the additional information given by the internal

Markov chain $(J_n)_n$, then $(J_n, M_n)_n$ is a Markov additive process (see Çinlar 1972) as it will appear in section 1.5.

Here, $(J_n)_n$ is positive recurrent but this does not imply the recurrence of $(S_n)_n$ or $(M_n)_n$. Moreover, $(S_n)_n$ and $(M_n)_n$ may have different behaviors. Explicit, necessary and sufficient conditions for the recurrence of $(M_n)_n$ in terms of characteristic functions and convergence of suitable series are given in Cénac *et al.* (2020, theorem 2.1). The following proposition states a dichotomy between some recurrence versus transience phenomenon.

THEOREM 1.1.– Under non-nullness assumption, the following dichotomy holds:

i) the series $\sum_n \mathbb{P}(M_n = 0)$ diverges if and only if the process $(M_n)_n$ is recurrent in the following sense:

$$\exists r > 0, \;\; \mathbb{P}\left(\liminf_{n\to\infty} \|M_n\| < r\right) = 1.$$

ii) the series $\sum_n \mathbb{P}(M_n = 0)$ converges if and only if the process $(M_n)_n$ is transient in the following sense:

$$\mathbb{P}\left(\lim_{n\to\infty} \|M_n\| = \infty\right) = 1.$$

Does the recurrence (respectively, the transience) of $(M_n)_n$ and $(S_n)_n$ occur at the same time? The answer to this 20-year-old question is no:

THEOREM 1.2 (Definitive invalidation of the conjecture in Mauldin *et al.* 1996).– There exist recurrent PRWs $(S_n)_n$ having an associated transient skeleton $(M_n)_n$.

Supposing that the persistence time distributions are horizontally and vertically symmetric is a natural necessary condition for the random walk $(S_n)_n$ to be recurrent. One example is given by the DRRW, originally introduced in Mauldin *et al.* (1996) (see Figure 1.7). Some particular values of the transition probabilities $q_{\alpha^n\beta}$ provide counterexamples. It is shown in Cénac *et al.* (2020) that the corresponding distributions of the persistence times must be non-integrable. In section 1.5, this non-integrability will be related to the non-existence of any invariant probability measure for the driving VLMC.

1.4. VLMC: existence of stationary probability measures

Consider a VLMC denoted by $U = (U_n)_{n\geq 0}$, defined by a pair $(\mathcal{T}, q)$ where $\mathcal{T}$ is a context tree on an alphabet $\mathcal{A}$ and $q = (q_c)_{c\in\mathcal{C}}$ a family of probability measures on $\mathcal{A}$, indexed by the contexts of $\mathcal{T}$. A probability measure π on $\mathcal{R}$ is *stationary* or *invariant* (with regard to U) whenever π is the distribution of every U_n as soon as it is the distribution of U_0. The question of interest consists here of finding conditions on

$(\mathcal{T}, q)$ for the process to admit at least one – or a unique one – stationary probability measure. The heuristic presentation aims to show how combinatoric objects, namely the α-*LIS* of contexts, and conditional probabilities, the *cascades*, naturally emerge.

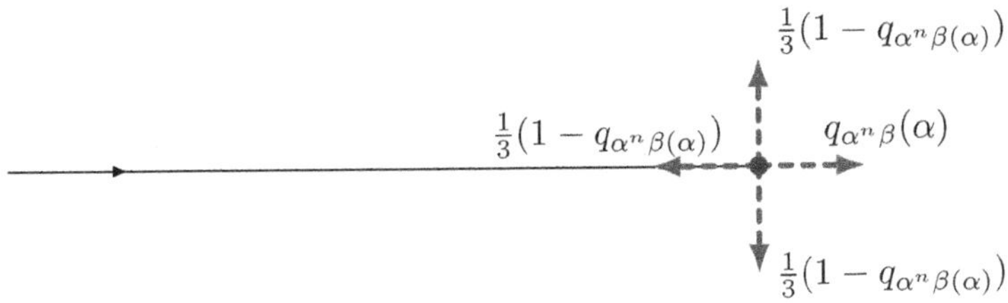

Figure 1.7. *The original directionally reinforced random walk (DRRW). For a color version of this figure, see www.iste.co.uk/barbu/data.zip*

Assume that π is a stationary probability measure on $\mathcal{R}$:

– First step: finite words. Since $\mathcal{R}$ is endowed with the cylinder σ-algebra, π is determined by its values $\pi(w\mathcal{R})$ on the cylinders $w\mathcal{R}$, where w runs over all finite words on $\mathcal{A}$.

– Second step: longest internal suffixes of words. Assume that e is a finite non-internal word and take $a \in \mathcal{A}$. Then, its pref is well defined and, because of formula [1.1], since π is stationary,

$$\pi(\alpha e\mathcal{R}) = q_{\text{pref}(e)}(\alpha) \times \pi(e\mathcal{R}). \tag{1.15}$$

Iterating this formula as far as possible leads to the following definitions. Consider any non-empty finite word w. It is uniquely decomposed as $w = p\alpha s = \beta_1\beta_2\beta_3\cdots\beta_\ell\alpha s$, where the α and the β are letters and s is the *longest internal suffix* of w. The integer ℓ is non-negative and $p = \beta_1\beta_2\cdots\beta_\ell$ is a prefix of w that may be empty – in which case $\ell = 0$.

DEFINITION 1.5 (Lis and α-LIS).– *With these notations, the longest internal suffix s is shortened as the* lis *of w. The word αs is called the* α-LIS *of w.*

DEFINITION 1.6 (Cascade).– *With the notation above, the* cascade *of w is the product*

$$\text{casc}(w) = q_{\text{pref}(\beta_2\cdots\beta_\ell\alpha s)}(\beta_1)q_{\text{pref}(\beta_3\cdots\beta_\ell\alpha s)}(\beta_2)\cdots q_{\text{pref}(\alpha s)}(\beta_\ell). \tag{1.16}$$

Note that this definition makes sense because all the $\beta_k\cdots\beta_\ell\alpha s$ are non-internal words, $k \geq 2$. Moreover, if $w = \alpha s$ where s is internal, then $\ell = 0$ and $\text{casc}(w) = 1$.

With these definitions, iterating formula [1.15] leads to the following equality, known as the *cascade formula*: for every non-empty finite word w having αs as an α-LIS,

$$\pi\left(w\mathcal{R}\right) = \mathrm{casc}(w) \times \pi\left(\alpha s\mathcal{R}\right). \qquad [1.17]$$

This shows that π is determined by its values on words of the form αs where s is internal and $\alpha \in \mathcal{A}$.

– Third step: finite contexts. Assume that s is an internal word and that $\alpha \in \mathcal{A}$. It is shown in Cénac *et al.* (2018a) that a stationary probability measure never charges infinite words so that, by disjoint union,

$$\pi\left(\alpha s\mathcal{R}\right) = \sum_{\substack{c:\ \text{finite context} \\ c=s\cdots}} \pi\left(\alpha c\mathcal{R}\right) = \sum_{\substack{c:\ \text{finite context} \\ c=s\cdots}} q_c(\alpha)\pi\left(c\mathcal{R}\right). \qquad [1.18]$$

Note that the set of indices may be infinite but the family is summable because π is a finite measure. This shows that π is entirely determined by its values $\pi\left(c\mathcal{R}\right)$ on the finite contexts.

– Fourth step: α-LIS of finite contexts. Cascade formula [1.16] applied to any finite context c (contexts are non-empty words) is written as $\pi\left(c\mathcal{R}\right) = \mathrm{casc}(c)\pi\left(\alpha_c s_c\mathcal{R}\right)$, where $\alpha_c s_c$ is the α-LIS of c. Denote by $\mathcal{S} = \mathcal{S}\left(\mathcal{T}\right)$ the set of finite context α-LIS:

$$\mathcal{S} = \left\{\alpha_c s_c :\ c \text{ finite context}\right\}.$$

If s is an internal word and if $\alpha \in \mathcal{A}$, then formula [1.18] leads to

$$\pi\left(\alpha s\mathcal{R}\right) = \sum_{\substack{c:\ \text{finite context} \\ c=s\cdots}} \mathrm{casc}\left(\alpha c\right)\pi\left(\alpha_c s_c\mathcal{R}\right), \qquad [1.19]$$

showing that π is determined by its values $\pi\left(\alpha_c s_c\mathcal{R}\right)$ on $\mathcal{S}$.

– Fourth step: a (generally infinite) linear system. When w and v are finite words and when $\alpha s \in \mathcal{S}$, the notation

$$w = v\cdots = \cdots[\alpha s]$$

stands for: w has v as a prefix and αs as an α-LIS. Writing formula [1.19] for every $\alpha s \in \mathcal{S}$ and grouping in each of them the terms that arise from contexts having the same α-LIS leads to the following square system (at most countably many unknowns $\pi\left(\alpha s\mathcal{R}\right)$ and as many equations):

$$\forall \alpha s \in \mathcal{S},\ \pi\left(\alpha s\mathcal{R}\right) = \sum_{\beta t\in\mathcal{S}} \pi\left(\beta t\mathcal{R}\right)\left(\sum_{\substack{c:\ \text{finite context} \\ c=s\cdots=\cdots[\beta t]}} \mathrm{casc}\left(\alpha c\right)\right). \qquad [1.20]$$

DEFINITION 1.7 (Matrix Q).– *When $\mathcal{T}$ is a context tree having $\mathcal{S}$ as a context α-LIS set, $Q = Q(\mathcal{T})$ is the $\mathcal{S}$-indexed square matrix defined by:*

$$\forall \alpha s, \beta t \in \mathcal{S},\ Q_{\beta t, \alpha s} = \sum_{\substack{c:\ \text{finite context} \\ c = s \cdots = \cdots [\beta t]}} \operatorname{casc}(\alpha c) \in [0, +\infty]. \quad [1.21]$$

Thus, system [1.20] tells us that, when π is a stationary measure, the row-vector $(\pi(\alpha s \mathcal{R}))_{\alpha s \in \mathcal{S}}$ appears as a left-fixed vector of the matrix Q.

DEFINITION 1.8 (Cascade series).– *For every $\alpha s \in \mathcal{S}$, denote*

$$\kappa_{\alpha s} = \sum_{\substack{c:\ \text{finite context} \\ c = \cdots [\alpha s]}} \operatorname{casc}(c) \in [0, +\infty].$$

When this series is summable, one says that the cascade series of αs converges. *Whenever the cascades series of all $\alpha s \in \mathcal{S}$ converge, one says that* the cascade series (of the VLMC) converges.

Note that the convergence of (all) the cascade series is sufficient to guarantee the finiteness of Q's entries. Actually, for a general VLMC, as it is made precise in Cénac *et al.* (2018a), the convergence of the cascade series appears as a pivot condition when dealing with existence and unicity of a stationary probability measure. In this chapter, we state a necessary and sufficient condition for a special kind of VLMC: the *stable* ones that have a finite $\mathcal{S}$. The following proposition is proven in Cénac *et al.* (2018a).

PROPOSITION 1.2.– Let $\mathcal{T}$ be a context tree. The following conditions are equivalent:

i) $\forall \alpha \in \mathcal{A}, \forall w \in \mathcal{W}, \alpha w \in \mathcal{T} \Longrightarrow w \in \mathcal{T}$;

ii) if c is a finite context and $\alpha \in \mathcal{A}$, then αc is non-internal;

iii) $\mathcal{T} \subseteq \mathcal{A}\mathcal{T} = \{\alpha w,\ \alpha \in \mathcal{A},\ w \in \mathcal{T}\}$;

iv) for any VLMC $(U_n)_n$ associated with $\mathcal{T}$, the process $(\operatorname{pref}(U_n))_{n \in \mathbb{N}}$ is a Markov chain that has the set of contexts as a state space.

The context tree is referred to as *stable* whenever one of these conditions is fulfilled.

It turns out that the stability of $\mathcal{T}$ together with the non-nullness of the VLMC imply both stochasticity and irreducibility of the matrix Q. Consequently, in the simple case where Q is a finite-dimensional matrix, there exists (because of stochasticity) a unique (because of irreducibility) left-fixed vector for Q. As a result of a much more general result proven in Cénac *et al.* (2018a), this implies the existence and unicity of a stationary probability measure for the VLMC, as stated below.

THEOREM 1.3.– Let $(\mathcal{T}, q)$ be a non-null stable probabilized context tree. If $\#\mathcal{S} < \infty$, then the following are equivalent:

1) the VLMC associated with $(\mathcal{T}, q)$ has a unique stationary probability measure;

2) the cascade series converge (see definition 1.8).

Note that in the non-stable case, the matrix Q is generally not stochastic nor is it even substochastic. Note also that, even in the stable case, when $\#\mathcal{S} = \infty$, the matrix Q may be stochastic, irreducible and positive recurrent, while the VLMC does not admit any stationary probability measure. One can find such an example in remark 3.16, page 20 in Cénac *et al.* (2020), built with a "left comb of left comb".

1.5. Where VLMC and PRW meet

On the one hand, a VLMC is defined by its context tree and its transition probability distributions q_c – in particular the double and the quadruple combs that are stable trees with finitely many context α-LIS.

Necessary and sufficient conditions of existence and uniqueness of stationary probability measures are given in terms of cascade series. On the other hand, for PRW (defined from VLMC), recurrence properties are written in terms of persistence times. Our aim is to build a bridge between these two families of objects and properties. The meeting point turns out to be the semi-Markov processes of α-LIS and bends.

1.5.1. *Semi-Markov chains and Markov additive processes*

Semi-Markov chains are defined following Barbu and Limnios (2008) because of so-called Markov renewal chains (MRCs).

DEFINITION 1.9 (Markov renewal chain).– *A Markov chain $(J_n, T_n)_{n\geq 0}$ with state space $\mathcal{E} \times \mathbb{N}$ is called a (homogeneous) MRC whenever the transition probabilities satisfy: $\forall n \in \mathbb{N}$, $\forall a, b \in \mathcal{E}$, $\forall j, k \in \mathbb{N}$,*

$$\mathbb{P}\left(J_{n+1} = b, T_{n+1} = k \middle| J_n = a, T_n = j\right) = \mathbb{P}\left(J_{n+1} = b, T_{n+1} = k \middle| J_n = a\right) =: p_{a,b}(k)$$

and $\forall a, b \in \mathcal{E}$, $p_{a,b}(0) = 0$. For such a chain, the family $p = \left(p_{a,b}(k)\right)_{a,b\in\mathcal{A},k\geq 1}$ is referred to as its semi-Markov kernel.

DEFINITION 1.10 (Semi-Markov chain).– *Let* $(J_n, T_n)_{n\geq 0}$ *be an MRC with state space* $\mathcal{E} \times \mathbb{N}$. *Assume that* $T_0 = 0$. *For any* $n \in \mathbb{N}$, *let* B_n *be defined by*

$$B_n = \sum_{i=0}^{n} T_i.$$

The semi-Markov chain *associated with* $(J_n, T_n)_{n\geq 0}$ *is the* $\mathcal{E}$*-valued process* $(Z_j)_{j\geq 0}$ *defined by*

$$\forall j \quad \text{such that} \quad B_n \leq j < B_{n+1}, \qquad Z_j = J_n.$$

Note that the sequence $(B_n)_{n\geq 0}$ is a.s. increasing because of the assumption $p_{a,b}(0) = 0$ (instantaneous transitions are not allowed) that guarantees that $T_n \geq 1$ a.s. for any $n \geq 1$.

The B_n are *jump times*, the T_n are *sojourn times* in a given state and Z_j stagnates at a same state between two successive jump times. The process $(J_n)_n$ is called the *internal (underlying) chain* of the semi-Markov chain $(Z_n)_n$.

The previous definitions make transitions to the same state between time n and time $n+1$ possible. Nevertheless, one can boil down to the case where $p_{a,a}(k) = 0$ for all $a \in \mathcal{E}, k \in \mathbb{N}$ (see the details in Cénac *et al.* 2018a).

A close notion, *Markov additive processes*, can be found in Çinlar (1972).

1.5.2. *PRWs induce semi-Markov chains*

Let us start with one-dimensional PRW, as defined in section 1.3.1. In this case, at each time $j, j \geq 0$, the increment X_j of the walk S takes d or u as a value (see Figure 1.5). Let us see that $(X_j)_{j\geq 0}$ is a semi-Markov chain, starting from $X_0 = d$. Remember that B_n denotes the nth jump times – see equation [1.5]. Define then $(J_n)_n$ by

$$J_n := X_{B_n}. \qquad [1.22]$$

Moreover, let T_n be the nth waiting time, namely $T_0 = 0$ and, for $n \geq 1$,

$$T_n = B_n - B_{n-1}.$$

These waiting times are related to the persistence times τ by the following formulas: for all $k \geq 1$,

$$T_{2k} := \tau_k^u \text{ and } T_{2k-1} := \tau_k^d. \qquad [1.23]$$

With these notations, $(J_n, T_n)_{n\geq 0}$ is an MRC and its semi-Markov kernel is written as: $\forall \alpha, \beta \in \{u, d\}, \alpha \neq \beta, \forall k \geq 1$,

$$p_{\alpha,\beta}(k) = \left(\prod_{j=1}^{k-1} q_{\alpha^j\beta}(\alpha) \right) q_{\alpha^k\beta}(\beta), \qquad [1.24]$$

as can be straightforwardly checked. Moreover, assumption 1.1 guarantees that the T_n are a.s. finite. Besides, formulas [1.7] are written as

$$\mathbb{E}\left(T_{2k}\right) = \Theta_u \text{ and } \mathbb{E}\left(T_{2k+1}\right) = \Theta_d.$$

The situation in dimension one is summarized by the following proposition.

PROPOSITION 1.3.– For a PRW in dimension one, defined by a VLMC associated with a double comb, the sequence $(X_j)_j$ of the increments is an $\mathcal{A}$-valued semi-Markov chain with MRC $(J_n, T_n)_n$ as defined in [1.22] and [1.23] and its semi-Markov kernel is given by equation [1.24].

Let us now deal with the two-dimensional PRW defined in section 1.3.2. At each time $j, j \geq 0$, the increment X_j of the walk S takes $\mathtt{n}, \mathtt{e}, \mathtt{w}$ or $\mathtt{s}$ as a value. But, as already noted, changing direction depends on the time spent in the current direction but also, contrary to the one-dimensional PRWs, on the previous direction. In other words, the bends play the main role. This gives rise to the process $\left(Z_j\right)_j$, valued in the set of bends $\{\alpha\beta : \alpha, \beta \in \mathcal{A}, \alpha \neq \beta\}$, defined in the following manner: $Z_0 = X_{-1}X_0 = \mathtt{ne}$ and, for $j \geq 1$, $Z_j = \alpha\beta$ if and only if $X_j = \beta$ and the first letter distinct from β in the sequence $X_{j-1}, X_{j-2}, X_{j-3}, \cdots$ is α. Let us see that $(Z_j)_{j\geq 0}$ is a semi-Markov chain. Use here notations $(J_n)_n$, $(B_n)_n$ and $(T_n)_n$ of section 1.3.2.

Note that, contrary to the one-dimensional case, the waiting times T_n are not independent. Nevertheless, $(J_n, T_n)_{n\geq 0}$ is an MRC with semi-Markov kernel

$$p_{\beta\alpha,\alpha\gamma}(k) := \left(\prod_{j=1}^{k-1} q_{\alpha^j\beta}(\alpha) \right) q_{\alpha^k\beta}(\gamma), \qquad [1.25]$$

as can be straightforwardly checked. Summarizing, the following proposition holds.

PROPOSITION 1.4.– For a PRW in dimension two, defined by a VLMC associated with a quadruple comb, the sequence $(Z_j)_j$ of the bends is a semi-Markov chain with MRC $(J_n, T_n)_n$ as defined in section 1.3.2. Its semi-Markov kernel is given by equation [1.25]. In addition, $(J_n, B_n)_n$ is a Markov additive process.

1.5.3. *Semi-Markov chain of the α-LIS in a stable VLMC*

In this section, let us consider a more general case than a double comb or a quadruple comb, namely a stable VLMC. In this case, there is always a semi-Markov chain induced by the process $(U_n)_n$, as described in the following.

Let $(U_n)_{n\geq 0}$ be a stable non-null VLMC such that the series of cascades converge (see definition 1.8). Recall that $\mathcal{S}$ denotes the set of context α-LIS of the VLMC. Let $(C_n)_{n\geq 0}$ be the sequence of contexts, and for $n \geq 0$, let Z_n be the α-LIS of C_n:

$$C_n = \text{pref}(U_n) \quad \text{and} \quad Z_n = \alpha_{C_n} s_{C_n}.$$

PROPOSITION 1.5.– Let $(B_n)_{n\geq 0}$ be the increasing sequence of times defined by $B_0 = 0$ and for any $n \geq 1$,

$$B_n = \inf\{k > B_{n-1}, |C_k| \leq |C_{k-1}|\} = \inf\{k > B_{n-1}, C_k \in \mathcal{S}\}$$

and let $T_n = B_n - B_{n-1}$ for $n \geq 1$ and $T_0 = 0$. For any $n \geq 0$, let $J_n = Z_{B_n}$. Then

i) B_n and T_n are a.s. finite and for $\alpha s \in \mathcal{S}$, $\mathbb{E}\left(T_n \middle| J_n = \alpha s\right) = \kappa_{\alpha s}$;

ii) $(Z_n)_{n\geq 0}$ is an $\mathcal{S}$-valued semi-Markov chain associated with the MRC $(J_n, T_n)_{n\geq 0}$;

iii) the associated semi-Markov kernel writes: $\forall \alpha s, \beta t \in \mathcal{S}, \forall k \geq 1$,

$$p_{\alpha s,\beta t}(k) = \sum_{\substack{c\in\mathcal{C},\ c=t\cdots \\ c=\cdots[\alpha s] \\ |c|=|\alpha s|+k-1}} \text{casc}\,(\beta c)\,.$$

The proof is detailed in Cénac *et al.* (2018a). It relies on the way the VLMC grows between two jump times: at the beginning, letters are added to the current context C_n, the α-LIS does not change and the length of the current context increases one by one. At a certain time (a.s. finite), adding a letter to the current context does not provide a context any more but an external node. At this moment, it happens (it is not trivial and only holds for a stable context tree) that

i) the α-LIS of the current context is renewed;

ii) the length of the current context does not grow;

iii) the current context begins by a lis.

These mechanisms explain the expressions of B_n and the formula giving the semi-Markov kernel.

REMARK 1.3.– In the very particular case of the double or quadruple comb, the semi-Markov chain $(Z_n)_n$ contains as much information as the chain $(U_n)_n$. But in general, the semi-Markov chain $(Z_n)_n$ contains less information than the chain $(U_n)_n$. To illustrate this, here is an example with a finite context tree.

α-LIS αs	contexts having αs as an α-LIS
10	10,010,110,0010,0110
000	000
111	111,0111
0011	0011

In this example, 0010 and 0110 are two contexts of the same length, with the same α-LIS 10 and beginning by the same lis 0. Hence, if we know that $J_n = 10$, $B_{n+1} - B_n = 3$ and $J_{n+1} = 10$, then Z_j is uniquely determined between the two successive jump times, whereas there are two possibilities to reconstruct the VLMC $(U_n)_n$. With the notations of proposition 1.5, there are two cascade terms in $p_{10,10}(3)$:

$$\begin{aligned} p_{10,10}(3) &= \mathbb{P}\left(C_{B_n+1} = 010, C_{B_n+2} = 0010, C_{B_n+3} = 10010 | C_{B_n} = 10\right) \\ &\quad + \mathbb{P}\left(C_{B_n+1} = 110, C_{B_n+2} = 0110, C_{B_n+3} = 10110 | C_{B_n} = 10\right) \\ &= q_{10}(0) q_{010}(0) q_{0010}(1) + q_{10}(1) q_{110}(0) q_{0110}(1) \\ &= \mathrm{casc}(10010) + \mathrm{casc}(10110). \end{aligned}$$

1.5.4. *The meeting point*

Summing up, the announced close encounter can be done with the following (commutative) diagram, together with the following explanations.

$$\begin{array}{ccc} \text{VLMC } (U_n)_n & \xrightarrow{\;D\;} & \text{PRW } (S_n)_n \\ \downarrow L & & \downarrow B \\ \text{MRC } \left(J_n^V, T_n\right)_n & \xleftarrow{\;N\;} & \text{MAP } \left(J_n^W, M_n\right)_n \\ \downarrow S_V & & \downarrow S_W \\ \text{Semi-Markov } \left(Z_n^V\right)_n & \xleftrightarrow{\;R\;} & \text{Semi-Markov } \left(Z_n^W\right)_n \end{array} \qquad [1.26]$$

The mapping D consists of defining the PRW from the VLMC: the random increments of the PRW are the initial letters of a VLMC. With the notations above, $S_n = \sum_{0 \le k \le n} X_k$, where X_k is the initial letter of U_k.

The mapping L associates with a VLMC the process of its successive different α-LIS that turns out to be an MRC when considered together with its jump times T_n (see section 1.5.3). Here, J_n^V is the nth distinct α-LIS of the successive right-infinite words $U_0, U_1, U_2, \cdots$ and T_n is the length of the nth run of identical letters in the sequence $X_0, X_1 X_2, \cdots$ The power V refers to the VLMC.

The mapping B associates with a PRW $(S_n)_n$ the process of its successive different bends (changes of directions). With our notations, J_n^W is the nth distinct bend and M_n is the value of S at the precise moment when the nth bend J_n^W occurs (see section 1.5.2). The power W refers to the PRW.

The mapping S_V only consists of defining a semi-Markov process from an MRC, as stated in section 1.5.1. The mapping S_W is defined in the same manner: it maps a MAP $\left(J_n^W, M_n\right)_n$ to the semi-Markov chain of the MRC $\left(J_n^W, M_n - M_{n-1}\right)_n$.

The mapping N acts on the first coordinate by reversing words: $J_n^V = \overline{J_n^W}$. The notation $\overline{w}$ stands for the reversed word of w: $\overline{ab} = ba$. For the second coordinate, remark first that $M_n - M_{n-1}$ is always of the form $k\alpha$, where k is a positive integer and α an increment vector. The integer T_n is this k.

Finally, the mapping R is simply the reversing of words: $Z_n^V = \overline{Z_n^W}$.

In fact, in these particular situations (double and quadruple combs), the composition $S_V \circ L$ is a bijection (see remark 1.3). Therefore, all these mappings are also one-to-one, showing that all these processes are essentially equivalent.

Now when our different processes are related, let us translate the parameters, properties and assumptions that come from the VLMC world in terms of PRW and vice versa.

Dimension one

The PRW in dimension one is driven by a VLMC based on the so-called double comb, as it was defined in example 1.2. The contexts of this tree are the $u^k d$, which have ud as an α-LIS, and the $d^k u$, which have du as an α-LIS ($k \geq 1$ for both families of contexts). The cascades of the contexts write

$$\operatorname{casc}\left(u^k d\right) = \prod_{j=1}^{k-1} q_{u^j d}(u) \quad \text{and} \quad \operatorname{casc}\left(d^k u\right) = \prod_{j=1}^{k-1} q_{d^j u}(d)$$

and there are two cascade series

$$\kappa_{ud} = \sum_{k \geq 1} \operatorname{casc}\left(u^k d\right) \quad \text{and} \quad \kappa_{du} = \sum_{k \geq 1} \operatorname{casc}\left(d^k u\right).$$

Theorem 1.3 guarantees that, under non-nullness assumption, this VLMC admits an invariant probability measure if and only if $\kappa_{ud} < \infty$ and $\kappa_{du} < \infty$. Since the double comb is a very simple context tree, one can also make a direct computation that leads to the following result: *a non-null double comb VLMC admits a σ-finite stationary measure if and only if* $\mathrm{casc}\left(u^k d\right) \to 0$ *and* $\mathrm{casc}\left(d^k u\right) \to 0$ *when k tends to infinity.*

It turns out that, on the side of the one-dimensional PRW, assumption 1.1 and the expectations of the persistence times τ_1^u and τ_1^d are functions of these cascades so that one can relate the above properties of the VLMC to the results of section 1.3.1 on one-dimensional PRW. The expectations of the waiting times are exactly the sums of cascades: $\kappa_{ud} = \Theta_u$ and $\kappa_{du} = \Theta_d$.

Finally, one can assert:

$$\begin{array}{ccc}
\begin{pmatrix} \mathrm{casc}\left(u^k d\right) \underset{k\to\infty}{\longrightarrow} 0 \\ \text{and} \\ \mathrm{casc}\left(d^k u\right) \underset{k\to\infty}{\longrightarrow} 0 \end{pmatrix} & \Longleftrightarrow & \text{assumption 1.1} \\
\Updownarrow & & \Updownarrow \\
\begin{pmatrix} \text{The VLMC admits} \\ \text{a } \sigma\text{-finite} \\ \text{invariant measure} \end{pmatrix} & \Longleftrightarrow & \begin{pmatrix} \tau_1^u \text{ and } \tau_1^d \\ \text{are a.s. finite} \end{pmatrix}
\end{array}$$

and

$$\begin{array}{ccc}
\begin{pmatrix} \displaystyle\sum_{k\geq 1} \mathrm{casc}\left(u^k d\right) < \infty \\ \text{and} \\ \displaystyle\sum_{k\geq 1} \mathrm{casc}\left(d^k u\right) < \infty \end{pmatrix} & \Longleftrightarrow & \begin{pmatrix} \tau_1^u \text{ and } \tau_1^d \\ \text{are integrable} \end{pmatrix} \\
\Updownarrow & & \\
\begin{pmatrix} \text{The VLMC admits} \\ \text{a unique probability} \\ \text{invariant measure} \end{pmatrix} & &
\end{array}$$

The link between recurrence or transience of the PRW and the behavior of the VLMC is only partial. For instance, the PRW may be recurrent while there is no invariant probability measure for the VLMC. The PRW may even be transient while the VLMC admits an invariant probability measure (see Table 1.1).

Dimension two

The PRW in dimension two is driven by a VLMC based on the so-called quadruple comb, as it is defined in example 1.2. Here, the contexts are the $\alpha^k\beta$, where $\alpha, \beta \in \mathcal{A} = \{\mathtt{n}, \mathtt{e}, \mathtt{w}, \mathtt{s}\}$, $\alpha \neq \beta$, $k \geq 1$. The α-LIS of the context $\alpha^k\beta$ is $\alpha\beta$, and its cascade is written as

$$\mathrm{casc}\left(\alpha^k\beta\right) = \prod_{i=1}^{k-1} q_{\alpha^i\beta}(\alpha).$$

Therefore, there are 12 cascade series, namely

$$\kappa_{\alpha\beta} = \sum_{k=1}^{\infty} \mathrm{casc}\left(\alpha^k\beta\right), \ \alpha, \beta \in \mathcal{A}, \ \alpha \neq \beta. \qquad [1.27]$$

As in dimension one, since the quadruple comb is a stable context tree having a finite set of context α-LIS, the non-null VLMC that drives the two-dimensional PRW admits a unique stationary probability measure if and only if the 12 cascade series [1.27] converge. This is a consequence of theorem 1.3 and, here again, due to the simplicity of the quadruple comb, one can directly check that *a non-null quadruple-comb VLMC admits a σ-finite stationary measure if and only if* $\mathrm{casc}\left(\alpha^k\beta\right) \to 0$ *when k tends to infinity for every $\alpha, \beta \in \mathcal{A}$, $\alpha \neq \beta$.*

The transition matrix of the Markov process $\left(J_n\right)_n$ of the PRW bends, denoted by P in formula [1.13], is also written as

$$P(\beta\alpha, \alpha\gamma) = \sum_{n\geq 1} \mathrm{casc}\left(\gamma\alpha^n\beta\right)$$

and all other entries vanish. Relating this expression to definition [1.21] of the Q-matrix of the VLMC leads to the following:

$$P(\beta\alpha, \alpha\gamma) = Q_{\alpha\beta,\gamma\alpha} \qquad [1.28]$$

so that, up to the re-ordering that consists of reversing the indices $\alpha\beta \rightsquigarrow \beta\alpha$, the stochastic matrices P and Q are the same ones. Note that, since the quadruple comb is stable, the process of the α-LIS of the VLMC is Markovian and Q is its transition matrix. Referring to the commutative diagram [1.26], formula [1.28] amounts to saying that the Markov chains $\left(J_n^V\right)_n$ and $\left(\overline{J_n^W}\right)_n$ are identical.

In terms of persistence times of the PRW versus stationary measures for the VLMC, the properties stated in section 1.3.2 show that the following equivalences hold:

$$\begin{pmatrix} \text{for all } \alpha,\beta\in\mathcal{A},\alpha\neq\beta, \\ \operatorname{casc}\left(\alpha^k\beta\right) \underset{k\to\infty}{\longrightarrow} 0 \end{pmatrix} \iff \quad \text{assumption 1.2}$$

$$\Updownarrow \qquad\qquad\qquad\qquad \Updownarrow$$

$$\begin{pmatrix} \text{The VLMC admits} \\ \text{a } \sigma\text{−finite} \\ \text{invariant measure} \end{pmatrix} \iff \left(\forall n,\ T_n \text{ is a.s. finite}\right)$$

and

$$\begin{pmatrix} \text{for all } \alpha,\beta\in\mathcal{A},\alpha\neq\beta, \\ \sum_{k\geq 1}\operatorname{casc}\left(\alpha^k\beta\right) < \infty \end{pmatrix} \iff (\forall n,\ T_n \text{ is integrable})$$

$$\Updownarrow$$

$$\begin{pmatrix} \text{The VLMC admits} \\ \text{a unique probability} \\ \text{invariant measure} \end{pmatrix}$$

The counterexample cited in theorem 1.2 is supported by these equivalences: an example of recurrent two-dimensional PRW having a transient skeleton $(M_n)_n$ cannot be found without assuming that the T_n are a.s. finite but non-integrable, as shown in Cénac *et al.* (2020). Reading the above equivalences shows that such a PRW must be driven by a VLMC the series of cascades of which diverge, while their general terms tend to zero at infinity.

1.6. References

Barbu, V.S., and Limnios, N. (2008). *Semi-Markov Chains and Hidden Semi-Markov Models Toward Applications*. Springer, New York.

Bejerano, G., and Yona, G. (2001). Variations on probabilistic suffix trees: Statistical modeling and prediction of protein families. *Bioinformatics*, 17(1), 23–43.

Busch, J., Ferrari, P., Flesia, A., Fraiman, R., Grynberg, S., and Leonardi, F. (2009). Testing statistical hypothesis on random trees and applications to the protein classification problem. *Ann. Appl. Stat.*, 3(2), 542–563.

Çinlar, E. (1972). Markov additive processes I, II, *Z. Wahrscheinlichkeitstheorie und Verw. Gebiete*, 24(85–93), 95–121.

Cénac, P., Chauvin, B., Paccaut, F., and Pouyanne, N. (2012). Variable length Markov chains and dynamical sources. *Séminaire de Probabilités XLIV, Lecture Notes in Mathematics*, 2046, 1–39.

Cénac, P., Chauvin, B., Herrmann, S., and Vallois, P. (2013). Persistent random walks, variable length Markov chains and piecewise deterministic Markov processes. *Markov Processes and Related Fields*, 19(1), 1–50.

Cénac, P., Chauvin, B., Paccaut, F., and Pouyanne, N. (2018a). Characterization of stationary probability measures for Variable Length Markov Chains [Online]. Available at: https://hal.archives-ouvertes.fr/hal-01829562

Cénac, P., Le Ny, A., De Loynes, B., and Offret, Y. (2018b). Persistent random walks I: Recurrence versus transience. *J. Theoret. Probab.*, 31(1), 232–243.

Cénac, P., Chauvin, B., Noûs, C., Paccaut, F., Pouyanne, N. (2020). Variable Length Memory Chains: Characterization of stationary probability measures [Online]. Available at: https://arxiv.org/abs/004.07893.

Csiszár, I., and Talata, Z. (2006). Context tree estimation for not necessarily finite memory processes, via BIC and MDL. *IEEE Transactions on Information Theory*, 52(3), 1007–1016.

Doeblin, W., and Fortet, R. (1937). Sur des chaînes à liaisons complètes. *Bull. Soc. Math. France*, 65, 132–148.

Erickson, K.B. (1973). The strong law of large numbers when the mean is undefined. *Trans. Amer. Math. Soc.*, 185, 371–381.

Galves, A., and Leonardi, F. (2008). Exponential inequalities for empirical unbounded context trees. *In and Out of Equilibrium 2: Progress in Probability*, 60, 257–269.

Garivier, A., and Leonardi, F. (2011). Context tree selection: A unifying view. *Stochastic Proc. Appl.*, 121, 2488–2506.

Harris, T.E. (1955). On chains of infinite order. *Pacific J. Math.*, 5, 707–724.

Mauldin, R.D., Monticino, M., and von Weizsäcker, H. (1996). Directionally reinforced random walks. *Adv. Math.*, 117(2), 239–252.

Rissanen, J. (1983). A universal data compression system. *IEEE Trans. Inform. Theory*, 29(5), 656–664.

2

Bootstraps of Martingale-difference Arrays Under the Uniformly Integrable Entropy

In this chapter, we consider the uniform central limit theorem for a bootstrapped martingale-difference array of a function-indexed stochastic process under the uniformly integrable entropy condition. We prove the consistency of the bootstrap. We finally apply our results for the bootstrap of the non-parametric semi-Markov kernel estimator.

2.1. Introduction and motivation

This chapter is mainly concerned with estimating the limiting distribution using the weighted bootstrap of sums of the form $\sum_{j\leq j(n)} V_{nj}(f)$, where the real-valued stochastic processes $\{V_{nj}(f) : f \in \mathcal{F}, 1 \leq j \leq j(n)\}$ for $n \geq 1$ are martingale-difference arrays on the probability space $(\Omega, \mathcal{E}, \mathbb{P})$ for arbitrary index set $\mathcal{F}$. Recall that the bootstrap technique, which is a form of resampling procedures for statistical inference, was introduced in Efron's seminal paper (1979). The bootstrap may be described briefly as follows. Let $T(F)$ be a functional of an unknown distribution function $F(\cdot)$, $X_1, \ldots, X_n$ a sample from $F(\cdot)$ and $X_1^*, \ldots, X_n^*$ an independent and identically distributed (i.i.d.) sample with common distribution given by the empirical distribution $F_n(\cdot)$ of the original sample. The distribution of $\{T(F_n) - T(F)\}$ is then approximated by that of $\{T(F_n^*) - T(F_n)\}$ conditionally upon $X_1, \ldots, X_n$, with $F_n^*(\cdot)$ being the empirical distribution of $X_1^*, \ldots, X_n^*$. The key idea behind the bootstrap is that if a sample is representative of the underlying population, then one can make inferences about the population characteristics by resampling from the current sample. Roughly speaking, it is known that the bootstrap works in the i.i.d. case if and only if the central limit theorem holds for the random

---

Chapter written by Salim BOUZEBDA and Nikolaos LIMNIOS.

variable under consideration. For further discussion, we refer the reader to the landmark paper by Giné and Zinn (1990). Note that the limiting distributions of the processes in which we are interested, or their functionals, are rather complicated, which does not permit explicit computation in practice. To overcome that difficulty, we shall propose a general bootstrap methodology and study some of its asymptotic properties by means of the modern empirical processes theory.

The remainder of this chapter is organized as follows. In section 2.2, we provide some necessary background and state the functional central limit theorem that we are interested in, where the notation and definitions are consistent with the work of Bae *et al.* (2010). The main theoretical results for the weighted bootstraps are given in section 2.3. The paper concludes in section 2.5, where all mathematical developments are given.

2.2. Some preliminaries and notation

Bae *et al.* (2010) introduced the following general setup for studying the uniform central limit theorem for a specific function-indexed process based on a martingale-difference array under the assumption of uniformly integrable entropy. Consider an array of sub-σ-fields $\{\mathcal{E}_{nj} : 0 \leq j \leq j(n), n \in \mathbb{N}\}$ on a given probability space $(\Omega, \mathcal{E}, \mathbb{P})$, satisfying $\mathcal{E}_{n0} \subset \mathcal{E}_{n1} \subset \cdots \subset \mathcal{E}_{nj(n)}$ for $n \in \mathbb{N}$. Given a collection $\mathcal{F}$ of real-valued functions defined on a measurable space $(\mathbb{X}, \mathcal{X})$, we consider an array $\{V_{nj}(f) : j \leq j(n), n \in \mathbb{N}, f \in \mathcal{F}\}$ of martingale difference of $\mathcal{L}_2$-process indexed by $\mathcal{F}$ with respect to the σ-fields $\{\mathcal{E}_{nj} : 0 \leq j \leq n, n \in \mathbb{N}\}$, which means that, for each $f \in \mathcal{F}$, $\{V_{nj}(f) : j \leq j(n), n \in \mathbb{N}\}$ is an array of random variables satisfying $\mathbb{E}(V_{nj}(f) \mid \mathcal{E}_{n,j-1}) = 0$ and $V_{nj}(f)$ is $\mathcal{E}_{nj}$-measurable. We simply denote $\mathbb{E}_{n,j-1}f$ to mean $\mathbb{E}(f \mid \mathcal{E}_{n,j-1})$, the conditional expectation of the random element f given the σ-field $\mathcal{E}_{n,j-1}$. Define a conditional variance process $v_{nj}(f) := \mathbb{E}_{n,j-1}(V_{nj}(f))^2$ for $f \in \mathcal{F}$. Notice that $v_{nj}(f)$ is also an $\mathcal{E}_{n,j-1}$-measurable random variable. For a given class $\mathcal{F}$ defined above, the covering number $N(\epsilon, \mathcal{F}, \|\cdot\|)$, simply denoted $N(\epsilon)$ when there is no risk of ambiguity, is the minimum number of balls $\{g : \|g - h\| < \epsilon\}$ of radius ϵ needed to cover $\mathcal{F}$. Let F be an envelope of $\mathcal{F}$. That is, F is a measurable function from $\mathbb{X}$ to $[0, \infty)$ such that

$$\sup_{f \in \mathcal{F}} |f(x)| \leq F(x), \quad \text{for all} \quad x \in \mathbb{X}.$$

Let $M(\mathbb{X}, F)$ be the set of all measures γ on $(\mathbb{X}, \mathcal{X})$ with

$$\gamma(F) := \int_{\mathbb{X}} F^2 d\gamma < \infty.$$

Given random measures γ on $(\mathbb{X}, \mathcal{X})$, we define the distance

$$d_\gamma^{(2)}(f,g) := [\gamma(f-g)^2]^{1/2} = \left[\int_{\mathbb{X}} (f-g)^2 d\gamma\right]^{1/2}.$$

Say that $\mathcal{F}$ has uniformly integrable entropy, with respect to L_2-norm, if

$$\int_0^\infty \sup_{\gamma\in M(\mathbb{X},F)} \left[\log\left(N\left(\epsilon[\gamma(F^2)]^{1/2}, \mathcal{F}, d_\gamma^{(2)}\right)\right)\right]^{1/2} d\epsilon < \infty.$$

When the class $\mathcal{F}$ has uniformly integrable entropy, $(\mathcal{F}, d_\gamma^{(2)})$ is totally bounded for any measure γ. Many important classes of functions, such as VC graph classes, have uniformly integrable entropy. See section 2.6 of van der Vaart and Wellner (1996), and we may refer also to Kosorok (2008). The one-dimensional CLT (Central Limit Theorem) for a martingale-difference array deals with convergence in distribution to a normal random variable of row sums of the martingale-difference array. See theorem 1 in Chapter 8 of Pollard (1984), among others. Throughout the chapter, events are identified with their indicator functions and $\mathbb{E}^*$ denotes the upper expectation with respect to the outer probability $\mathbb{P}^*$ (refer to van der Vaart and Wellner (1996) for definition). We are interested in estimating the limiting distribution of the process $\{S_n(f) : f \in \mathcal{F}\}$ defined by

$$S_n(f) := \sum_{j\leq j(n)} V_{nj}(f), \quad \text{for } f \in \mathcal{F}.$$

We define

$$\sigma_n^2(f,g) := \sum_{j\leq j(n)} \mathbb{E}_{n,j-1}[V_{nj}(f) - V_{nj}(g)]^2, \quad \text{for all } f, g \in \mathcal{F}.$$

We now state the following result of Bae *et al.* (2010).

THEOREM 2.1 (Bae *et al.* 2010).– Let $\{V_{nj}(f) : j \leq j(n), n \in \mathbb{N}, f \in \mathcal{F}\}$ be a martingale-difference array of $\mathcal{L}_2$-process indexed by a class $\mathcal{F}$ of measurable functions with an envelope $F(\cdot)$ on a measurable space $(\mathbb{X}, \mathcal{X})$. Suppose that $\mathcal{F}$ has uniformly integrable entropy. Let $\mu_n, n \in \mathbb{N}$, be random measures on $(\mathbb{X}, \mathcal{X})$, such that

$$\mathbb{P}^*\left\{\sup_{f,g\in\mathcal{F}} \frac{\sigma_n^2(f,g)}{(d_{\mu_n}^{(2)}(f,g))^2} \geq L\right\} \to 0, \quad \text{as } n\to\infty \text{ for a constant } L.$$

Suppose

$$L_n(\sigma) := \frac{6}{\delta} \sum_{j\leq j(n)} \mathbb{E}[V_{nj}(F)^2\{V_{nj}(F) > \delta\}] \to 0, \quad \text{for every } \delta > 0.$$

Then, given $\epsilon > 0$ and $\gamma > 0$, there exists an $\eta > 0$ for which

$$\limsup_{n\to\infty} \mathbb{P}^* \left(\sup_{d^{(2)}_{\mu_n}(f,g)\leq\eta} |S_n(f) - S_n(g)| > 5\gamma \right) \leq 3\epsilon.$$

Establishing a uniform CLT essentially means showing that $\{\mathcal{L}(S_n(f)) : f \in \mathcal{F}\} \to \{\mathcal{L}(Z(f) : f \in \mathcal{F}\}$, where the processes are indexed by $\mathcal{F}$ and are considered as random elements of the Banach space

$$B(\mathcal{F}) := \left\{ z : \mathcal{F} \to \mathbb{R} : \|z\|_{\mathcal{F}} := \sup_{f\in\mathcal{F}} |z(f)| < \infty \right\},$$

the space of the bounded real-valued functions on $\mathcal{F}$, taken with the sup norm. The limiting process $Z = \{Z(f) : f \in \mathcal{F}\}$ is a Gaussian process whose sample paths are contained in

$$U_B(\mathcal{F}, \rho) := \{z \in B(\mathcal{F}) : z \text{ is uniformly continuous with respect to } \rho\},$$

where ρ is a metric on $\mathcal{F}$. Note that $(B(\mathcal{F}), \|\cdot\|_{\mathcal{F}})$ is a Banach space and $U_B(\mathcal{F}, \rho)$ is a closed subspace of $(B(\mathcal{F}), \|\cdot\|_{\mathcal{F}})$ and hence it is also a Banach space. In particular, $U_B(\mathcal{F}, \rho)$ is separable if and only if $(\mathcal{F}, \rho)$ is totally bounded. Write

$$d(f,g) := d^{(2)}_{\gamma}(f,g) := \left[\int_{\mathbb{X}} (f-g)^2 d\gamma \right]^{1/2},$$

where γ is a fixed measure on $(\mathbb{X}, \mathcal{X})$ with

$$\gamma(F^2) := \int_{\mathbb{X}} F^2 d\gamma < \infty.$$

We equip the space $\mathcal{F}$ with the pseudometric d so that $(\mathcal{F}, d)$ is totally bounded. We use the following definition of weak convergence due to Hoffmann-Jørgensen (1991).

DEFINITION 2.1.– *A sequence of $B(\mathcal{F})$-valued random functions $\{T_n : n \geq 1\}$ converges in law to a $B(\mathcal{F})$-valued Borel measurable random function T whose law concentrates on a separable subset of $B(\mathcal{F})$, denoted $T_n \rightsquigarrow T$, if,*

$$\mathbb{E}g(T) = \lim_{n\to\infty} \mathbb{E}^* g(T_n), \;\; \textit{for any} \;\; g \in C(B(\mathcal{F}), \|\cdot\|_{\mathcal{F}}),$$

where $C(B(\mathcal{F}), \|\cdot\|_{\mathcal{F}})$ is the set of all bounded, continuous functionals from $(B(\mathcal{F}), \|\cdot\|_{\mathcal{F}})$ into $\mathbb{R}$.

We now give the following result of Bae *et al.* (2010).

THEOREM 2.2 (Bae *et al.* 2010).– Let $\{V_{nj}(f) : j \leq j(n), n \in \mathbb{N}, f \in \mathcal{F}\}$ be an array of martingale difference of $\mathcal{L}_2$-process indexed by a class $\mathcal{F}$ of measurable functions with an envelope F on a measurable space $(\mathbb{X}, \mathcal{X})$. Suppose that F has uniformly integrable entropy. Assume that there exists a constant L such that

$$\mathbb{P}^* \left\{ \sup_{f,g \in \mathcal{F}} \frac{\sigma_n^2(f,g)}{(d_\gamma^{(2)}(f,g))} \geq L \right\} \to 0, \quad \text{as } n \to \infty. \qquad [2.1]$$

Suppose that, as $n \to \infty$, in probability

$$\sum_{j \leq j(n)} v_{nj}(f) \to \sigma^2(f), \quad \text{for any } f \in \mathcal{F},$$

where $\sigma^2(f)$ are positive constants, and for every $\epsilon > 0$, in probability

$$\sum_{j \leq j(n)} \mathbb{E}_{n,j-1}((V_{nj}(F)^2)\mathbb{1}\{V_{nj}(F) > \epsilon\}) \to 0.$$

Suppose there exists a Gaussian process Z such that finite dimensional distributions of S_n converge to those of Z. Then

$$\{S_n(f), \quad f \in \mathcal{F}\} \rightsquigarrow \{Z(f), f \in \mathcal{F}\} \text{ as random elements of } B(\mathcal{F}).$$

The limiting process $Z = (Z(f) : f \in \mathcal{F})$ is mean zero Gaussian with covariance structure $\mathbb{E}Z(f)Z(g)$ and the sample paths of Z belong to $U_B(\mathcal{F}, d)$.

Note that a sufficient assumption to the Lipschitz condition [2.1] is that

$$\mathbb{E}^* \sup_{f,g \in \mathcal{F}} \frac{\sigma_n^2(f,g)}{(d_\gamma^{(2)}(f,g))} < \infty.$$

Bae *et al.* (2010) obtained, as a particular case of the last theorem, the uniform CLT for a sequence of martingale difference in Bae and Choi (1999) by applying theorem 2.2 with $V_{nj}(f) = n^{-1/2}D_j(f)$.

Let $\mathcal{F}$ be a class of measurable functions on a measurable space $(\mathbb{X}, \mathcal{X})$. Let $\{D_j(f) : j \in \mathbb{N}, f \in \mathcal{F}\}$ be a sequence of martingale difference in $\mathcal{L}_2$, indexed by a class $\mathcal{F}$, with respect to an increasing sequence of σ-fields $\{\mathcal{E}_j : j \in \mathbb{N}\}$. Suppose that $\mathcal{F}$ has uniformly integrable entropy and that there exists a constant L, such that

$$\mathbb{P}^* \left\{ \sup_{f,g \in \mathcal{F}} \sum_{j=1}^{n} \frac{\mathbb{E}_{n,j-1}[D_j(f) - D_j(g)]^2}{nd_\gamma^{(2)}(f,g)} \geq L \right\} \to 0, \quad \text{as } n \to \infty.$$

Suppose that, as $n \to \infty$, in probability

$$\frac{1}{n}\sum_{j=1}^{n} \mathbb{E}_{j-1}(D_j(f))^2 \to \sigma^2(f), \quad \text{for each } f \in \mathcal{F},$$

where $\sigma^2(f)$ are positive constants, and for every $\epsilon > 0$, in probability

$$\frac{1}{n}\sum_{j=1}^{n} \mathbb{E}_{j-1}((D_j(F)^2)\mathbb{1}\{D_j(F) > \epsilon\sqrt{n}\}) \to 0.$$

Suppose there exists a Gaussian process Z such that the finite-dimensional distributions of S_n converge to those of Z. Then

$$\left\{\frac{1}{\sqrt{n}}\sum_{j=1}^{n} D_j(f),\ f \in \mathcal{F}\right\} \rightsquigarrow \{Z(f), f \in \mathcal{F}\} \text{ as random elements of } B(\mathcal{F}).$$

We give, in the sequel, some examples of classes of functions.

EXAMPLE 2.1.– The set $\mathscr{F}$ of all indicator functions $\mathbb{I}_{\{(-\infty,t]\}}$ of cells in $\mathbb{R}$ satisfies

$$\mathcal{N}\left(\epsilon, \mathscr{F}, d_{\mathbb{P}}^{(2)}\right) \leq \frac{2}{\epsilon^2},$$

for any probability measure $\mathbb{P}$ and $\epsilon \leq 1$. Note that

$$\int_0^1 \sqrt{\log\left(\frac{1}{\epsilon}\right)}d\epsilon \leq \int_0^\infty u^{1/2}\exp(-u)du \leq 1.$$

For more details and discussion on this example refer to example 2.5.4 of van der Vaart and Wellner (1996), and Kosorok (2008, p. 157). The covering numbers of the class of cells $(-\infty, t]$ in higher dimension satisfy a similar bound, but with a higher power of $(1/\epsilon)$ (see theorem 9.19 of Kosorok 2008).

EXAMPLE 2.2 (*Classes of functions that are Lipschitz in a parameter*, section 2.7.4 in van der Vaart and Wellner (1996)).– Let $\mathscr{F}$ be the class of functions $x \mapsto f(t, x)$ that are Lipschitz in the index parameter $t \in T$. Suppose that

$$|f(t_1, x) - f(t_2, x)| \leq d(t_1, t_2)\kappa(x)$$

for some metric d on the index set T, the function $\kappa(\cdot)$ defined on the sample space $\mathcal{X}$ and all x. According to theorem 2.7.11 of van der Vaart and Wellner (1996) and lemma 9.18 of Kosorok (2008), it follows, for any norm $\|\cdot\|_{\mathscr{F}}$ on $\mathscr{F}$, that

$$\mathcal{N}(\epsilon\|F\|_{\mathscr{F}}, \mathscr{F}, \|\cdot\|_{\mathscr{F}}) \leq \mathcal{N}(\epsilon/2, T, d).$$

Hence, if (T, d) satisfy $J(\infty, T, d) = \int_0^\infty \sqrt{\log \mathcal{N}(\epsilon, T, d)} d\epsilon < \infty$, then the conclusion holds for $\mathscr{F}$.

EXAMPLE 2.3.– Let us consider as example the classes of functions that are smooth up to order α defined as follows (see section 2.7.1 of Vaart and Wellner 1996). For $0 < \alpha < \infty$, let $\lfloor \alpha \rfloor$ be the greatest integer strictly smaller than α. For any vector $k = (k_1, \ldots, k_d)$ of d integers, define the differential operator

$$D^{k.} := \frac{\partial^{k.}}{\partial^{k_1} \ldots \partial^{k_d}},$$

where

$$k_. := \sum_{i=1}^{d} k_i.$$

Then, for a function $f : \mathcal{X} \to \mathbb{R}$, let

$$\|f\|_\alpha := \max_{k. \leq \lfloor \alpha \rfloor} \sup_x |D^k f(x)| + \max_{k. = \lfloor \alpha \rfloor} \sup_x \frac{D^k f(x) - D^k f(y)}{\|x - y\|^{\alpha - \lfloor \alpha \rfloor}},$$

where the suprema are taken over all x, y in the interior of $\mathcal{X}$ with $x \neq y$. Let $C_M^\alpha(\mathcal{X})$ be the set of all continuous functions $f : \mathcal{X} \to \mathbb{R}$ with

$$\|f\|_\alpha \leq M.$$

Note that for $\alpha \leq 1$, this class consists of bounded functions f that satisfy a Lipschitz condition. Kolmogorov and Tihomirov (1961) computed the entropy of the classes of $C_M^\alpha(\mathcal{X})$ for the uniform norm. As a result, there exists a constant K depending only on α, d and the diameter of $\mathcal{X}$ such that for every measure γ and every $\epsilon > 0$,

$$\log \mathcal{N}_{[\,]}(\epsilon M \gamma(\mathcal{X}), C_M^\alpha(\mathcal{X}), L_2(\gamma)) \leq K \left(\frac{1}{\epsilon}\right)^{d/\alpha},$$

$\mathcal{N}_{[\,]}$ is the bracketing number; refer to definition 2.1.6 of van der Vaart and Wellner (1996) and theorem 2.7.1 of van der Vaart and Wellner (1996) for a variant of the last inequality. By lemma 9.18 of Kosorok (2008), we have

$$\log \mathcal{N}(\epsilon M \gamma(\mathcal{X}), C_M^\alpha(\mathcal{X}), L_2(\gamma)) \leq K \left(\frac{1}{2\epsilon}\right)^{d/\alpha}.$$

2.3. Main results

In this section, we shall establish the consistency of bootstrapping under general conditions in the framework of martingale-difference arrays. Let $\{z_i\}$ be a sequence of random variables satisfying:

A) the $\{z_i\}$ are bounded independent and identically distributed, on the probability space $(\Omega_z, \mathcal{A}_z, P_z)$ with mean zero and variance 1.

The boundedness assumption of the $\{z_i\}$ can be replaced by a finite moment assumption, but this will add much extra complexity to the proofs. Throughout the paper, we assume that the bootstrap weights z_i's are independent from the data $\{V_{nj}(f) : j \leq j(n), n \in \mathbb{N}, f \in \mathcal{F}\}$. Define, for a measurable function $f \in \mathcal{F}$,

$$\widehat{S}_n(f) := \sum_{i \leq j(n)} z_i V_{ni}(f).$$

THEOREM 2.3.– Assume that the conditions of theorem 2.2 and (A) hold. Then we have

$$\{\widehat{S}_n(f) : f \in \mathcal{F}\} \rightsquigarrow \{Z(f) : f \in \mathcal{F}\} \text{ as random elements of } B(\mathcal{F}).$$

We regain the uniform CLT for a bootstrapped sequence of martingale-difference analogue to that in Bae and Choi (1999) by applying theorem 2.3 with $V_{nj}(f) = n^{-1/2} D_j(f)$ in a similar way as in Bae *et al.* (2010).

COROLLARY 2.1.– Assume that the conditions of theorem 2.3 hold for $\{D_j(f)/\sqrt{n} : f \in \mathcal{F}\}$. Then we have,

$$\left\{ \frac{1}{\sqrt{n}} \sum_{i=1}^{n} z_i D_i(f) : f \in \mathcal{F} \right\} \rightsquigarrow \{Z(f) : f \in \mathcal{F}\} \text{ as random elements of } B(\mathcal{F}).$$

2.4. Application for the semi-Markov kernel estimators

In this section, we will use the notation from Bouzebda *et al.* (2018). Consider an infinite countable set, say E, and an E-valued càdlàg time-homogeneous semi-Markov process Z_t, $t \in \mathbb{R}_+$,with embedded Markov renewal process (J_k, S_k), for $k \in \mathbb{N}$, where (J_k) is the E-valued embedded Markov chain (EMC) of the successive visited states, and $0 = S_0 \leq S_1 \leq \cdots \leq S_k \leq S_{k+1} \leq \cdots$ are the jump times of (Z_t). Define also $X_k := S_k - S_{k-1}, k \geq 1$, the inter-jump times, and the process $N(t)$, $t \in \mathbb{R}_+$, which counts the number of jumps of (Z_t), in the time interval $(0, t]$, by $N(t) := \sup\{k \geq 0 : S_k \leq t\}$. Let us also define $N_i(t)$ as the number of visits of (Z_t) to state $i \in E$ up to time t, and $N_{ij}(t)$ the number of direct jumps of (Z_t) from state i to state j up to time t, (see, for example, Limnios and Oprişan 2001). More precisely, we have

$$N_i(t) := \sum_{k=1}^{N(t)} \mathbb{1}_{\{J_{k-1}=i\}} \text{ and } N_{ij}(t) := \sum_{k=1}^{N(t)} \mathbb{1}_{\{J_{k-1}=i, J_k=j\}},$$

here and elsewhere, $\mathbb{1}_A$ stands for the indicator function of the event A. Consider the renewal process $(S_n^i)_{n\in\mathbb{N}}$ of the successive times of visits to state i, and $N_i(t)$ is its counting process of jumps. Let us define the following observation in the interval $(0,t]$, for $t>0$,

$$\mathcal{H}_t := \{Z_u, 0 \leq u \leq t\} = \begin{cases} \{J_0, X_1, \ldots, J_{N(t)}, U_t\} & \text{if } N(t) > 0, \\ \{J_0, U_t = t\} & \text{if } N(t) = 0, \end{cases} \qquad [2.2]$$

where $U_t := t - S_{N(t)}$ is the backward recurrence time. Let us denote by $Q(t) = \{Q_{ij}(t) : i,j \in E\}, t \geq 0$ the semi-Markov kernel, which is defined by

$$\begin{aligned} Q_{ij}(t) &:= \mathbb{P}(J_{k+1} = j, X_{k+1} \leq t | J_k = i) \\ &= P(i,j)F_{ij}(t), \quad t \geq 0, \quad i,j \in E, \end{aligned} \qquad [2.3]$$

where

$$P(i,j) := \mathbb{P}(J_{k+1} = j | J_k = i)$$

is the transition kernel of the EMC (J_k), and

$$F_{ij}(t) := \mathbb{P}(X_{k+1} \leq t | J_k = i, J_{k+1} = j)$$

is the conditional distribution function of the sojourn time in the state i, given that the next visited state is j, $j \neq i$. The empirical estimator $\widehat{Q}(x,t) := \{\widehat{Q}_{ij}(x,t) : i,j \in E\}$, $x \in (0,t]$, for $t > 0$, of the semi-Markov kernel $Q(x)$ is defined by the following formula:

$$\widehat{Q}_{ij}(x,t) := \frac{1}{N_i(t)} \sum_{k=1}^{N(t)} \mathbb{1}_{\{J_{k-1}=i, J_k=j, X_k \leq x\}}, \quad 0 \leq x \leq t, \quad i,j \in E. \qquad [2.4]$$

The empirical estimator $\widehat{F}(x,t) := \{\widehat{F}_{ij}(x,t) : i,j \in E\}$, $x \in (0,t]$, for $t > 0$, of the sojourn-time distribution function $F_{ij}(x)$ is given by

$$\widehat{F}_{ij}(x,t) := \frac{1}{N_{ij}(t)} \sum_{k=1}^{N(t)} \mathbb{1}_{\{J_{k-1}=i, J_k=j, X_k \leq x\}}, \quad 0 \leq x \leq t, \quad i,j \in E. \qquad [2.5]$$

Let us denote by μ_{ii} the mean recurrence times of (S_n^i), i.e.

$$\mu_{ii} := \mathbb{E}[S_2^i - S_1^I],$$

and by $\nu = (\nu)_{i\in E}$ the stationary distribution of the embedded Markov chain J. Let us define the mean sojourn time of the semi-Markov process Z as

$$\overline{m} := \sum_{i\in E} \nu_i m_i$$

with m_i is defined to be the mean sojourn time in state $i \in E$, i.e.

$$m_i := \int_0^{\infty} [1 - \sum_{i \in E} Q_{ij}(t)]dt.$$

We also recall the property $\mu_{ii} = \overline{m}/\nu_i$. Empirical estimation of the stationary distribution of the embedded Markov chain and for the mean sojourn time are presented in Limnios *et al.* (2005). In the continuous time case, we can define the filtration $\mathbf{F} = (\mathcal{F}_t)_{t \in \mathbb{R}}$ with respect to a non-empty space Ω contained in $\mathcal{F}$, which is an increasing and right-continuous family of sub-σ-algebras of $\mathcal{F}$. Moreover, we assume that $\mathcal{F}_0$ contains all $\mathbb{P}$-null sets of $\mathcal{F}$. Consider the continuous-time filtered probability space with respect to filtration $\mathbf{F}$ as the quadruple $\mathcal{B} = (\Omega, \mathcal{F}, \mathbf{F}, \mathbb{P})$. For each $n \in \mathbb{N}$, let $\mathcal{B}^n = (\Omega^n, \mathcal{F}^n, \mathbf{F}^n, \mathbb{P}^n)$, $\mathbf{F}^n = (\mathcal{F}_t^n)_{t \in \mathbb{R}_+}$, be a stochastic basis. In the discrete time case, the filtration is denoted as $\widetilde{\mathbf{F}} = (\mathcal{F}_k)_{k \in \mathbb{N}}$. Therefore, the discrete-time filtered probability space with respect to filtration $\widetilde{\mathbf{F}}$ is, accordingly, the quadruple $\widetilde{\mathcal{B}} = (\Omega, \mathcal{F}, \widetilde{\mathbf{F}}, \mathbb{P})$. For all $n \in \mathbb{N}$, consider a discrete-time stochastic basis $\widetilde{\mathcal{B}}^n = (\Omega^n, \mathcal{F}^n, \widetilde{\mathbf{F}}^n, \mathbb{P}^n)$. In our case, assume that, for any $n \in \mathbb{N}^*$, it holds $\mathcal{F}_k^n = \mathcal{F}_k = \sigma(J_0, J_i, X_i; 0 \leq i \leq k)$ with $X_0 = 0$, and consider the continuous-time filtration $\mathcal{F}_t^n = \mathcal{F}_{N(nt)}$.

To establish our results, the following conditions are tacitly assumed for all asymptotic results presented in this chapter:

1) the Markov chain $\{J_k : k \geq 0\}$ is ergodic with stationary distribution ν;

2) the mean sojourn times $\{m_i : i \in E\}$ are finite;

3) the distribution functions of the sojourn times are not degenerate, i.e. not equal to δ_0 and Dirac distributions concentrated at 0.

REMARK 2.1.– The first and second assumptions (1) and (2) are used in order to obtain the asymptotic result, since we are working on a single trajectory of the semi-Markov process $\{Z_t : t \geq 0\}$. In particular, the stationarity of the semi-Markov process $\{Z_t : t \geq 0\}$ is guaranteed by assumption (2). Finally, assumption (3) is used to avoid the explosion of the semi-Markov process $\{Z_t : t \geq 0\}$. However, the Dirac distribution concentrated at the point $a > 0$, these distributions are useful in order to introduce fixed durations.

From now on, we will consider a finite set $E = \{1, \ldots, d\}$. For any states $i, j \in E$ and any fixed time $x \in \mathbb{R}_+$, we set $Y_k = \{Y_k^{i,j} : i, j \in E\} \in \mathbb{R}^{d^2}$, where $(Y_k^{ij})_{k \in \mathbb{N}}$ are random sequences on $\widetilde{\mathcal{B}}$ defined by

$$Y_k^{ij} := \mathbb{1}_{\{J_{k-1}=i, J_k=j, X_k \leq x\}} - \mathbb{1}_{\{J_{k-1}=i\}} Q_{ij}(x).$$

Let us define the double sequence $(Y_k^n)_{k\in\mathbb{N}^*}$, $n \in \mathbb{N}^*$, on $\widetilde{\mathcal{B}}^n$, by

$$Y_k^n = \frac{Y_k}{\sqrt{n}}. \qquad [2.6]$$

Also, for any $n \in \mathbb{N}^*$, the set of non-null integers, let us define the stochastic processes

$$S_t^{ij,n} := \sum_{k=1}^{N(nt)} Y_k^{ij,n} = \frac{1}{\sqrt{n}} \sum_{k=1}^{N(nt)} Y_k^{ij}, \quad t \in \mathbb{R}_+^*.$$

Making use of the above notation, we have the sequence of stochastic processes on $\mathcal{B}^n$

$$S_t^n = \{S_t^{ij,n} : i,j \in E\}, \quad t \in \mathbb{R}_+^*, n \in \mathbb{N}^*.$$

Let $D[0,\infty)$ be the Skorohod space (the space of all right-continuous real-valued functions defined on $[0,\infty)$, which have left-hand limits, equipped with the Skorohod topology), by "$\rightsquigarrow$" we denote the weak convergence in this space. The process $W_t := \{W_t^{ij} : i,j \in E\}$ is the d^2-dimensional standard Brownian motion on $\mathcal{B}$. We have, for any fixed $x \in \mathbb{R}_+$, the following weak convergence holds, as n tends to infinity

$$\{S_t^n : t \in \mathbb{R}_+^*\} \rightsquigarrow \{\mathcal{C}^{1/2} W(t) : t \in \mathbb{R}_+^*\}, \qquad [2.7]$$

where

$$\mathcal{C} = (\mathcal{C}^{i,j,\ell,r})_{(i,j),(\ell,r)\in E\times E} = \left(\delta_{i\ell} \frac{1}{\mu_{ii}} Q_{ij}(x)(\delta_{jr} - Q_{ir}(x))\right)_{(i,j),(\ell,r)\in E\times E}$$

is the covariance matrix and δ_{ij} stands for the Kronecker symbol. For more details on the convergence in [2.7], we may refer to Georgiadis and Limnios (2012). For each $i,j \in E$ and $k \in \mathbb{N}^*$ fixed, we have

$$\begin{aligned}\mathbb{E}\left(Y_k^{ij} \mid \mathcal{F}_{k-1}\right) &= \mathbb{E}\left(\{\mathbb{1}_{\{J_{k-1}=i, J_k=j, X_k\le x\}} - \mathbb{1}_{\{J_{k-1}=i\}} Q_{ij}(x)\} \mid \mathcal{F}_{k-1}\right) \\ &= \mathbb{E}\left(\mathbb{1}_{\{J_{k-1}=i, J_k=j, X_k\le x\}} - \mathbb{1}_{\{J_{k-1}=i\}} Q_{ij}(x) \mid \mathcal{F}_{k-1}\right) \\ &= \left\{\mathbb{1}_{\{J_{k-1}=i\}} \mathbb{P}(J_k = j, X_k \le x \mid J_{k-1}) - \mathbb{1}_{\{J_{k-1}=i\}} Q_{ij}(x)\right\} \\ &= 0. \end{aligned} \qquad [2.8]$$

This, in turn, implies that $(Y_k^n)_{k\in\mathbb{N}^*}$ is a $(\mathcal{F}_k)$-martingale difference for any $n \in \mathbb{N}^*$. Then the weighted bootstrap of $S_t^{ij,n}$ is given by

$$\widetilde{S}_t^{ij,n} := \sum_{k=1}^{N(nt)} z_k Y_k^{ij,n} = \frac{1}{\sqrt{n}} \sum_{k=1}^{N(nt)} z_k Y_k^{ij}, \quad t \in \mathbb{R}_+^*.$$

THEOREM 2.4.– For any fixed $x \in \mathbb{R}_+$, the following unconditional weak convergence holds, as n tends to infinity

$$\{\widetilde{S}_t^n : t \in \mathbb{R}_+^*\} \rightsquigarrow \{\mathcal{C}^{1/2}W(t) : t \in \mathbb{R}_+^*\}.$$

For any states $i, j \in E$ and any fixed time $x \in \mathbb{R}_+$, we set $V_k = \{V_k^{ij} : i, j \in E\} \in \mathbb{R}^{d^2}$, where $(V_k^{ij})_{k\in\mathbb{N}^*}$ are random sequences on $\widetilde{\mathcal{B}}$ defined by

$$V_k := \left\{\mathbb{1}_{\{J_{k-1}=i, J_k=j, X_k\leq x\}} - \mathbb{1}_{\{J_{k-1}=i\}}F_{ij}(x)\right\}.$$

For each $i, j \in E$ and $k \in \mathbb{N}^*$ fixed, we have, as in [2.8],

$$\mathbb{E}\left(V_k^{ij} \mid \mathcal{F}_{k-1}\right) = 0.$$

This, in turn, implies that $(V_k^n)_{k\in\mathbb{N}^*}$ is a $(\mathcal{F}_k)$-martingale difference for any $n \in \mathbb{N}^*$. Let us define the double sequence $(V_k^n)_{k\in\mathbb{N}^*}$, $n \in \mathbb{N}^*$, on $\widetilde{\mathcal{B}}^n$, by

$$V_k^n = \frac{V_k}{\sqrt{n}}. \qquad [2.9]$$

Also, for any $n \in \mathbb{N}^*$, let us define the bootstrapped stochastic processes

$$\widetilde{T}_t^{ij,n} := \sum_{k=1}^{N(nt)} z_k V_k^{ij,n} = \frac{1}{\sqrt{n}} \sum_{k=1}^{N(nt)} z_k V_k^{ij}, \quad t \in \mathbb{R}_+^*.$$

Making use of the notation above, we have the sequence of stochastic processes on $\mathcal{B}^n$

$$\widetilde{T}_t^n = \{\widetilde{T}_t^{ij,n} : i, j \in E\}, \quad t \in \mathbb{R}_+^*, n \in \mathbb{N}^*.$$

THEOREM 2.5.– We have, for any fixed $x \in \mathbb{R}_+$, the following unconditional weak convergence holds, as n tends to infinity,

$$\{\widetilde{T}_t^n : t \in \mathbb{R}_+^*\} \rightsquigarrow \{\mathcal{D}^{1/2}W(t) : t \in \mathbb{R}_+^*\},$$

where

$$\mathcal{D} = (\mathcal{D}^{i,j,\ell,r})_{(i,j),(\ell,r)\in E\times E} = \left(\delta_{i\ell}\delta_{jr}\frac{1}{\mu_{ii}}Q_{ij}(x)(1 - F_{ij}(x))\right)_{(i,j),(\ell,r)\in E\times E}.$$

A more general versions of theorems 2.4 and 2.5 are discussed in Bouzebda *et al.* (2018) and Bouzebda and Limnios (2013a), Bouzebda and Limnios (2013b) and Bouzebda and Limnios (2014).

2.5. Proofs

This section is devoted to the proofs of our results. The previously defined notation continues to be used below. For a metric space $\{D, d\}$, let $BL_1(D)$ be the space of real-valued functions on D with Lipschitz norm bounded by 1, i.e. for any $f \in BL_1(D)$, $\sup_{x \in D} |f(x)| \leq 1$ and $|f(x) - f(y)| \leq d(x, y)$ for all $x, y \in D$.

LEMMA 2.1.– Let $\{Y_{ni} : i = 1, \ldots, j(n), n \geq 1\}$ be a martingale-difference array of vectors in $\mathbb{R}^d$, and we let $\{z_i\}$ to be a sequence of random variables satisfying condition (A). Suppose also that

1) $\lim_{n\to\infty} \sum_{i=1}^{j(n)} \mathbb{E} Y_{ni} Y_{ni}^\top = V_0$, where superscript $\top$ denotes transpose and entries of matrix V_0 are finite;

2) for every $\eta > 0$,

$$\limsup_{n\to\infty} \sum_{i=1}^{j(n)} \mathbb{E}\|Y_{ni}\|^2 \mathbb{1}\{\|Y_{ni}\| > \eta\} = 0,$$

where $\|\cdot\|$ is the Euclidean norm on $\mathbb{R}^d$.

Then

i) $\sum_{i=1}^{j(n)} Y_{ni}$ converges weakly to $Y_0 \stackrel{d}{=} N_d(0, V_0)$;

ii) $$\sup_{h \in BL_1(\mathbb{R}^d)} \left| \mathbb{E} h\left(\sum_{i=1}^{j(n)} z_i Y_{ni}\right) - \mathbb{E}(h(Y_0)) \right| \to 0,$$

in probability, as $n \to \infty$, where $\mathbb{R}^d$ is endowed with the uniform metric.

PROOF OF LEMMA 2.1.– Part (i) follows from theorem 4.1 of Pauly (2011). For part (ii), we work as follows.

From independence of sequences (z_i) and (Y_{ni}), we can write:

$$\begin{aligned} \mathbb{E}\left(\sum_{i=1}^{j(n)} z_i^2 Y_{ni} Y_{ni}^\top\right) &= \sum_{i=1}^{j(n)} \mathbb{E}(Y_{ni} Y_{ni}^\top) \mathbb{E}\left(z_i^2\right) \\ &= \sum_{i=1}^{j(n)} \mathbb{E}(Y_{ni} Y_{ni}^\top) \to V_0. \end{aligned}$$

Fix $\eta > 0$. Since, for a matrix norm $\|\cdot\|_m$,

$$\left\| \mathbb{E}\left(\sum_{i=1}^{j(n)} z_i^2 Y_{ni} Y_{ni}^{\top} \mathbb{1}\left\{ |z_i| \times \|Y_{ni}\| > \eta \right\} \right) \right\|_m$$

$$\leq \mathbb{E}\left(z_i^2 \mathbb{1}\left\{ |z_i| > k \right\} \right) \sum_{i=1}^{j(n)} \|\mathbb{E}(Y_{ni} Y_{ni}^{\top})\|_m$$

$$+ k^2 \sum_{i=1}^{j(n)} \|\mathbb{E}(Y_{ni} Y_{ni}^{\top})\|_m \mathbb{1}\{ \|Y_{ni}\| > \eta/k \},$$

for any positive $k < \infty$, we have that the left-hand side converges to zero in probability since we can choose k to make the first expectation on the right-hand side arbitrarily small. Since this is true for every $\eta > 0$, we can now replace η with a sequence $\{\eta_n\}$ going to zero. Thus for every subsequence n', there exists a further subsequence n'' such that

$$\limsup_{n'' \to \infty} \mathbb{E}\left(\sum_{i=1}^{j(n'')} z_i^2 Y_{n''i} Y_{n''i}^{\top} \mathbb{1}\left\{ |z_i| \times \|Y_{n''i}\| > \eta_{n''} \right\} \right) = 0,$$

almost surely. Thus, by the Lindeberg–Feller theorem combined with theorem 1.12.2 of van der Vaart and Wellner (1996), we have, with probability 1, that

$$\lim_{n'' \to \infty} \sup_{h \in BL_1(\mathbb{R}^d)} \left| \mathbb{E}h\left(\sum_{i=1}^{j(n'')} z_i Y_{n''i} \right) - \mathbb{E}(h(Y_0)) \right| = 0.$$

Since this is true for every subsequence n', part (ii) follows by lemma 1.9.2 in van der Vaart and Wellner (1996). □

PROOF OF THEOREM 2.3.– We will first apply theorem 2.1 to $\{z_j V_{nj}(f, \omega) : j \leq j(n), n \in \mathbb{N}, f \in \mathcal{F}\}$. For each $j \in \mathbb{N}$,

$$\mathbb{E}(z_j V_{nj}(f) \mid \mathcal{E}_{n,j-1}) = 0$$

and $z_i V_{nj}(f)$ is $\mathcal{A}_z \times \mathcal{E}_{nj}$-measurable. This, in turn, implies that $\{z_j V_{nj}(f, \omega) : j \leq j(n), n \in \mathbb{N}, f \in \mathcal{F}\}$ is a $\mathcal{E}_{nj}$-martingale difference for any $j \in \mathbb{N}$. Let us introduce, for $f \in \mathcal{F}$,

$$v_{nj}^*(f) := \mathbb{E}((z_i V_{nj}(f))^2 \mid \mathcal{E}_{n,j-1}) := \mathbb{E}_{n,j-1}((z_i V_{nj}(f))^2).$$

First, we show that, as $n \to \infty$, in probability

$$\sum_{j \leq j(n)} \mathbb{E} v_{nj}^*(f) \to \sigma^2(f), \quad \text{for each } f \in \mathcal{F},$$

where $\sigma^2(f)$ are positive constants, and for every $\epsilon > 0$, in probability

$$\sum_{j \le j(n)} \mathbb{E}\mathbb{E}_{n,j-1}[(z_j V_{nj}(F))^2 \mathbb{1}\{|z_j| V_{nj}(F) > \epsilon\}] \to 0.$$

We have

$$\sum_{j \le j(n)} \mathbb{E}\mathbb{E}[(z_j V_{nj}(f))^2 \mid \mathcal{E}_{n,j-1}] = \sum_{j \le j(n)} \mathbb{E}(z_j^2)\mathbb{E}((V_{nj}(f))^2 \mid \mathcal{E}_{n,j-1})$$
$$= \sum_{j \le j(n)} \mathbb{E}((V_{nj}(f))^2 \mid \mathcal{E}_{n,j-1}) \to \sigma^2(f).$$

Fix $\eta > 0$. Since

$$\mathbb{E}\left(\sum_{i=1}^{j(n)} \mathbb{E}_{n,j-1}[(z_j V_{nj}(F))^2 \mathbb{1}\{|z_j| V_{nj}(F) > \eta\}]\right)$$
$$\le \mathbb{E}\left(z_i^2 \mathbb{1}\left\{|z_i| > k\right\}\right) \sum_{i=1}^{j(n)} \mathbb{E}_{n,j-1}((V_{nj}(F))^2 \mathbb{1}\{V_{nj}(F) > \eta/k\})$$
$$+ k^2 \sum_{i=1}^{j(n)} \mathbb{E}_{n,j-1}[(V_{nj}(F))^2 \mathbb{1}\{V_{nj}(F) > \eta/k\}],$$

for any positive $k < \infty$, we have that the left-hand side converges to zero in probability since we can choose k to make the first expectation on the right-hand side arbitrarily small. Let us define

$$\sigma_n^{2*}(f,g) := \sum_{j \le j(n)} \mathbb{E}_{n,j-1}[z_j V_{nj}(f) - z_j V_{nj}(g)]^2, \quad \text{for all } f, g \in \mathcal{F}.$$

For $L' = L_1 \times L > 0$, we want to show that

$$\mathbb{P}^* \left\{ \sup_{f,g \in \mathcal{F}} \frac{\sigma_n^{2*}(f,g)}{(d_{\mu_n}^{(2)}(f,g))^2} \ge L' \right\} \to 0, \quad \text{as } n \to \infty.$$

Notice that

$$\mathbb{P}^* \left\{ \sup_{f,g \in \mathcal{F}} \frac{\sigma_n^{2*}(f,g)}{(d_{\mu_n}^{(2)}(f,g))^2} \ge L' \right\}$$
$$\le \mathbb{P}^* \left\{ \left(\max_{j \le j(n)} z_j \right) \sup_{f,g \in \mathcal{F}} \frac{\sigma_n^{2}(f,g)}{(d_{\mu_n}^{(2)}(f,g))^2} \ge L' \right\}$$

$$\leq \mathbb{P}\left\{ \max_{j \leq j(n)} z_j \geq L_1 \right\} + \mathbb{P}^* \left\{ \sup_{f,g \in \mathcal{F}} \frac{\sigma_n^2(f,g)}{(d_{\mu_n}^{(2)}(f,g))^2} \geq L \right\}$$

$$= \mathbb{P}^* \left\{ \sup_{f,g \in \mathcal{F}} \frac{\sigma_n^2(f,g)}{(d_{\mu_n}^{(2)}(f,g))^2} \geq L \right\} + o(1) \to 0, \text{ as } n \to \infty.$$

Note that the value of z_i in the process $\sum_{i \leq j(n)} z_i V_{ni}(f, \omega)$ does not affect the geometry of $\mathscr{F}$. Thus, one can take the envelope $\{|z_i|F(\cdot)\}$, which is exactly the argument used in the proof of lemma 1 of Ma and Kosorok (2005). It is easy to see that

$$\int_{\mathbb{R}_+} \int_{\mathbb{X}} t^2 F^2(x) d\gamma(x) dP_z(t) = \left\{ \int_{\mathbb{X}} F^2(x) d\gamma(x) \right\} \left\{ \int_{\mathbb{R}_+} t^2 dP_z(t) \right\}$$

$$= \left\{ \int_{\mathbb{X}} F^2(x) d\gamma(x) \right\} < \infty.$$

The rest of the proof is similar to Ma and Kosorok (2005), which we state for reader convenience. Since this is true for every $\eta > 0$, we can now replace η with a sequence $\{\eta_n\}$ going to zero. Thus, as in the proof of van der Vaart and Wellner (1996, p. 357), for every subsequence n', there exists a further subsequence n'' such that

$$\limsup_{n'' \to \infty} \mathbb{E} \left(\sum_{i=1}^{j(n'')} \mathbb{E}_{n'',j-1}((z_j V_{n''j}(F))^2 \mathbb{1}\{|z_j| V_{n''j}(F) > \eta\}) \right) = 0,$$

almost surely. Thus, by the Lindeberg–Feller theorem combined with theorem 1.12.2 of van der Vaart and Wellner (1996), we have with probability 1 that

$$\lim_{n'' \to \infty} \sup_{h \in BL_1(\mathbb{R}^d)} \left| \mathbb{E} h \left(\sum_{i=1}^{j(n'')} z_i Y_{n''i} \right) - \mathbb{E}(h(Y_0)) \right| = 0.$$

Since this is true for every subsequence n', part (ii) follows by lemma 1.9.2 of van der Vaart and Wellner (1996).

For each $\delta > 0$, let M_δ assign to each $f \in \mathcal{F}$ a closest element of a given finite δ-net for $\mathcal{F}$ with respect to $d_\gamma^{(2)}$. Thus, by the fact that $Z(f)$ concentrates on $U_B(\mathcal{F}, d_\gamma^{(2)})$ we have that

$$\lim_{\delta \to 0} \sup_{h \in BL_1(\ell^\infty(\mathcal{F}))} |h(Z(M_\delta(\cdot))) - h(Z(\cdot))| = 0.$$

By lemma 2.1, we next have that

$$\sup_{h \in BL_1(\ell^\infty(\mathcal{F}))} |\mathbb{E} h(\widehat{S}_n(M_\delta(\cdot))) - \mathbb{E} h(S_n(M_\delta(\cdot)))| \to 0,$$

in probability as $n \to \infty$. Now,

$$\sup_{h \in BL_1(\ell^\infty(\mathcal{F}))} |\mathbb{E}h(\widehat{S}_n(M_\delta(\cdot))) - \mathbb{E}h(\widehat{S}_n(\cdot))|$$

$$\leq \sup_{h \in BL_1(\ell^\infty(\mathcal{F}))} \mathbb{E}|h(\widehat{S}_n(M_\delta(\cdot))) - h(\widehat{S}_n(\cdot))|$$

$$\leq \mathbb{E}^* \left(\sup_{f,g \in \mathcal{F}, \rho(f,g) \leq \delta} |\widehat{S}_n(f) - \widehat{S}_n(g)| \right),$$

but the $\limsup_{n\to\infty}$ of the expectation of this last term goes to zero, as $\delta \to 0$, by the previously established unconditional weak convergence of $\widehat{S}_n(\cdot)$. Thus, the desired results follow. □

2.6. References

Bae, J., and Choi, M. J. (1999). The uniform CLT for martingale difference of function-indexed process under uniformly integrable entropy. *Commun. Korean Math. Soc.*, 14(3), 581–595.

Bae, J., Jun, D., and Levental, S. (2010). The uniform CLT for martingale difference arrays under the uniformly integrable entropy. *Bull. Korean Math. Soc.*, 47(1), 39–51.

Bouzebda, S., and Limnios, N. (2013a). Exchangeably weighted bootstraps of empirical estimators of a semi-Markov kernel. *C. R. Math. Acad. Sci. Paris*, 351(13–14), 569–573.

Bouzebda, S., and Limnios, N. (2013b). On general bootstrap of empirical estimator of a semi-Markov kernel with applications. *J. Multivariate Anal.*, 116, 52–62.

Bouzebda, S., and Limnios, N. (2014). Some asymptotic results for exchangeably weighted bootstraps of the empirical estimator of a semi-Markov kernel with applications. In *Statistical Models and Methods for Reliability and Survival Analysis*, Couallier, V., Gerville-Réache, L., Huber-Carol, C., Limnios, N., and Mesbah, M. (eds), 195–212, ISTE Ltd, London and John Wiley & Sons, New York.

Bouzebda, S., Papamichail, C., and Limnios, N. (2018). On a multidimensional general bootstrap for empirical estimator of continuous-time semi-Markov kernels with applications. *J. Nonparametr. Stat.*, 30(1), 49–86.

Efron, B. (1979). Bootstrap methods: Another look at the jackknife. *Ann. Statist.*, 7(1), 1–26.

Georgiadis, S., and Limnios, N. (2012). A multidimensional functional central limit theorem for an empirical estimator of a continuous-time semi-Markov kernel. *J. Nonparametr. Stat.*, 24(4), 1007–1017.

Giné, E., and Zinn, J. (1990). Bootstrapping general empirical measures. *Ann. Probab.*, 18(2), 851–869.

Hoffmann-Jørgensen, J. (1991). *Stochastic Processes on Polish Spaces*. Aarhus Universitet, Matematisk Institut, Aarhus.

Kolmogorov, A.N., and Tihomirov, V.M. (1961). ε-entropy and ε-capacity of sets in functional space. *Amer. Math. Soc. Transl.*, 17(2), 277–364.

Kosorok, M.R. (2008). *Introduction to Empirical Processes and Semiparametric Inference*. Springer, New York.

Limnios, N., and Oprişan, G. (2001). *Semi-Markov Processes and Reliability*. Birkhäuser, Boston.

Limnios, N., Ouhbi, B., and Sadek, A. (2005). Empirical estimator of stationary distribution for semi-Markov processes. *Comm. Statist. Theory Methods*, 34(4), 987–995.

Ma, S., and Kosorok, M.R. (2005). Robust semiparametric M-estimation and the weighted bootstrap. *J. Multivariate Anal.*, 96(1), 190–217.

Pauly, M. (2011). Weighted resampling of martingale difference arrays with applications. *Electron. J. Stat.*, 5, 41–52.

Pollard, D. (1984). *Convergence of Stochastic Processes*. Springer, New York.

van der Vaart, A.W., and Wellner, J.A. (1996). *Weak Convergence and Empirical Processes*. Springer, New York.

3

A Review of the Dividend Discount Model: From Deterministic to Stochastic Models

This chapter introduces a comprehensive overview of the dividend discount models, with a focus on the modeling of the dividend growth process. The analysis starts with the basic Gordon growth model and its extensions to more advanced models based on the Markov chain, leading the reader to the latest frontiers of the stock valuation.

3.1. Introduction

This chapter presents a review of the dividend discount models starting from the basic models (Williams 1938; Gordon and Shapiro 1956) to more recent and complex models (Ghezzi and Piccardi 2003; Barbu *et al.* 2017; D'Amico and De Blasis 2019) with a focus on the modeling of the dividend process rather than the discounting factor, that is assumed constant in most of the models. The chapter starts with an introduction to the basic valuation model with some general aspects to consider when performing the computation. Then, section 3.3 presents the Gordon growth model (Gordon 1962) with some of its extensions (Malkiel 1963; Fuller and Hsia 1984; Molodovsky *et al.* 1965; Brooks and Helms 1990; Barsky and De Long 1993) and reports some empirical evidences. Extended reviews of the Gordon stock valuation model and its extensions can be found in Kamstra (2003) and Damodaran (2012). In section 3.4, the focus is directed to more recent advancements in which the dividend process is modeled as a Markov chain (Hurley and Johnson 1994; Yao 1997; Hurley and Johnson 1998; Ghezzi and Piccardi 2003; Barbu *et al.* 2017; D'Amico and De Blasis 2019). The advantage of these models is the possibility to obtain a different valuation that depends on the state of the dividend series, allowing the model to be closer to reality. In addition, these models allow one to obtain a measure of the risk of the single stock or a portfolio of stocks. Finally, in section 3.5 some concluding remarks are made.

Chapter written by Guglielmo D'AMICO and Riccardo DE BLASIS.

3.2. General model

Stock valuation is one of the basic aspects of financial markets. Discussions about the fair price of a stock, or its overpricing and underpricing, have always been of paramount importance to investors. Williams (1938) was the first to recognize that market prices and fundamental values are "separate and distinct things not to be confused". In his work, he states that an asset's intrinsic long-term value is the present value of all future cash flows, i.e. dividends and future selling price.

Let $P(t)$ be the random variable giving the fundamental value of a stock at time $t \in \mathbb{N}$. Let $D(t)$ be the dividend at time $t \in \mathbb{N}$, also assumed to be a random variable, and denote by $k_e(t)$ the required rate of return on the stock at time t. If we buy a stock at time t and plan to sell it at time $t+1$, the price $p(t) := \mathbb{E}_{(t)}[P(t)]$ that we pay is the expected value of the stock price at time $t+1$ plus the cash flows distributed by the company, all discounted at an appropriate measure of risk $k_e(t)$, given as

$$p(t) = \mathbb{E}_{(t)}\left[\frac{P(t+1)+D(t+1)}{1+k_e(t)}\right]. \qquad [3.1]$$

If we buy and hold the stock indefinitely, and assuming (see, for example, Samuelson 1973)

$$\lim_{i\to+\infty} \mathbb{E}_{(t)}\left[\frac{P(t+i)}{\prod_{j=0}^{i}\left[1+k_e(t+j)\right]}\right] = 0, \qquad [3.2]$$

then the price we pay is the expected value of all future cash flows in the form of dividends,

$$p(t) = \sum_{i=0}^{+\infty} \mathbb{E}_{(t)}\left[\frac{D(t+i+1)}{\prod_{j=0}^{i}\left[1+k_e(t+j)\right]}\right]. \qquad [3.3]$$

If condition [3.2] is not assumed, then Blanchard and Watson (1982) proved that there could exist different solutions of the fundamental equation, i.e. there is a presence of bubbles in the stock market.

To solve equation [3.3], we have to identify two inputs, namely future dividends and the required measure of risk. When estimating future dividends, because of the impossibility of making predictions through to infinity, many models make assumptions about the dividend growth. The basic Gordon model (Gordon 1962) is based on a constant dividend growth rate, while multistage models are advanced by

Brooks and Helms (1990) and Barsky and De Long (1993) to better describe the dividend growth series. Donaldson and Kamstra (1996) generalize the Gordon growth model to allow for arbitrary dividend growth and discount rates using a Monte Carlo simulation. On the contrary, other models apply specific stochastic processes to forecast dividends. Gutiérrez and Vázquez (2004) propose a model that allows a regime switching in the dividend process, and the Korn and Rogers (2005) model dividends as a deterministic transformation of a Lévy process. Hurley (2013) introduces dividends modeled as a Bernoulli process with a continuous set of values, while Eisdorfer and Giaccotto (2014) model the time series behavior of dividend growth rates with a first-order autoregressive process. In this chapter, we focus on how the various models make assumptions about the dividend process, with particular attention paid to the Markov chain based models.

The second input of the equation is the discount factor $k_e(t)$, or the cost of equity, that represents a measure of the asset's riskiness. In most of the dividend discount models, it is assumed to be constant, k_e. Traditionally, the estimation of k_e has been performed using the capital asset pricing model (CAPM). This model originates from the idea of the mean-variance efficient portfolio of Markowitz (1952), and it is formalized by Sharpe (1964) and Lintner (1965) and extended by Black (1972). The rationale of the model is that risky investments R_i, e.g. stocks in financial markets, are expected to be more remunerating than the risk-free assets

$$\mathbb{E}[R_i] = R_f + \beta_{im}(\mathbb{E}[R_m] - R_f), \quad [3.4]$$

$$\beta_{im} = \frac{Cov[R_i, R_m]}{Var[R_m]}, \quad [3.5]$$

where R_m is the return on the market portfolio, and R_f is the return on the risk-free asset. Black's (1972) version substitutes the risk-free rate with a zero-beta portfolio uncorrelated with the market. The coefficient β_{im} represents the correlation of the stock with the market and can be estimated as the slope coefficient of the ordinary least squares (OLS) regression

$$Z_{it} = \alpha_{im} + \beta_{im} Z_{mt} + \epsilon_{it}, \quad [3.6]$$

where Z_{it} is the excess return of the stock on the risk-free asset, or equity premium, and Z_{mt} is the market risk premium, $\mathbb{E}[R_i] - R_f$. In practical applications, the market return and the risk-free rate are proxied by market indices, e.g. the S&P 500 Index, and government treasury bonds, respectively. The estimation is based on a period of time that generally extends to about 5 years of historical data (Campbell *et al.* 1997).

Many authors provide empirical evidence on the CAPM application (see, for example, Jensen *et al.* 1972; Fama and MacBeth 1973; Blume and Friend 1973; Basu 1977; Fama and French 1992, 1993), while Roll (1977) criticizes it because the market portfolio is not observable and therefore the model is not testable. For a comprehensive description of the CAPM models and its variations with econometrics analysis, see, for example, Campbell *et al.* (1997) and Cochrane (2009).

In general, the dividend discount model is a very attractive model because it is intuitive and easy to implement. Nevertheless, it encounters much criticism because of the limits it poses. The main argument is the applicability of the model only to certain firms with stable, high-paying dividend policy. Moreover, the firms' recent practice of performing share buybacks instead of paying dividends, for obvious tax reasons, reduces the dividend cash flow and the application of the dividend discount model results in an underestimation of the value of the firm. The same principle applies to other assets that are ignored in the model, e.g. the value of brand names. However, share buybacks and values of other assets can be included in the dividends flow and treated as such with adequate adjustments (see, for example, Damodaran 2012).

3.3. Gordon growth model and extensions

In this section, we introduce the basic Gordon growth model that represents the cornerstone of the stock valuation. Then, we discuss some extensions, mainly based on the modeling of the dividend process, starting from the two-stage model up to the geometric or arbitrary dividend growth model.

3.3.1. *Gordon model*

Equation [3.3] can be rewritten in terms of dividend growth, defining

$$g(t) = \frac{D(t+1) - D(t)}{D(t)}, \qquad [3.7]$$

as the growth rate of dividends from time t to time $t+1$, so that $D(t+1) = D(t)(1+g(t))$ and $D(t+2) = D(t)(1+g(t))(1+g(t+1))$. Then, the price becomes

$$p(t) = D(t) \sum_{i=0}^{+\infty} \mathbb{E}_{(t)} \left[\prod_{j=0}^{i} \frac{1+g(t+j)}{1+k_e(t+j)} \right]. \qquad [3.8]$$

Assuming a constant dividend growth rate $g(t+j) = g$ and a constant discounting factor $k_e(t+j) = k_e$, equation [3.8] becomes

$$p(t) = D(t) \sum_{i=0}^{+\infty} \frac{(1+g)^i}{(1+k_e)^i}, \qquad [3.9]$$

and summing the geometric progression, we obtain the *Gordon fundamental price estimate* (Gordon 1962)

$$p^G(t) = D(t)\frac{1+g}{k_e - g}, \quad \text{or} \quad p^G(t) = \frac{D(t+1)}{k_e - g}, \qquad [3.10]$$

with the constraint $g < k_e$ to obtain a finite price.

The Gordon model is straightforward because it requires only estimates of the dividend growth rate and discount rate that are both easily obtained from a company's historical data. Nevertheless, it has some limitations. The model can result in incorrect estimations of the price when the growth rate approaches the discount rate, as the price tends to grow up to infinity. Therefore, this model is more suitable for companies with a stable dividend policy with a growth that is less than the growth of the economy. Moreover, empirical applications of the Gordon model show that dividends tend to grow exponentially, meaning that a linear growth model is not suitable for the stock valuation (see, for example, Campbell and Shiller 1987; West 1988).

3.3.2. *Two-stage model*

The assumption that the dividends will experience constant growth forever is not realistic. To relax this assumption, Malkiel (1963) introduces a two-stage model, with the first period of n years of extraordinary growth followed by a stable growth forever. The value of a stock can be obtained as the sum of first years' values, calculated from the general model plus a discounted value of the Gordon growth model at year n:

$$p^{2st}(t) = \mathbb{E}_{(t)}\left[\sum_{i=0}^{n} \frac{D(t+i+1)}{\prod_{j=0}^{i}\left[1+k_e(t+j)\right]} + \frac{P^G(n)}{\prod_{j=0}^{n}\left[1+k_e(t+j)\right]}\right], \qquad [3.11]$$

where $P^G(n)$ is the Gordon growth fundamental price estimate [3.10] at year n.

A further assumption of constant growth in the first phase, g_h, and constant discount rate, $k_{e,h}$, simplifies equation [3.11] to

$$p^{2st}(t) = \frac{D(t)(1+g_h)\left[1 - \frac{(1+g_h)^n}{(1+k_{e,h})^n}\right]}{k_{e,h} - g} + \frac{P^G(n)}{(1+k_{e,h})^n}. \qquad [3.12]$$

This model is suitable for valuing companies that expect to have an initial growth period higher than normal, because of a specific investment or a patent right, that will result in higher profits. At the same time, it presents some limits. First, the growth rate is expected to drop drastically from high to normal level, and second, it is hard to define the length of the high growth period in practical terms.

3.3.3. *H model*

To avoid the sharp drop from high to stable growth rate, Fuller and Hsia (1984) propose a linear decline of the growth in their "H" model. The high growth phase with decline is assumed to last $2H$ periods up to the stable growth phase g_n, with an initial growth rate g_a. The model assumes that the discount rate k_e is constant over time, as well as the dividend payout ratio, and it is given as follows:

$$p^H(t) = \frac{D(t)(1+g_a)}{k_e - g_n} + \frac{D(t)H(g_a - g_n)}{k_e - g_n}. \qquad [3.13]$$

A constant payout ratio assumption poses some limits to this model. Generally, a company is expected to have lower payout ratios in high growth phases and higher payout ratios in the stable growth phase, as shown in Figure 3.1.

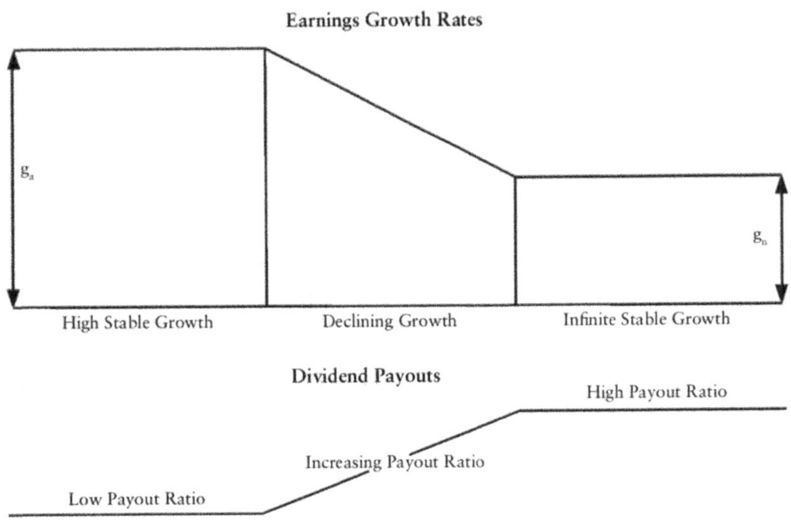

Figure 3.1. *Expected growth in a three-stage dividend discount model (figure from Damodaran 2012, p. 341)*

3.3.4. *Three-stage model*

A three-stage model, initially formulated by Molodovsky *et al.* (1965) and derived from a combination of the H model and the two-stage model, with the inclusion of a variable payout policy and different discount factors for the various phases, overcomes the limits of previous models, but it requires a larger number of inputs. Let $k_{e,h}$, $k_{e,d}$ and $k_{e,st}$ be the discount factors for high, declining, and stable phases, respectively. Let g_a and g_n be the growth rate at the beginning and the end of the period. Let EPS

be the earnings per share, and Π_a and Π_n the payout ratios at the beginning and end of the period, respectively. The stock valuation for the three-stage model is

$$p^{3st}(t) = \sum_{i=0}^{n1} \frac{EPS(t)\Pi_a(1+g_a)^i}{(1+k_{e,h})^i} + \sum_{i=n1+1}^{n2} \frac{D(t+i)}{(1+k_{e,d})^i} + \frac{EPS(t+n2)\Pi_n(1+g_n)}{(k_{e,st}-g_n)(1+k_{e,h})^{n1}(1+k_{e,d})^{n2-n1}}. \quad [3.14]$$

An empirical comparison of the Gordon model and its variations is in Sorensen and Williamson (1985). The authors analyze the intrinsic value of a random sample of 150 firms from the S&P 400 using data available in 1981 from four different valuation models, i.e. price/earning, constant growth, two-period and three-period models. They base the analysis on normalized earnings and a dividend payout ratio of approximately 45%. The discount factor is calculated using the CAPM model for the growth period, according to the beta of the stock and the high growth period is assumed to last 5 years for all the stock. Then, based on the assumption that all mature firms look alike, an equal risk measure of 8% among all the stocks is adopted for the stable phase.

For every model, the authors generate five portfolios of 30 stocks each, ordered from undervalued to overvalued securities, estimating returns for 2 years. Results show that the increased complexity of the model improves the annualized returns. As well as looking at the risk characteristics of the portfolios, the three-stage model outperforms the other model.

3.3.5. *N-stage model*

Brooks and Helms (1990) generalize the two-stage model from Malkiel (1963). They propose an N-stage model, with quarterly dividends and fractional periods. Within each stage, dividends growth is assumed as constant, and the discount rate is based on quarterly compounding $r_e = (1 - k_e)^{\frac{1}{4}} - 1$. They test the model on the case of the Commonwealth Edison Company (CWE), an electricity supplier, estimating the required rate of return for three cases: (a) annual dividends, no fractional period; (b) quarterly dividends, no fractional periods; and (c) quarterly dividends, fractional periods. They show that ignoring quarterly compounding and fractional periods, the results present a downward bias.

3.3.6. *Other extensions*

Another extension of the Gordon growth models is given in Barsky and De Long (1993). The authors propose to model the permanent dividend growth as a geometric average of past dividend changes:

$$g(t) = (1-\theta)\sum_{i=0}^{t}\theta^i \Delta D(t-i) + \theta^t g(0), \qquad [3.15]$$

with $g(t)$ following a random walk process and, thus, change in dividends following an IMA(1,1).

Donaldson and Kamstra (1996) generalize the Gordon growth model, allowing for arbitrary dividend growth and discount rates. Their methodology involves a Monte Carlo simulation and numerical integration of the random joint process of dividend growth and discount rates

$$y(t+1) = \frac{1+g(t+j)}{1+k_e(t+j)}. \qquad [3.16]$$

They forecast a range of possible evolution of the process $y(t+1)$ up to a certain point in the future, $t+I$, and calculate the average of several estimations of the present stock value

$$p(t) = D(t)\sum_{i=0}^{I}\prod_{j=0}^{i} y(t+1). \qquad [3.17]$$

3.4. Markov chain stock models

According to equation [3.8], the stock valuation is obtained through two inputs, namely the dividend growth and the discount factor. The idea of the Markov chain stock models is to describe the dividend growth rate as a sequence of independent, identically distributed discrete random variables, and model it as a Markov process. In all these models, the discount factor k_e is kept constant.

3.4.1. *Hurley and Johnson model*

Hurley and Johnson (1994) model the dividend growth as a *Markov dividend stream*. They assume that in each period the dividend can increase with probability q, and can be constant with probability $1-q$, to resemble a step pattern in the long term. Moreover, they include the possibility for the firm to go bankrupt, with probability q_B. They propose two variations of the model, an additive model and a geometric model, both giving an estimation of the value, along with a lower bound estimation for each of these values.

In the *additive model*, the dividend at time $t+1$ increases by the amount Δ with probability q, and assuming a constant discount rate k_e, the value of the firm is

$$p(t) = \begin{cases} D(t) + \Delta + p(t+1)\frac{D(t)+\Delta}{1+k_e} & \text{with prob } q \\ D(t) + p(t+1)\frac{D(t)}{1+k_e} & \text{with prob } 1-q-q_B \\ 0 & \text{with prob } q_B \end{cases}, \qquad [3.18]$$

and the closed-form solutions for the value and the lower bound are

$$p^A(t) = \frac{D(t)}{k_e} + \left[\frac{1}{k_e} + \frac{1}{k_e^2}\right] q\Delta, \qquad [3.19]$$

and

$$p^A_{low}(t) = \frac{D(t)(1-q_B)}{k+q_B} + \left[\frac{1}{k+q_B} + \frac{1}{(k+q_B)^2}\right] q\Delta. \qquad [3.20]$$

Note that, when $q_b = 0$, $p^A_{low} = p^A$.

The *geometric model* assumes that the dividend increases with a growth rate g and with a probability q

$$D(t+1) = \begin{cases} D(t)(1+g) & \text{with prob } q \\ D(t) & \text{with prob } 1-q-q_B \end{cases}. \qquad [3.21]$$

The closed-form solutions for the value and the lower bound become

$$p^G(t) = \frac{D(t)(1+qg)}{k_e - qg}, \qquad [3.22]$$

and

$$p^G_{low}(t) = D(t)\left[\frac{1+qg-q_B}{k_e - (qg - q_B)}\right]. \qquad [3.23]$$

It is worth noting that the geometric model reduces to the Gordon model, setting the expected growth rate to $qg - q_B$, or, if we exclude the possibility of bankruptcy, setting the expected growth rate to qg.

An empirical application to three stocks, provided in Hurley and Johnson (1994), shows that the geometric method performs well when the dividend series is erratic and does not always show increases. The model gives an estimation that is very close to the actual stock prices.

Hurley and Johnson (1998) formulate a generalized version of their model to include the possibility of a decrease in the dividends, so the dividend at time t is $D(t) = D(t-1) + \Delta_i$ for the additive model, and $D(t) = D(t-1)(1+g_i)$ with probability q_i for the geometric model. Both Δ_i and g_i include the possibility of dividends reduction or suspensions. Under the condition $q_0 + \sum_{i=1}^{n} q_i = 1$, the closed-form solutions for both models are

$$p^A(t) = \frac{D(t)}{k_e} + \left[\frac{1}{k_e} + \frac{1}{k_e^2}\right] \sum_{i=1}^{n} q_i \Delta_i, \qquad [3.24]$$

and

$$p^G(t) = D(t) \frac{1 + \sum_{i=1}^{n} q_i g_i}{k_e - \sum_{i=1}^{n} q_i g_i}. \qquad [3.25]$$

When $n = 1$, the models reduce to Hurley and Johnson (1994) models.

3.4.2. *Yao model*

Yao (1997) advances the same proposal of a dividend reduction extending Hurley and Johnson (1994) models. The author introduces a trinomial dividend valuation model and extends the additive model, where the dividend at time $t + 1$ is

$$D(t+1) = \begin{cases} D(t) + \Delta & \text{with prob } q^u \\ D(t) - \Delta & \text{with prob } q^d \\ D(t) & \text{with prob } q^c = 1 - q^u - q^d \end{cases}, \qquad [3.26]$$

with closed solution for the stock value

$$p^A(t) = \frac{D(t)}{k_e} + \left[\frac{1}{k_e} + \frac{1}{k_e^2}\right] (q^u - q^d)\Delta. \qquad [3.27]$$

Then, the geometric model, with

$$D(t+1) = \begin{cases} D(t)(1 + g) & \text{with prob } q^u \\ D(t)(1 - g) & \text{with prob } q^d \\ D(t) & \text{with prob } q^c = 1 - q^u - q^d \end{cases}, \qquad [3.28]$$

and closed solution

$$p^G(t) = D(t) \frac{1 + (q^u - q^d)g}{k_e - (q^u - q^d)g}. \qquad [3.29]$$

Lower bounds for both models are also given by the author. Moreover, a practical application on five firms, provided in Yao (1997), shows that the model produces better estimates than Hurley and Johnson (1994) models.

3.4.3. *Markov stock model*

Ghezzi and Piccardi (2003) start from the previous Markov models to formulate a more general Markov chain stock model. The authors begin with a description of the simple model for the dividend growth rate using a two-state discrete Markov chain, and a constant discount rate $r = 1 + k_e$. Finally, they extend the model to an n-state Markov chain and define a vector of price–dividend ratios as the solution of a system of linear equations.

In previous models, Hurley and Johnson (1994, 1998) and Yao (1997) assume that the dividend growth rates are independent, identically distributed (i.i.d), discrete random variables, thus obtaining one closed-form solution irrespective of the state of the dividend. On the contrary, Ghezzi and Piccardi (2003) relax the i.i.d. assumption and obtain a different price–dividend solution for each state of the dividends. This variety allows the Markov chain stock model to be closer to reality.

The dividend series obeys the difference equation

$$D(k+1) = G(k+1)D(k), \quad k = t, t+1, \ldots, \qquad [3.30]$$

where $G(k+1)$ is the dividend growth factor described by a Markov chain.

The dividend series relation [3.30] states that given the initial dividend value $D(0) = d \in \mathbb{R}$, we can compute the next random dividend $D(1) = G(1)D(0) = G(1)d$, $D(2) = G(2)D(1) = G(2)G(1)d$ and so on. Generally, $D(n) = \prod_{i=1}^{n} G(i)d$.

The combination of the dividend discount model equation [3.8] and [3.30], with a constant discount factor r, i.e. one plus the required rate of return, yields

$$p(k) = d(k) \sum_{i=1}^{+\infty} \frac{\mathbb{E}_{(k)}[\prod_{j=1}^{i} G(k+j)]}{r^i} =: d(k)\psi_1(g(k)), \qquad [3.31]$$

where $d(k)$ and $g(k)$ are the values at time k of the dividend process and of the growth–dividend process, respectively, and $\psi_1(g(k))$ is the price–dividend ratio.

The simple case is modeled with a two-state Markov chain taking values in the state space $E = \{g_1, g_2\}$. Let $\mathbf{P} = (p_{ij})_{i,j \in E}$ be the one-step transition probability matrix of this Markov chain, and let

$$\mathbf{A1} : \overline{g} := \max(p_{11}g_1 + p_{12}g_2, p_{21}g_1 + p_{22}g_2) < r, \qquad [3.32]$$

be the largest one step conditional expectation on the dividend growth rate.

If A1 holds true, then the series $p(k) = \sum_{i=1}^{+\infty} \frac{\mathbb{E}_{(k)}[D(k+i)]}{r^i}$ converges and satisfies the asymptotic condition in [3.2], and the pair $(\psi_1(g_1), \psi_1(g_2))$ is the unique and non-negative solution of the linear system

$$\begin{aligned} \psi_1(g_1) &= p_{11}\frac{\psi_1(g_1)g_1 + g_1}{r} + p_{12}\frac{\psi_1(g_2)g_2 + g_2}{r} \\ \psi_1(g_2) &= p_{21}\frac{\psi_1(g_1)g_1 + g_1}{r} + p_{22}\frac{\psi_1(g_2)g_2 + g_2}{r}. \end{aligned} \tag{3.33}$$

Assuming that for any given $D(k)$, we obtain the same $\mathbb{E}[D(t+1)]$, irrespective of the initial states g_1, g_2, then $p_{11} = q$ and $p_{22} = 1 - q$, therefore the solution to [3.33] becomes

$$\psi_1(g_1) = \psi_1(g_2) = \frac{qg_1 + (1-q)g_2}{r - 1g_1 - (1-q)g_2}, \tag{3.34}$$

thus implying that the same price–dividend ratio is attached to each state, sharing the same results as Hurley and Johnson (1994, 1998) and Yao (1997) assume.

Results can be easily extended to the case of an s-state Markov chain with state space $E = \{g_1, g_2, \ldots, g_s\}$, where assumption A1 becomes

$$\overline{g} := \max_{i \in E} \left(\sum_{j=1}^{s} p_{ij} g_j \right) < r. \tag{3.35}$$

If $\overline{g} < r$, the series [3.31] converges and the unique and non-negative solution to the linear system is

$$\psi(g_i) = \sum_{j=1}^{s} p_{ij} \frac{\psi(g_j)g_j + g_j}{r}, \quad i = 1, 2, \ldots, s. \tag{3.36}$$

This model has the advantage of assigning a different price–dividend ratio to each value of the states that does not depend on the time. Forecasts on the dividend growth rate are updated based on the previous value of the state, according to the Markov property, thus the price of the stock is updated according to the state of the dividend process. On the contrary, all previous models make fixed assumptions on forecasts and obtain a unique valuation.

Agosto and Moretto (2015) complement the model calculating a closed-form expression for the variance of random stock prices in a multinomial setting. The authors argue that for proper investment decisions, a measure of risk should be taken into consideration. Thus applying the standard mean-variance analysis, an investor can deal with financial decisions under uncertainty. In their model, they relate the variance of stock prices with the variance of the dividend rate of growth, obtaining a measure of the stock riskiness.

An extension of the Markov stock model is given in Barbu *et al.* (2017). The authors provide a formula for the computation of the second-order moment of the fundamental price process in the case of a two-state Markov chain,

$$\begin{aligned} p^{(2)}(k) &= \sum_{i=1}^{+\infty} \frac{\mathbb{E}_k[D^2(k+i)]}{r^{2i}} + \\ & 2\sum_{i=1}^{+\infty}\sum_{j>i} \frac{\mathbb{E}_k[D(k+i)D(k+j)]}{r^{i+j}} = \psi_2(g(k))\, d^2(k). \end{aligned} \quad [3.37]$$

To obtain the convergence of the series [3.37] and to satisfy the asymptotic condition in [3.2] and

$$lim_{N\to+\infty} \sum_{i=1}^{N} \frac{\mathbb{E}_k[D(k+i)P(k+N)]}{r^{i+N}} = 0,$$

the authors introduce a further assumption that avoids the presence of speculative bubbles,

$$\mathbf{A2} : \overline{g}^{(2)} := max(p_{11}g_1^2 + p_{12}g_2^2, p_{21}g_1^2 + p_{22}g_2^2) < r^2, \quad [3.38]$$

where $\overline{g}^{(2)}$ is the largest one-step second-order moment of the dividend growth rate.

If assumptions A1 and A2 hold true, the pair $(\psi_2(g_1), \psi_2(g_2))$ is the unique and non-negative solution of the linear system

$$\begin{aligned} &\psi_2(g_1)(r^2 - p_{11}g_1^2) - \psi_2(g_2)p_{12}g_2^2 \\ &\quad = p_{11}g_1^2(1+2\psi_1(g_1)) + p_{12}g_2^2(1+2\psi_1(g_2)) \\ &\psi_2(g_2)(r^2 - p_{22}g_2^2) - \psi_2(g_1)p_{21}g_1^2 \\ &\quad = p_{21}g_1^2(1+2\psi_1(g_1)) + p_{22}g_2^2(1+2\psi_1(g_2)). \end{aligned} \quad [3.39]$$

To extend the results to an s-state Markov chain with state space $E = \{g_1, g_2, \dots, g_s\}$, assumptions A1 should be formulated as [3.35] and A2 as follows:

$$\overline{g}^{(2)} := \max_{i\in E} \left(\sum_{j=1}^{s} p_{ij} g_j^2 \right) < r^2. \quad [3.40]$$

In this general case, systems [3.33] and [3.39] can be conveniently represented in the matrix form,

$$(\mathbf{I}_r - \mathbf{P}\cdot\mathbf{I_g})\cdot\mathbf{\Psi}_1 = \mathbf{P}\cdot\mathbf{g}, \quad [3.41]$$

$$\left(\mathbf{I}_r^2 - \mathbf{P} \cdot \mathbf{I}_{\mathbf{g}}^2\right) \cdot \boldsymbol{\Psi}_2 = \mathbf{P} \cdot \left((\mathbf{g} \diamond \mathbf{g}) + 2\boldsymbol{\Psi}_1 \diamond (\mathbf{g} \diamond \mathbf{g})\right), \qquad [3.42]$$

where:

– $\boldsymbol{\Psi}_1 = (\psi_1(g_1), \ldots, \psi_1(g_n))^\top$ and $\boldsymbol{\Psi}_2 = (\psi_2(g_1), \ldots, \psi_2(g_n))^\top$;

– $\mathbf{I}_r := r\mathbf{I}$, for any $r \in \mathbb{R}^* := \mathbb{R} - \{0\}$, and, more generally, $\mathbf{I}_r^n = \mathbf{I}_{r^n}$;

– $\mathbf{I}_{\mathbf{g}} = (I_g(i,j))_{i,j \in E}$, $I_{\mathbf{g}}(i,j) = \begin{cases} g_i, & \text{if } i = j \\ 0, & \text{if } i \neq j \end{cases}$;

– $\mathbf{I}$ is the identity matrix of dimension $s \times s$;

– $\cdot$ denotes the row by column matrix product and $\diamond$ denotes the Hadamard element by element product.

The matrix $(\mathbf{I}_r - \mathbf{P} \cdot \mathbf{I}_g)$ is invertible, therefore system [3.41] has a unique solution,

$$\boldsymbol{\Psi}_1 = (\mathbf{I}_r - \mathbf{P} \cdot \mathbf{I}_{\mathbf{g}})^{-1} \cdot \mathbf{P} \cdot \mathbf{g}. \qquad [3.43]$$

Similarly, the matrix $\left(\mathbf{I}_r^2 - \mathbf{P} \cdot \mathbf{I}_g^2\right)$ is invertible and the solution to system [3.42] is

$$\begin{aligned} \boldsymbol{\Psi}_2 &= \left(\mathbf{I}_r^2 - \mathbf{P} \cdot \mathbf{I}_{\mathbf{g}}^2\right)^{-1} \cdot \mathbf{P} \cdot \left((\mathbf{g} \diamond \mathbf{g}) + 2\boldsymbol{\Psi}_1 \diamond (\mathbf{g} \diamond \mathbf{g})\right) \\ &= \left(\mathbf{I}_{r^2} - \mathbf{P} \cdot \mathbf{I}_{\mathbf{g}^2}\right)^{-1} \cdot \mathbf{P} \cdot \left((\mathbf{g} \diamond \mathbf{g}) + 2\boldsymbol{\Psi}_1 \diamond (\mathbf{g} \diamond \mathbf{g})\right). \end{aligned} \qquad [3.44]$$

Relation [3.44] represents an explicit formula for the second-order price–dividend ratio that multiplied by $d^2(t)$ results in the second moment of the price process that is expressed as a function of the model parameters $\mathbf{P}$ and $\mathbf{g}$.

Barbu *et al.* (2017) completed the Markov stock model framework developing non-parametric statistical techniques for the inferential analysis of the model where they propose estimators of price, risk and forecasted prices. For each estimator, they demonstrate that they are strongly consistent and that, after proper centralization and normalization, they converge in distribution to normal random variables. Finally, they give the interval estimators.

A further generalization of Ghezzi and Piccardi (2003) is available in D'Amico (2013). The author models the dividend growth rate as a semi-Markov chain. In this setting, prices become duration dependent. Therefore, they are influenced by the current state of the dividend growth process and by the elapsed time in the state. The same author proposes another extension of the model describing the dividend growth series via a continuous state space semi-Markov model (D'Amico 2017).

3.4.4. *Multivariate Markov chain stock model*

The previous analysis of the dividend discount model focused on the valuation of a single firm based on its dividend process. In this section, we analyze the problem of valuating multiple stocks when they constitute a financial portfolio. When dealing with more than one price series, it is important to consider the possible dependencies that characterize the pool of stocks. In a recent paper, Agosto *et al.* (2019) compute the covariance between two stocks that may be held in a portfolio. They consider a Markov chain with state space equal to the set of possible couples of the growth–dividend values for both stocks. However, this strategy cannot be easily implemented in real applications, especially when we introduce dependencies between more than two stocks as the number of parameters to estimate increase drastically.

D'Amico and De Blasis (2019) propose an extension of the Markov stock model to a multivariate setting, computing the first and the second-order price–dividend ratios. Moreover, the authors provide a formula for the computation of the variances and covariances between stocks in a portfolio. The model belongs to the class of mixture transition distribution models originated by Raftery (1985) in a high-order Markov chain setting and is further extended in Ching and Ng (2006) to a multivariate Markov chain setting. This approach makes it possible for one to overcome the limitations of Agosto *et al.* (2019) because it reduces the number of parameters to estimate.

With a portfolio of multiple stocks, $\alpha = 1, 2, \ldots, \gamma$, the dividend series expressed in [3.30] becomes

$$D^{(\alpha)}(k+1) = G^{(\alpha)}(k+1) \cdot D^{(\alpha)}(k), \qquad [3.45]$$

where $\{G^{(\alpha)}\}_{k\in\mathbb{N}}$ is the growth–dividend random process for stock α, and the multivariate Markov chain model follows the relationship,

$$\mathbf{A}^{(\alpha)}(k+1) = \sum_{\beta=1}^{\gamma} \mathbf{A}^{(\beta)}(k) \cdot \lambda_{\beta,\alpha} \cdot \mathbf{P}^{(\beta,\alpha)}, \qquad [3.46]$$

where:

– $\mathbf{A}^{\alpha}(k) := [A_1^{(\alpha)}, \ldots, A_m^{(\alpha)}]$ is a probability distribution vector with $A_i^{(\alpha)}(k) := \mathbb{P}[G^{(\alpha)}(k) = i]$ being the probability of growth–dividend of stock α to be at time k in state i;

– $\lambda_{\beta,\alpha} \in [0,1]$, $\sum_{\beta=1}^{\gamma} \lambda_{\beta,\alpha} = 1$;

– $\mathbf{P}^{(\beta,\alpha)}$ is the transition probability matrix of stock α given the state occupied one time step before by stock β, i.e.

$$\mathbf{P}_{i,j}^{(\beta,\alpha)} = \mathbb{P}[G^{(\alpha)}(k+1) = j \mid G^{(\beta)}(k) = i]. \qquad [3.47]$$

According to equation [3.46], the probability distribution function of the growth–dividend process at time $k+1$ for the stock α depends on the state of the growth–dividend process of the same stock at time k, and, at the same time, on the set of states visited by each stock in the portfolio at time k.

To extend the model to a multivariate setting, the price process series [3.31] and [3.37] can be rewritten as

$$\begin{aligned} p^{(\alpha)}(\boldsymbol{g}(k)) &= \sum_{i=1}^{+\infty} \frac{\mathbb{E}_{(k)}[D^{(\alpha)}(k+i)]}{r_\alpha^i} \\ &= \sum_{i=1}^{+\infty} \left(\frac{\mathbb{E}_{(k)}[\prod_{j=1}^{i} G^{(\alpha)}(k+j)]}{r_\alpha^i} \right) d^{(\alpha)}(k), \end{aligned} \qquad [3.48]$$

$$\begin{aligned} p_2^{(\alpha,\beta)}(k) &:= \sum_{i=1}^{+\infty} \mathbb{E}_{(k)} \left[\frac{D^{(\alpha)}(k+i) D^{(\beta)}(k+i)}{r_\alpha^i \cdot r_\beta^i} \right] \\ &+ \sum_{i=1}^{+\infty} \sum_{j>i} \mathbb{E}_{(k)} \left[\frac{D^{(\alpha)}(k+i) D^{(\beta)}(k+j)}{r_\alpha^i \cdot r_\beta^j} \right] \\ &+ \sum_{i=1}^{+\infty} \sum_{j>i} \mathbb{E}_{(k)} \left[\frac{D^{(\alpha)}(k+j) D^{(\beta)}(k+i)}{r_\alpha^j \cdot r_\beta^j} \right], \end{aligned} \qquad [3.49]$$

respectively. Equation [3.49] represents the fundamental formula of the price-product and reduces to the second-order moment of the price process when considering the same price series, $\alpha = \beta$.

To guarantee the convergence of the series [3.48] and [3.49] in the multivariate setting, D'Amico and De Blasis (2019) extend the transversality conditions in [3.35] and [3.40],

$$\overline{g}^{(\alpha;1)} := \max_{e^{(1)},\ldots,e^{(\gamma)}} \left(\sum_{j=1}^{m} \sum_{\beta=1}^{\gamma} \sum_{h=1}^{m} e_h^{(\beta)} \lambda_{\beta,\alpha} \mathbf{P}_{h,j}^{(\beta,\alpha)} g_j \right) < r_\alpha, \qquad [3.50]$$

$$\overline{g}^{(\alpha;2)} := \max_{e^{(1)},\ldots,e^{(\gamma)}} \left(\sum_{j=1}^{m} \sum_{\beta=1}^{\gamma} \sum_{h=1}^{m} e_h^{(\beta)} \lambda_{\beta,\alpha} \mathbf{P}_{h,j}^{(\beta,\alpha)} (g_j)^2 \right) < r_\alpha^2. \qquad [3.51]$$

If assumptions in [3.50] holds true, then the first-order price–dividend ratio, $\psi_1^{(\alpha)}(\boldsymbol{g}(k))$, can be computed as a linear system of m^γ equations in m^γ unknown that admits a unique solution,

$$\psi_1^{(\alpha)}(g_{a_1}^{(1)},\ldots,g_{a_\gamma}^{(\gamma)}) = \frac{1}{r_\alpha}\Big\{\sum_{j_\alpha=1}^{m}\sum_{\beta=1}^{\gamma}\sum_{h=1}^{m} e_h^{(\beta)}(k)\lambda_{\beta,\alpha}\mathbf{P}_{h,j_\alpha}^{(\beta,\alpha)}g_{j_\alpha}^{(\alpha)}$$

$$+\sum_{j_1,\ldots,j_\gamma=1}^{m}\psi_1^{(\alpha)}(g_{j_1}^{(1)},\ldots,g_{j_\gamma}^{(\gamma)})\cdot g_{j_\alpha}^{(\alpha)}\cdot\prod_{f=1}^{\gamma}\sum_{w=1}^{\gamma}\sum_{c=1}^{m}e_c^{(w)}(k)\lambda_{w,f}\mathbf{P}_{c,j_f}^{(w,f)}\Big\}. \quad [3.52]$$

Correspondingly, if assumptions in [3.50] and [3.51] hold true, then the second-order price–dividend ratio, $\psi_2^{(\alpha)}(\boldsymbol{g}(k))$, can be computed as a linear system of m^γ equations in m^γ unknown that admits a unique solution,

$$r_\alpha^2\psi_2^{(\alpha)}(g_{a_1}^{(1)},\ldots,g_{a_\gamma}^{(\gamma)}) -$$

$$\sum_{j_1,\ldots,j_\gamma=1}^{m}\psi_2^{(\alpha)}(g_{j_1}^{(1)},\ldots,g_{j_\gamma}^{(\gamma)})(g_{j_\alpha}^{(\alpha)})^2\left(\prod_{f=1}^{\gamma}\sum_{w=1}^{\gamma}\sum_{c=1}^{m}e_c^{(w)}(k)\lambda_{w,f}\mathbf{P}_{c,j_f}^{(w,f)}\right)$$

$$= 2\sum_{j_1,\ldots,j_\gamma=1}^{m}\psi_1^{(\alpha)}(g_{j_1}^{(1)},\ldots,g_{j_\gamma}^{(\gamma)})(g_{j_\alpha}^{(\alpha)})^2\left(\prod_{f=1}^{\gamma}\sum_{w=1}^{\gamma}\sum_{c=1}^{m}e_c^{(w)}(k)\lambda_{w,f}\mathbf{P}_{c,j_f}^{(w,f)}\right)$$

$$+\sum_{j=1}^{m}\sum_{\beta=1}^{\gamma}\sum_{h=1}^{m}e_h^{(\beta)}(k)\lambda_{\beta,\alpha}\mathbf{P}_{h,j}^{(\beta,\alpha)}(g_j^{(\alpha)})^2. \quad [3.53]$$

The solutions of the first- and second-order price–dividend ratio in [3.52] and [3.53] present a different price–dividend ratio attached to each combination of states of the growth process of each stock.

Finally, considering the possible correlation between the stocks and holding assumptions in [3.50] and [3.51], it is possible to compute the product price–dividend ratio, $\psi_2^{(\alpha;\beta)}(\boldsymbol{g}(k))$,

$$r_\alpha r_\beta\psi_2^{(\alpha,\beta)}(g_{a_1}^{(1)},\ldots,g_{a_\gamma}^{(\gamma)})$$

$$= \sum_{j_\alpha,j_\beta=1}^{m}g_{j_\alpha}^{(\alpha)}g_{j_\beta}^{(\beta)}\left(\prod_{f\in\{\alpha,\beta\}}\sum_{w=1}^{\gamma}\sum_{c=1}^{m}e_c^{(w)}(k)\lambda_{w,f}\mathbf{P}_{c,j_f}^{(w,f)}\right)$$

$$+\sum_{j_1,\ldots,j_\gamma=1}^{m}\psi_1^{(\beta)}(g_{j_1}^{(1)},\ldots,g_{j_\gamma}^{(\gamma)})(g_{j_\alpha}^{(\alpha)})(g_{j_\beta}^{(\beta)})\left(\prod_{f=1}^{\gamma}\sum_{w=1}^{\gamma}\sum_{c=1}^{m}e_c^{(w)}(k)\lambda_{w,f}\mathbf{P}_{c,j_f}^{(w,f)}\right)$$

$$+\sum_{j_1,\ldots,j_\gamma=1}^{m}\psi_1^{(\alpha)}(g_{j_1}^{(1)},\ldots,g_{j_\gamma}^{(\gamma)})(g_{j_\alpha}^{(\alpha)})(g_{j_\beta}^{(\beta)})\left(\prod_{f=1}^{\gamma}\sum_{w=1}^{\gamma}\sum_{c=1}^{m}e_c^{(w)}(k)\lambda_{w,f}\mathbf{P}_{c,j_f}^{(w,f)}\right)$$

$$+\sum_{j_1,\dots,j_\gamma=1}^{m}(g_j^{(\alpha)})(g_j^{(\beta)})\psi_1^{(\alpha)}(g_{j_1}^{(1)},\dots,g_{j_\gamma}^{(\gamma)})\psi_1^{(\beta)}(g_{j_1},\dots,g_{j_\gamma})$$
$$\cdot\left(\prod_{f=1}^{\gamma}\sum_{w=1}^{\gamma}\sum_{c=1}^{m}e_c^{(w)}(k)\lambda_{w,f}\mathbf{P}_{c,j_f}^{(w,f)}\right). \quad [3.54]$$

Knowing the product price–dividend ratio for any couple (α, β) of stocks, it is simple to compute the covariance function between the prices of two stocks:

$$\begin{aligned}&Cov(\mathcal{P}^{(\alpha)}(\boldsymbol{g}(k)),\mathcal{P}^{(\beta)}(\boldsymbol{g}(k)))\\&=\mathbb{E}_{(k)}[\mathcal{P}^{(\alpha)}(\boldsymbol{g}(k))\cdot\mathcal{P}^{(\beta)}(\boldsymbol{g}(k))]-\mathbb{E}_{(k)}[\mathcal{P}^{(\alpha)}(\boldsymbol{g}(k))]\cdot\mathbb{E}_{(k)}[\mathcal{P}^{(\beta)}(\boldsymbol{g}(k))]\\&=d^{(\alpha)}(k)d^{(\beta)}(k)\left(\psi_2^{(\alpha,\beta)}(\boldsymbol{g}(k))-\psi_1^{(\alpha)}(\boldsymbol{g}(k))\psi_1^{(\beta)}(\boldsymbol{g}(k))\right).\end{aligned} \quad [3.55]$$

The authors apply the model to a portfolio of three U.S. stock with a stable dividend policy with a long history and compare results with other valuation models. Finally, they show how to obtain the risk of the portfolio for different combinations of the stocks.

3.5. Conclusion

This chapter presented a review of the dividend discount model from its basic formulation to more recent and advanced stochastic models based on the Markov chain modeling of the dividend process. As the fundamental valuation of the firms represents an important function in the financial markets, especially for long-term investments, the Markov stock model clearly shows some advantages over the Gordon model and its extensions. In particular, the Markov stock model makes it possible for one to obtain a different valuation depending on the state of the growth–dividend process, or on a combination of the states of the various series in the multivariate case.

However, the Markov stock model presents some limitations that are shared with the other cited models. First, the valuation is based on the dividend process, therefore it is only applicable to companies that pay dividend and with a long history of payments. Second, the discounting factor, k_e, is considered constant, thus it is not a realistic assumption when considering the very long timeframe.

Future extensions of the Markov stock model could consider the inclusion of some restrictions on the estimation of the transition probability matrix to reduce the number of parameters to estimate and permit the use of shorter dividend series.

Moreover, the cost of equity could be modeled as a stochastic process interdependent with the dividend process. Finally, the model could be extended to companies without a dividend policy, perhaps using the earnings or similar cash flows.

3.6. References

Agosto, A., and Moretto, E. (2015). Variance matters (in stochastic dividend discount models). *Annals of Finance*, 11(2), 283–295.

Agosto, A., Mainini, A., and Moretto, E. (2019). Stochastic dividend discount model: Covariance of random stock prices. *Journal of Economics and Finance*, 43(3), 552–568.

Barbu, V.S., D'Amico, G., and De Blasis, R. (2017). Novel advancements in the Markov chain stock model: Analysis and inference. *Annals of Finance*, 13(2), 125–152.

Barsky, R.B., and De Long, J.B. (1993). Why does the stock market fluctuate? *The Quarterly Journal of Economics*, 108(2), 291–311.

Basu, S. (1977). Investment performance of common stocks in relation to their price-earnings ratios: A test of the efficient market hypothesis. *The Journal of Finance*, 32(3), 663–682.

Black, F. (1972). Capital market equilibrium with restricted borrowing. *The Journal of Business*, 45(3), 444–455.

Blanchard, O.J., and Watson, M.W. (1982). Bubbles, rational expectations and financial markets. In *Crises in the Economic and Financial Structure*, Wachtel, P. (ed.). D.C. Heathand Company, Lexington.

Blume, M.E., and Friend, I. (1973). A new look at the capital asset pricing model. *The Journal of Finance*, 28(1), 19–34.

Brooks, R., and Helms, B. (1990). An n-stage, fractional period, quarterly dividend discount model. *Financial Review*, 25(4), 651–657.

Campbell, J.Y., and Shiller, R.J. (1987). Cointegration and tests of present value models. *Journal of Political Economy*, 95(5), 1062–1088.

Campbell, J.Y., Lo, A.W., and MacKinlay, A.C. (1997). *The Econometrics of Financial Markets*, vol. 2. Princeton University Press, NJ.

Ching, W.-K., and Ng, M.K. (2006). *Markov Chains: Models, Algorithms and Applications*. Springer, New York.

Cochrane, J.H. (2009). *Asset Pricing*, revised edition. Princeton University Press, NJ.

D'Amico, G. (2013). A semi-Markov approach to the stock valuation problem. *Annals of Finance*, 9(4), 589–610.

D'Amico, G. (2017). Stochastic dividend discount model: Risk and return. *Markov Processes and Related Fields*, 23(3), 349–376.

D'Amico, G., and De Blasis, R. (2019). A multivariate Markov chain stock model. *Scandinavian Actuarial Journal*, 2020(4), 272–291.

Damodaran, A. (2012). *Investment Valuation: Tools and Techniques for Determining the Value of Any Asset*. Wiley, New York.

Donaldson, R.G., and Kamstra, M. (1996). A new dividend forecasting procedure that rejects bubbles in asset prices: The case of 1929's stock crash. *The Review of Financial Studies*, 9(2), 333–383.

Eisdorfer, A., and Giaccotto, C. (2014). Pricing assets with stochastic cash-flow growth. *Quantitative Finance*, 14(6), 1005–1017.

Fama, E.F. and French, K.R. (1992). The cross-section of expected stock returns. *The Journal of Finance*, 47(2), 427–465.

Fama, E.F., and French, K.R. (1993). Common risk factors in the returns on stocks and bonds. *Journal of Financial Economics*, 33(1), 3–56.

Fama, E.F., and MacBeth, J.D. (1973). Risk, return, and equilibrium: Empirical tests. *Journal of Political Economy*, 81(3), 607–636.

Fuller, R.J., and Hsia, C.-C. (1984). A simplified common stock valuation model. *Financial Analysts Journal*, 40(5), 49–56.

Ghezzi, L.L., and Piccardi, C. (2003). Stock valuation along a Markov chain. *Applied Mathematics and Computation*, 141(2–3), 385–393.

Gordon, M.J. (1962). *The Investment, Financing, and Valuation of the Corporation*. RD Irwin, Homewood, IL.

Gordon, M.J., and Shapiro, E. (1956). Capital equipment analysis: The required rate of profit. *Management Science*, 3(1), 102–110.

Gutiérrez, M.-J., and Vázquez, J. (2004). Switching equilibria: The present value model for stock prices revisited. *Journal of Economic Dynamics and Control*, 28(11), 2297–2325.

Hurley, W.J. (2013). Calculating first moments and confidence intervals for generalized stochastic dividend discount models. *Journal of Mathematical Finance*, 3(2), 275–279.

Hurley, W.J., and Johnson, L.D. (1994). A realistic dividend valuation model. *Financial Analysts Journal*, 50(4), 50–54.

Hurley, W.J., and Johnson, L.D. (1998). Generalized Markov dividend discount models. *The Journal of Portfolio Management*, 25(1), 27–31.

Jensen, M.C., Black, F., and Scholes, M.S. (1972). The capital asset pricing model: Some empirical tests. In *Studies in the Theory of Capital Markets*, Jensen, M.C. (ed.). Praeger, New York.

Kamstra, M. (2003). Pricing firms on the basis of fundamentals. *Economic Review: Federal Reserve Bank of Atlanta*, 88(1), 49–66.

Korn, R., and Rogers, L. (2005). Stocks paying discrete dividends: Modeling and option pricing. *Journal of Derivatives*, 13(2), 44.

Lintner, J. (1965). Security prices, risk, and maximal gains from diversification. *The Journal of Finance*, 20(4), 587–615.

Malkiel, B.G. (1963). Equity yields, growth, and the structure of share prices. *The American Economic Review*, 53(5), 1004–1031.

Markowitz, H. (1952). Portfolio selection. *The Journal of Finance*, 7(1), 77–91.

Molodovsky, N., May, C., and Chottiner, S. (1965). Common stock valuation: Principles, tables and application. *Financial Analysts Journal*, 21(2), 104–123.

Raftery, A.E. (1985). A model for high-order Markov chains. *Journal of the Royal Statistical Society: Series B (Methodological)*, 47(3), 528–539.

Roll, R. (1977). A critique of the asset pricing theory's tests Part I: On past and potential testability of the theory. *Journal of Financial Economics*, 4(2), 129–176.

Samuelson, P.A. (1973). Proof that properly discounted present values of assets vibrate randomly. *The Bell Journal of Economics and Management Science*, 4(2), 369–374.

Sharpe, W.F. (1964). Capital asset prices: A theory of market equilibrium under conditions of risk. *The Journal of Finance*, 19(3), 425–442.

Sorensen, E.H., and Williamson, D.A. (1985). Some evidence on the value of dividend discount models. *Financial Analysts Journal*, 41(6), 60–69.

West, K.D. (1988). Dividend innovations and stock price volatility. *Econometrica*, 56(1), 37–61.

Williams, J.B. (1938). *The Theory of Investment Value*. Harvard University Press, Cambridge.

Yao, Y.F. (1997). A trinomial dividend valuation model. *The Journal of Portfolio Management*, 23(4), 99–103.

4

Estimation of Piecewise-deterministic Trajectories in a Quantum Optics Scenario

The manipulation of individual copies of quantum systems is one of the most groundbreaking experimental discoveries in the field of quantum physics. On both an experimental and a theoretical level, it has been shown that the dynamics of a single copy of an open quantum system is a trajectory of a piecewise-deterministic process (PDP). To the best of our knowledge, this application field has not been explored by the literature in applied mathematics, from both probabilistic and statistical perspectives. The objective of this chapter is to provide a self-contained presentation of this kind of model, as well as its specificities in terms of observations scheme of the system, and a first attempt to deal with a statistical issue that arises in the quantum world.

4.1. Introduction

4.1.1. *The postulates of quantum mechanics*

Quantum mechanics is one of the most successful theoretical frameworks in physics, its predictions having been tested and verified with remarkable precision in countless scenarios. Roughly speaking, and with the due exceptions, quantum mechanics deals with phenomena happening at the subatomic, atomic and molecular scales, and with how these systems interact with one another. Several new and counterintuitive phenomena arise at these scales, such as non-locality (Brunner *et al.* 2014), non-determinism (Bera *et al.* 2017), wave-particle duality (Wheeler and Zurek 1986), quantum entanglement (Horodecki *et al.* 2009), zero-point vacuum fluctuations (Jaekel and Reynaud 1997), to name just a few. One key point, central in quantum theory, is the fundamental and irreducible role played by measurements of a

Chapter written by Romain AZAÏS and Bruno LEGGIO.

quantum system on the quantum system itself (Wheeler and Zurek 1986). Quantum mechanics, rather than describing how a quantum system behaves in absolute terms, is in fact a theory of how a quantum system reacts when measured. This difference, which may appear almost semantic, is, on the contrary, of fundamental importance for the correct interpretation of the theory and lies at the core of many of the exotic phenomena observed in the quantum world. This bears two main consequences: first, on a practical level, the way measurements affect the observed system and the way the system interacts with its surroundings (thereby including the measuring apparata) is an essential point in the structure of the theory; second, as suggested by the theory and verified by experiments, measuring does not simply reveal a property of a system but, instead, contributes to determining the property itself (Wiseman and Milburn 2009). Quantum mechanics is thus an overcoming of classical mechanics but, at the same time, needs classical objects (measuring apparata) for the appropriate definition of physical properties to associate to a system. This intricate tie between quantum and classical physics is probably the element that, more than others, makes the interpretation of quantum mechanics a non-trivial task.

In the absence of an intuitive level of understanding, the mathematical structure of the theory is clear and well-developed (Von Neumann and Wheeler 2018). It depends on the following postulates, the ensemble of which is known as the Copenhagen interpretation of quantum mechanics.

Postulate 1 (System's state): The state of an isolated quantum system is described by a vector $|\psi\rangle$ in a complex projective Hilbert space $\mathcal{H}$. This vector contains the full information on any possible outcome of measurements performed on the system and, as such, represents the most complete description of the system's state. By definition of projective Hilbert spaces, state vectors have norm 1.

In the following, we focus on the special but important case of systems whose Hilbert space is finite dimensional.

Postulate 2 (System's properties): If the dimension of the system's Hilbert space is N, any measurable physical property $\mathcal{A}$ on a system is represented by a Hermitian matrix A of dimension $N \times N$. Conversely, any $N \times N$ Hermitian matrix represents a measurable property on the system. Hermitian matrices are often referred to as observables. By the properties of Hermitian matrices, observables have real eigenvalues and orthonormal eigenvectors, forming a basis in the Hilbert space.

Postulate 3 (Measurement outcomes): The measurement outcome of an observable A on a system in the state $|\psi\rangle$ is a stochastic variable x. The support of x is the set $\mathcal{E}_A$ of eigenvalues of A. For any $a \in \mathcal{E}_A$, $|a\rangle$ being the corresponding eigenvector, the probability function is given by $\mathbb{P}(x = a) = |\langle a|\psi\rangle|^2$, where $\langle v|u\rangle$ stands for the Hermitian inner product $\langle u, v\rangle$. In case of degeneracy, the probability is simply $\mathbb{P}(x = a) = \sum_i |\langle a_i|\psi\rangle|^2$, $\{|a_i\rangle\}$ being the subset of eigenvectors of A associated with eigenvalue a.

Postulate 3 is central in the interpretation of quantum theory as it specifies what kind of information the state vector provides and how this information can be experimentally accessed. The state vector is not directly measurable: what is measurable is the probabilistic structure it imposes on measurement outcomes.

Postulate 4 (Wave-function collapse): If property A is measured on a system in a state $|\psi\rangle$ yielding the outcome a, the state of the system changes *instantaneously* to the eigenstate $|a\rangle$ of A associated with the eigenvalue a. The discontinuous change $|\psi\rangle \rightarrow |a\rangle$ is referred to as wave-function collapse. In case of degeneracy, the state after measurement collapses to $\frac{1}{\mathcal{N}} \sum_i c_{a_i} |a_i\rangle$, $\{|a_i\rangle\}$ being the subset of eigenvectors of A associated with eigenvalue a, c_{a_i} being the coefficient of $|a_i\rangle$ in the expansion of $|\psi\rangle$ on the basis of eigenvectors of A (see Postulate 2), and $\mathcal{N} = \sqrt{\sum_i |c_{a_i}|^2}$ is a normalization factor.

Postulate 5 (Time evolution): The state of an isolated system between measures evolves in time according to the so-called Schrödinger equation

$$\frac{\mathrm{d}|\psi\rangle}{\mathrm{d}t} = -\frac{i}{\hbar} H|\psi\rangle, \qquad [4.1]$$

where $\hbar \simeq 1.05 \times 10^{-34}\,\mathrm{m^2\,kg\,s^{-1}}$ is the reduced Planck's constant and H, the system's Hamiltonian, is the observable of the energy of the system. For time-independent Hamiltonians, its formal solution is $|\psi(t)\rangle = U(t)|\psi(0)\rangle$, with the time-evolution operator $U(t) = e^{-\frac{i}{\hbar}Ht}$

4.1.2. *Dynamics of open quantum Markovian systems*

Section 4.1.1 gave the basic postulates of quantum mechanics for isolated systems (with the exception of their interaction with measuring apparata). However, in many situations of physical interest, the assumption of isolated systems breaks down due to the almost unavoidable coupling to the external world (Rotter and Bird 2015). It is a matter of both theoretical interest and practical significance to extend the five postulates to the case of an open quantum system, i.e. a system S allowed to exchange energy and information with another system E, considered as an external environment. One can always imagine to identify E in such a way that the composite system $S + E$ can be regarded as isolated. The state of the total system is thus represented by a vector $|\psi_{SE}\rangle$. Under the assumption however of only having access to the subsystem S rather than to the global system $S + E$, the distribution of measurement outcomes on S will reflect, at the same time, the intrinsic probabilistic nature of quantum systems (Postulate 3) and our subjective lack of knowledge on the global system state due to our ignorance on E. Because information on the state of the system is only partial, the description of S in terms of a state vector (Postulate 1) does not hold anymore[1]. To account for a lack of knowledge about the specific vector state of a system S, a new class of quantum states needs to be introduced (Breuer and Petruccione 2002) describing statistical ensembles of vector states. Given a set of (not necessarily orthogonal) vectors $\{|\psi_i\rangle\}$, let us introduce the matrix

$$\rho = \sum_i w_i |\psi_i\rangle\langle\psi_i|, \qquad [4.2]$$

with the weights $0 \leq w_i \leq 1$ and $\sum_i w_i = 1$. The notation $|u\rangle\langle v|$ stands for the complex outer product $u \otimes v^*$, v^* being the complex conjugate of v. By construction, ρ is a Hermitian matrix. In the case $w_k = 1$ for a certain k, the matrix [4.2] is referred to as a *pure state*, and a *mixed state* otherwise. A mixed state represents a classical statistical mixture of pure states $|\psi_i\rangle\langle\psi_i|$, each contributing with a weight w_i. The five postulates of section 4.1.1 generalize as follows:

Postulate 1b (Density matrices): The state of a (generally open) quantum system is described by a positive semidefinite Hermitian matrix ρ of the form [4.2] called its *density matrix*. The eigenvalues of this matrix are the probabilities that the system S is found in the corresponding eigenvector, which is a vector in the system's Hilbert space. This implies that $\text{Tr}\rho = 1$. A density matrix represents a pure state if and only if $\text{Tr}\rho^2 = 1$. A pure state is simply an alternative way of describing a vector state.

Postulate 2 does not change, since it does not depend on the state of the system.

Postulate 3b (Measurement outcomes on density matrices): Given any matrix M and any vector $|\psi\rangle$, the expression $\langle\psi|M|\psi\rangle = \Big\langle |\psi\rangle, M|\psi\rangle \Big\rangle$ is the inner product of the vector $M|\psi\rangle$ with the vector $|\psi\rangle$ and is often referred to as the average of M on $|\psi\rangle$. When measuring A on a state ρ, the probability of obtaining the specific eigenvalue a is $\mathbb{P}(x = a) = \langle a|\rho|a\rangle$. If ρ is a pure state $|\psi\rangle\langle\psi|$, $\mathbb{P}(x = a) = \langle a|\psi\rangle\langle\psi|a\rangle = |\langle a|\psi\rangle|^2$, as expected from Postulate 3. In the general case, one obtains $\mathbb{P}(x = a) = \sum_i w_i |\langle a|\psi_i\rangle|^2$, as expected from a statistical mixture of pure states.

Postulate 4b (Density matrix collapse): If the measurement of A on ρ yields the outcome a associated with the eigenvector $|a\rangle$, the density matrix instantaneously collapses to $\frac{|a\rangle\langle a|\rho|a\rangle\langle a|}{\text{Tr}(|a\rangle\langle a|\rho|a\rangle\langle a|)}$.

1 This point of view, although useful in providing an intuition on the need of extending the structure of quantum mechanics to a new class of states, only gives a partial view: due to the existence of a specifically quantum kind of correlations between two systems (entanglement), the state of a subsystem in a global entangled state cannot be described by a vector, even if full information on the global state is available. This aspect, however, goes beyond the scope of this brief overview.

More in general, thereby including the case of degeneracy in the spectrum of observables, consider the spectral decomposition of $A = \sum_i a_i A_i$. Then $\mathbb{P}(x = a_i) = \mathrm{Tr}(A_i \rho A_i)$ and the state collapses to $\frac{A_i \rho A_i}{\mathrm{Tr}(A_i \rho A_i)}$, i.e. to the normalized projection of ρ onto the subspace of A corresponding to the eigenvalue a_i.

There is no general closed extension of Postulate 5 to open quantum systems. For a closed system in a pure state $\rho = |\psi\rangle\langle\psi|$, one has $\frac{\mathrm{d}\rho}{\mathrm{d}t} = \frac{\mathrm{d}|\psi\rangle}{\mathrm{d}t}\langle\psi| + |\psi\rangle\frac{\mathrm{d}\langle\psi|}{\mathrm{d}t}$ which, using equation [4.1], can be recast as

$$\frac{\mathrm{d}\rho}{\mathrm{d}t} = -\frac{i}{\hbar}[H, \rho], \qquad [4.3]$$

with the commutator of two matrices $[A, B] = AB - BA$. Equation [4.3] is known as the von Neumann equation.

Assuming, now, that the global system $S + E$ is isolated, its pure state $\rho_{SE} = |\psi_{SE}\rangle\langle\psi_{SE}|$ evolves according to the von Neumann equation $\frac{\mathrm{d}\rho_{SE}}{\mathrm{d}t} = -\frac{i}{\hbar}[H_{SE}, \rho_{SE}]$, H_{SE} being the Hamiltonian of the global system, containing, in general, local Hamiltonians H_S and H_E of, respectively, S and E, and an interaction term V_{SE}. The density matrix ρ_S of subsystem S is obtained from ρ_{SE} by performing a partial trace $\mathrm{Tr}_E(\rho_{SE})$ over the degrees of freedom of E. Hence, the general extension of Postulate 5 is as follows:

Postulate 5b (Open quantum system's dynamics): The (generally mixed) density matrix of an open quantum system S that interacts with an environment E evolves as

$$\frac{\mathrm{d}\rho_S}{\mathrm{d}t} = \frac{\mathrm{d}\mathrm{Tr}_E(\rho_{SE})}{\mathrm{d}t} = \mathrm{Tr}_E\left(\frac{\mathrm{d}\rho_{SE}}{\mathrm{d}t}\right) = -\frac{i}{\hbar}\mathrm{Tr}_E\left([H_{SE}, \rho_{SE}]\right). \qquad [4.4]$$

This equation is not in closed form, i.e. it still explicitly depends on the state ρ_{SE} of the global system. No general way is known for obtaining a closed form for the right-hand side (r.h.s.) of equation [4.4] without any additional assumptions on the environment E and its interaction with S. However, in the case of weakly interacting systems and rapidly decaying correlation functions of E, the so-called Born-Markov approximation can be performed (Breuer and Petruccione 2002), which ultimately leads to a closed Markovian equation for the evolution of S (quantum master equation), which for an N-dimensional quantum system can always be cast in the so-called GKSL form as

$$\frac{\mathrm{d}\rho_S}{\mathrm{d}t} = -\frac{i}{\hbar}[\widetilde{H}_S, \rho_S] + \sum_{i=1}^{N^2-1} \gamma_i\Big(L_i \rho_S L_i^\dagger - \frac{1}{2}\{L_i^\dagger L_i, \rho_S\}\Big). \qquad [4.5]$$

Here, $\widetilde{H}_S$ is the system's Hamiltonian, modified to take into account the change in the system's energy due to its interaction with E. The first term on the r.h.s. accounts for the free dynamics of S, not triggered by its interaction with E. The second term on the r.h.s., on the other hand, describes the dynamics of S directly induced by its interaction with E: L_i (Lindblad operators) are non-Hermitian matrices ($L_i^\dagger$ stands here for Hermitian conjugation), each describing a specific dynamical process in S triggered by E and happening at a rate γ_i, and $\{A, B\}$ stands here for the anti-commutator $AB + BA$ of two matrices. The explicit form of matrices $\{L_i\}$ and of rates $\{\gamma_i\}$ depends on the specific problem at hand. In particular, the physical properties of E are encoded in the rates $\{\gamma_i\}$.

4.1.3. *Stochastic wave function: quantum dynamics as PDPs*

In section 4.1.2, mixed quantum states have been introduced to account for classical-like stochasticity in the state vector, in addition to the purely quantum one. As a result, the density matrix [4.2] has been introduced together with its Markovian time evolution [4.5]. The formalism in terms of density matrices describes the average response of a collection of identical quantum systems, the state vector of which has a certain amount of stochasticity. The density matrix can be regarded as an average of systems' state vectors, and its evolution [4.5] as the average time evolution of such an ensemble of quantum systems. One can, however, wonder whether it is possible to obtain a description of the evolution of a single element of this ensemble. In other words, whether it is possible to write down, for a single quantum system, an equation for its state vector extending equation [4.1] to the case of open system dynamics, in the same way as, for density matrices, equation [4.5] extends [4.3] to the case of an ensemble of open systems. Since, as discussed in section 4.1.2, the interaction with an environment generally "hides" part of the information about the state vector of a system S, its time evolution cannot be deterministic and must take into account the environmental-induced stochasticity. The two natural requirements for such a single-open-system dynamical equation are that it must reduce to the deterministic equation [4.1] for a closed system, and its average must reproduce equation [4.5]. This second requirement is fundamental if one wants the single-open-system equation to describe the dynamics of one element of the ensemble [4.2]. There are many possible formulations of equations respecting these two requirements (Dalibard *et al.* 1992; Dum *et al.* 1992; Wiseman and Milburn 1993; Wiseman 1994; Breuer and Petruccione 1995; Wiseman 1996; Plenio and Knight 1998; Breuer and Petruccione 2002), the full treatment of which goes well beyond the scope of this introduction. Here, we give a brief overview on what goes under the name of *Markovian stochastic wave-function approach* (Dalibard *et al.* 1992; Breuer and Petruccione 1995, 2002), which has the advantage of having a clear and direct experimental interpretation and has recently been employed to investigate quantum stochastic thermodynamics (Hekking and Pekola 2013; Leggio *et al.* 2013).

Skipping the derivation (the details of which can be found in (Breuer and Petruccione 2002)) and with the same L_i, γ_i and $\widetilde{H}_S$ as in equation [4.5], the equation for the evolution of the state vector $|\psi\rangle$ of an open quantum system can be given under the form of the stochastic differential equation

$$\mathrm{d}|\psi\rangle = -\frac{i}{\hbar}\mathcal{L}(|\psi\rangle)\mathrm{d}t + \sum_{i=1}^{N^2-1}\left(\frac{L_i|\psi\rangle}{|L_i|\psi\rangle|} - |\psi\rangle\right)\mathrm{d}N_i. \quad [4.6]$$

Equation [4.6] is the sum of a deterministic time evolution with generator $\mathcal{L}(|\psi\rangle)$ and a stochastic contribution in terms of inhomogeneous Poisson processes $N_i = N_i(t)$ satisfying

$$\mathbb{E}\left[\mathrm{d}N_i(t)\right] = \gamma_i|L_i|\psi\rangle|^2\mathrm{d}t \quad [4.7]$$

and $\mathrm{d}N_i\mathrm{d}N_j = \delta_{ij}\mathrm{d}N_i$, where δ_{ij} is the Kronecker delta. The generator of the deterministic evolution has the form

$$\mathcal{L}(|\psi\rangle) = \left(\widetilde{H}_S - \hbar\frac{i}{2}\sum_{i=1}^{N^2-1}\gamma_i\left(L_i^\dagger L_i - |L_i|\psi\rangle|^2\right)\right)|\psi\rangle.$$

This deterministic evolution has the form of a (nonlinear) Schrödinger equation, which can be shown, for time-independent Hamiltonian $\widetilde{H}_S$ and Lindblad operators L_i, to have the formal solution

$$|\psi(t)\rangle_{\mathrm{Det}} = e^{-\frac{i}{\hbar}\widehat{H}t}|\psi(0)\rangle, \quad [4.8]$$

through the effective non-Hermitian Hamiltonian $\widehat{H} = \widetilde{H}_S - \hbar\frac{i}{2}\sum_{i=1}^{N^2-1}\gamma_i L_i^\dagger L_i$. Note that, due to the non-Hermitian part $-\hbar\frac{i}{2}\sum_{i=1}^{N^2-1}\gamma_i L_i^\dagger L_i$, the norm of $|\psi(t)\rangle$ in equation [4.8] generally decreases in time.

On the other hand, the stochastic contribution in [4.6] produces, in the evolution of $|\psi(t)\rangle$, random instantaneous transitions. As evident from [4.6], there is a discrete and finite set of possible jumps for the state vector of the form

$$|\psi\rangle \to \frac{L_i|\psi\rangle}{|L_i|\psi\rangle|},\ i = 1, \ldots, N^2 - 1. \quad [4.9]$$

Each of these jumps, due to property [4.7], happens at a rate $\gamma_i|L_i|\psi\rangle|^2\mathrm{d}t$: the dynamics of the state vector is thus a Markovian PDP. It is possible to show that the waiting-time distribution during a deterministic evolution from the initial state $|\psi(0)\rangle$ is $f(t) = 1 - ||\psi(t)\rangle_{\mathrm{Det}}|^2$, which explains the meaning of a decreasing norm during deterministic evolutions. It is easy to see that equation [4.6] reduces to the deterministic Schrödinger equation when $\gamma_i = 0, \forall i$ (closed-system case). It is also possible to show rigorously that

$$\mathbb{E}\left[|\psi(t)\rangle\langle\psi(t)|\right] = \rho(t), \quad [4.10]$$

where$|\psi(t)\rangle$ is the solution of equation [4.6] and$\rho(t)$ is the solution of equation [4.5].

The above point proves that the trajectories given by equation [4.6] are indeed a valid single-system representation of the ensemble dynamics. In addition, these trajectories have a direct and interesting experimental interpretation. To make this interpretation clear, it is convenient to state here that the Markovian master equation [4.5] is invariant under unitary or inhomogeneous transformations (Breuer and Petruccione 2002). In the latter case, the transformations $L_i \to L_i'$ and $\widetilde{H}_S \to \widetilde{H}_S'$ read

$$L_i' = L_i + \beta_i \mathbb{I}, \text{ and } \widetilde{H}_S' = \widetilde{H}_S + \frac{1}{2i} \sum_{k=1}^{N^2-1} \gamma_k \left(\beta_k^* L_k - \beta_k L_k^\dagger \right) + \alpha \mathbb{I}, \qquad [4.11]$$

where $\beta_i \in \mathbb{C}$ and $\alpha \in \mathbb{R}$. Despite leaving equation [4.5] invariant, the transformation [4.11] changes equation [4.6] while still assuring that $\mathbb{E}[|\psi(t)\rangle\langle\psi(t)|] = \rho(t)$. This means that different single-system-dynamics decompositions of the same ensemble dynamics are possible. It turns out that these different decompositions correspond to the different ways of performing continuous measurements on the environment E the system S is in contact with. Intuitively, since a direct measurement of S would collapse its state vector and thus stop its dynamics, one must follow the dynamics of system S by indirect measures. Specifically, one looks for signatures, in the continuously monitored state of E, of possible transitions happening in S. Because of the fact that E and S interact and that it is, properly, this interaction that causes departures from the closed-system dynamics, monitoring S through E is enough to reconstruct its dynamics. Under this interpretation, the deterministic parts of the evolution of S correspond to the absence of detected transitions in the state of E, while jumps correspond to the detection of a particular transition i.

4.1.4. *Estimation for PDPs*

As explained above, the pure quantum state of an open system evolves as a piecewise-deterministic Markov process. In recent years, the literature has extensively investigated this class of stochastic processes, in particular for statistical purposes in a general setting (Azaïs *et al.* 2014; Azaïs and Muller-Gueudin 2016) or in the context of particular applications: while not exhaustive, applications considered are various, from cell cycle modeling (Doumic *et al.* 2011, 2015) to fatigue crack propagation analysis (Abdessalem *et al.* 2016) through insurance risk models (Kovacevic and Pflug 2011), food contaminants exposure (Bertail *et al.* 2008) and neurobiology (Buckwar and Riedler 2011). Nevertheless, to the best of our knowledge, no paper deals with statistical estimation in the particular context of quantum dynamics, whose specificities need to be taken into account. We think that three of them are particularly significant. First, the state space of quantum trajectories is the unit ball of a complex Hilbert space, which is a bit unusual. In addition, post-jump locations lie on the boundary so that the transition kernel does not admit a density with respect to the Lebesgue measure on the whole state space, but only on the unit circle. For instance, the techniques developed by Azaïs (2014) for estimating the transition distribution do not apply in this context. Last but not least, the direct observation of the trajectory is problematic due to the wave-function collapse phenomenon and can, at most, be performed only once at a specific time managed by the experimenter, while usual observation schemes assume repeated observations of the trajectory, either at the jump times or on a regular temporal grid.

4.2. Problem formulation

Let us imagine that we have n_s copies of a quantum system S, each in contact with its own copy of the same Markovian environment E. Assume that we know some properties of S (in particular, we know its Hamiltonian), but we do not know its initial state $|\psi(0)\rangle$. We also do not know the state of E, which is assumed not to evolve in time compatibly with the Born–Markov approximation. We assume here that we are allowed to continuously monitor E, such that the trajectory of each single quantum system can be followed.

We wonder whether information on additional properties of S, on the initial state of its n_s copies and on the state of E can be inferred by looking at the system's trajectories. On top of the piecewise-deterministic trajectories, we allow a single direct measurement to be performed on each copy of S. This instantaneously stops the dynamics of S and hence destroys its trajectory, which means that after each such measurement a new copy of S must be considered and followed in time.

4.2.1. *Atom-field interaction*

We consider, in particular, our system S to be an atom[2] with two distinct energy levels only, a situation often encountered in practical applications in the fields of quantum information and quantum computation (Nielsen and Chuang 2011). Let $|0\rangle$ and $|1\rangle$ be these two energy levels, and be $\varepsilon_g = \varepsilon_1 - \varepsilon_0 > 0$ their energy gap. The system's Hamiltonian, expressed in the $\{|0\rangle, |1\rangle\}$ basis, simply reads

$$H_S = \begin{pmatrix} 0 & 0 \\ 0 & \varepsilon_g \end{pmatrix},$$

while most general expression for the system's state vector is

$$|\psi\rangle = c_0|0\rangle + c_1|1\rangle, \qquad [4.12]$$

2 By "atom", we mean here any quantum system whose internal structure we can model as having a discrete set of energy levels, for example, real atoms, quantum dots, superconducting qubits and polarization states of photons.

with $c_0, c_1 \in \mathbb{C}$ and $|c_0|^2 + |c_1|^2 = 1$. The (Markovian) environment E considered here is a thermal electromagnetic field, at the unknown temperature T. In a semiclassical treatment of the problem (Cohen-Tannoudji *et al.* 1989), in which S is considered a quantum system but E is treated classically, and in the weak-coupling limit, the atom interacts with the field through the dipole moment of its $|0\rangle \leftrightarrow |1\rangle$ transition. Such an interaction leads to either the absorption of a photon of energy ε_g from E (transition $|0\rangle \to |1\rangle$) or to the emission of an excitation of energy ε_g into E (transition $|1\rangle \to |0\rangle$). The atomic dynamics is given, on an ensemble level, by the master equation

$$\begin{aligned}\frac{\mathrm{d}\rho}{\mathrm{d}t} &= -\frac{i}{\hbar}[H_S, \rho] \\ &\quad + \gamma_1\left(\sigma^- \rho \sigma^+ - \frac{1}{2}\{\sigma^+\sigma^-, \rho\}\right) + \gamma_2\left(\sigma^+ \rho \sigma^- - \frac{1}{2}\{\sigma^-\sigma^+, \rho\}\right), \end{aligned} \quad [4.13]$$

which is of the form [4.5] with $L_1 = \sigma^- = |0\rangle\langle 1|$ describing the emission of a photon to E, $L_2 = \sigma^+ = (\sigma^-)^\dagger = |1\rangle\langle 0|$ the absorption of a photon from E with rates, respectively, $\gamma_1 = \big(1 + n(\varepsilon_g, T)\big)\gamma_0$ and $\gamma_2 = n(\varepsilon_g, T)\gamma_0$. Here, γ_0 is the atomic vacuum spontaneous emission rate and

$$n(\varepsilon, T) = \frac{1}{-1 + e^{\varepsilon/k_B T}} \quad [4.14]$$

is the average number of photons of energy ε in a field at temperature T, $k_B \simeq 1.38 \times 10^{-23}\,\mathrm{J\,K^{-1}}$ being Boltzmann's constant.

It is possible to solve equation [4.13] analytically, obtaining

$$\rho(t) = \begin{pmatrix} p_0(t) & K(t) \\ K^*(t) & 1 - p_0(t) \end{pmatrix},$$

with

$$\begin{aligned} p_0(t) &= p_0(0)e^{-(\gamma_1+\gamma_2)t} + \frac{\gamma_1}{\gamma_1+\gamma_2}\left(1 - e^{-(\gamma_1+\gamma_2)t}\right), \\ K(t) &= K(0)e^{\frac{i\varepsilon_g t}{\hbar}} e^{-\frac{\gamma_1+\gamma_2}{2}t} \end{aligned} \quad [4.15]$$

being, respectively, the $|0\rangle$-state ensemble population and the so-called quantum coherence between states $|0\rangle$ and $|1\rangle$.

4.2.2. *Piecewise-deterministic trajectories*

On the other hand, when continuously monitoring the field, the dynamics of S can be followed under the form of a PDP. In the specific example at hand, as a particular case of the general setting discussed in section 4.1.3, more than one possibility exists for monitoring the field. Here, we consider two possible schemes, equivalent, respectively, to the so-called *direct photodetection* and the so-called *homodyne photodetection* (Wiseman and Milburn 1993; Breuer and Petruccione 2002). In the first case, the jump operators are simply σ^- and σ^+ with associated rates γ_1 and γ_2, respectively, the same as in equation [4.13]. This configuration corresponds to directly probing the field for photons emitted/absorbed by the atom. In the second case, four jump operators are introduced as $L_e^\pm = \sigma^- \pm i\beta$ and $L_a^\pm = \sigma^+ \pm i\beta^* = (L_e^\mp)^\dagger$ ($\beta \in \mathbb{C}$), with rates $\gamma_e^\pm = \frac{\gamma_1}{2}$ and $\gamma_a^\pm = \frac{\gamma_2}{2}$. This is a transformation of the kind [4.11] (it is easy to see that it does not change the Hamiltonian of the system) and, as such, produces a valid set of Lindblad operators. These operators correspond to monitoring the field E after it interferes with another local electromagnetic field of amplitude β, which is not directly coupled to S.

Note that the "direct-photodetection-like" scheme is obtained from the "homodyne-photodetection-like" scheme in the limit $\beta \to 0$. We will hence hereafter focus on the homodyne strategy alone. The deterministic part of the system's trajectories is governed by the non-Hermitian Hamiltonian

$$\widehat{H} = \begin{pmatrix} -\hbar\frac{i}{2}\left[\gamma_1|\beta|^2 + \gamma_2(1+|\beta|^2)\right] & 0 \\ 0 & \varepsilon_g - \hbar\frac{i}{2}\left[\gamma_2|\beta|^2 + \gamma_1(1+|\beta|^2)\right] \end{pmatrix} \qquad [4.16]$$

while, using equations [4.9] and [4.12], the four possible jumps are

$$J_e^\pm : \; c_0|0\rangle + c_1|1\rangle \to \frac{1}{\sqrt{N_e^\pm}}\Big((c_1 \pm i\beta c_0)|0\rangle \pm i\beta c_1|1\rangle\Big), \qquad [4.17]$$

$$J_a^\pm : \; c_0|0\rangle + c_1|1\rangle \to \frac{1}{\sqrt{N_a^\pm}}\Big(\pm i\beta^* c_0|0\rangle + (c_0 \pm i\beta^* c_1)|1\rangle\Big), \qquad [4.18]$$

where $N_e^\pm$ and $N_a^\pm$ are normalization factors. Following equation [4.7], the associated probabilities are

$$p_e^\pm = \frac{\gamma_e^\pm N_e^\pm}{\gamma_e^+ N_e^+ + \gamma_e^- N_e^- + \gamma_a^+ N_a^+ + \gamma_a^- N_a^-}, \qquad [4.19]$$

$$p_a^\pm = \frac{\gamma_a^\pm N_a^\pm}{\gamma_e^+ N_e^+ + \gamma_e^- N_e^- + \gamma_a^+ N_a^+ + \gamma_a^- N_a^-}. \qquad [4.20]$$

Finally, given that the last jump has brought the system to the state $h_0|0\rangle + h_1|1\rangle$, the waiting-time distribution for the next jump is

$$f(\tau) = 1 - |h_0|^2 e^{-\tau\left[\gamma_1|\beta|^2 + \gamma_2(1+|\beta|^2)\right]} - |h_1|^2 e^{-\tau\left[\gamma_2|\beta|^2 + \gamma_1(1+|\beta|^2)\right]}. \qquad [4.21]$$

4.2.3. *Measures*

As discussed in section 4.1.3, a quantum trajectory (or PDP) is generated when the environment of an open quantum system is continuously monitored. This allows us, in principle, to detect the transitions induced by the system–environment interaction and to determine whether the system has "jumped" to a new quantum state.

On top of that, here we assume that the analyst is allowed to directly measure the quantum system only once per repetition of the PDP. By virtue of Postulate 4 of section 4.1.1, this radically interferes with the system's dynamics and projects the state vector onto one eigenvector of the measured observable. This is the reason why only one direct measurement is allowed during the PDP: after measuring, the PDP is fundamentally altered and no further information on the system's original dynamics can be extracted. Measuring properties of a single-system's wave function is a challenging but possible task (Lundeen *et al.* 2011; Hinds and Blatt 2012). The quantity we choose to measure is, for the ith trajectory in the state $|\psi_i(t_i)\rangle$ at time t_i,

$$Y^i_{t_i} = |\langle 0|\psi_i(t_i)\rangle|^2, \tag{4.22}$$

i.e. the amplitude of the projection of the state onto the atomic ground state. Note that, in light of equation [4.10],

$$\begin{aligned}\mathbb{E}[Y^i_{t_i}] &= \mathbb{E}[\langle 0|\psi_i(t_i)\rangle\langle\psi_i(t_i)|0\rangle] = \langle 0|\,\mathbb{E}\left[|\psi_i(t_i)\rangle\langle\psi_i(t_i)|\right]\,|0\rangle \\ &= \langle 0|\,\rho(t_i)\,|0\rangle = p_0(t_i).\end{aligned} \tag{4.23}$$

A specific example of distribution of the outcomes $Y^i_{t_i}$ on 10^2 repetitions of the atomic PDP is shown in the upper panel of Figure 4.1. The bottom panel of Figure 4.1 shows the right-hand side (crosses) and the left-hand side (solid line) of equation [4.23].

4.3. Estimation procedure

4.3.1. *Strategy*

We are interested in the estimation of the unknown parameters of the system, namely the initial condition $p_0(0)$ and the parameters γ_1 and γ_2. As aforementioned, we observe a dataset composed of independent measures $Y^i_{t_i}$, $1 \le i \le n$, where the time of measurement t_i can be chosen by the analyst. In this chapter, we propose applying an ordinary least squares estimation method that only takes advantage of the average behavior of the measures expressed in equation [4.23]. Precisely, the estimator $(\hat{p}_0(0), \hat{\gamma}_1, \hat{\gamma}_2)$ is expected to be obtained by minimizing the function

$$V(p_0(0), \gamma_1, \gamma_2) = \sum_{i=1}^{n} \left(Y^i_{t_i} - p_0(t_i)\right)^2,$$

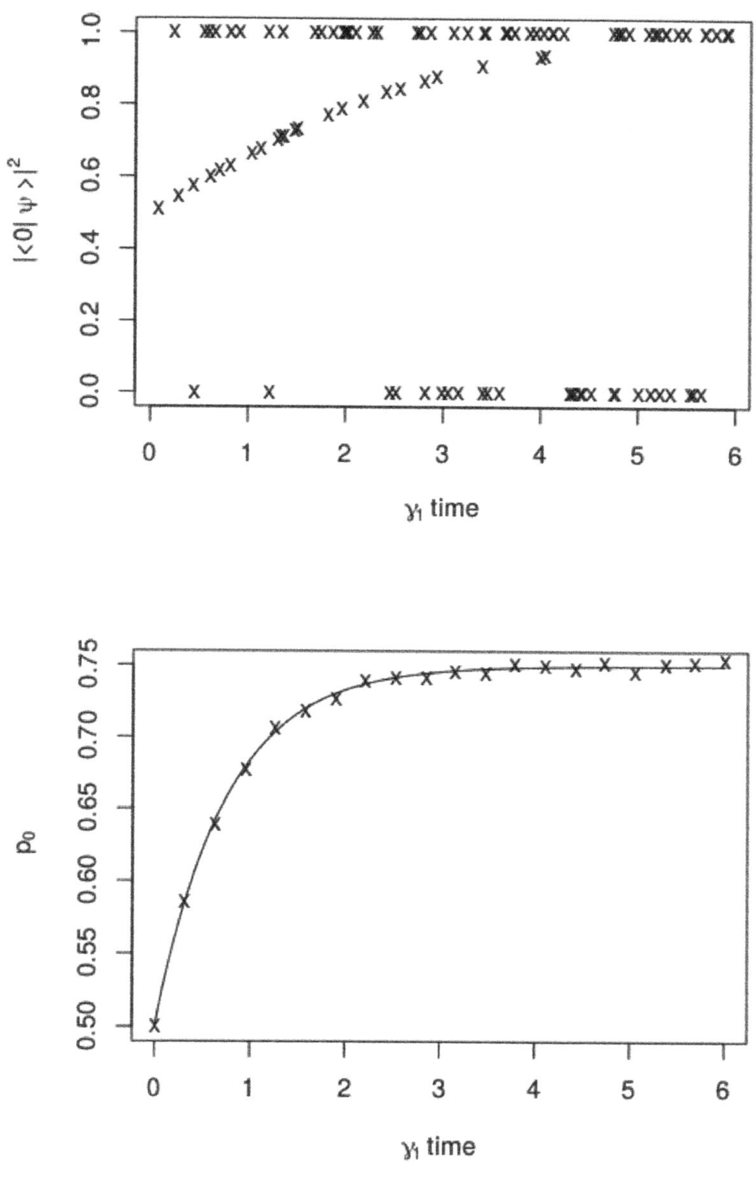

Figure 4.1. *Top: Distribution of measurement outcomes $Y^i_{t_i}$ (equation [4.23]) on 10^2 trajectories, all having the same initial state $\frac{1}{\sqrt{2}}(|0\rangle + |1\rangle)$ and measured at randomly chosen timepoints t_i. Bottom: Average value [4.23]. Each point at time t_x corresponds to the average of $|\langle 0|\psi_i(t_x)\rangle|^2$ over 10^4 trajectories, all having the same initial state $\frac{1}{\sqrt{2}}(|0\rangle + |1\rangle)$. The solid line is the analytical solution [4.15] of the ensemble dynamics. Other system parameters are $\varepsilon_g = 10^{14}$ rad $\cdot$ s^{-1} and $\beta = 0$*

where the dependency of p_0 on the three parameters is implicit. The expression of p_0 being rather complex, and in order to avoid gradient algorithms for minimizing the sum of squares function, we exploit the fact that the experimenter can manage the times of measurement to simplify the equations. First, as t goes to infinity,

$$p_0(t) \to \frac{\gamma_1}{\gamma_1 + \gamma_2}.$$

In addition, expanding around $t = 0$, we have

$$p_0(t) \sim p_0(0) + t\left[\gamma_1 - p_0(0)(\gamma_1 + \gamma_2)\right].$$

Consequently, if the $n/2$ first measurements have been made in long time while the $n/2$ last measurements have been made in short time, the parameters can be estimated by simultaneously minimizing the approximated functions

$$V^l(\gamma_1, \gamma_2) = \sum_{i \leq n/2} \left(Y^i_{t_i} - \frac{\gamma_1}{\gamma_1 + \gamma_2} \right)^2, \qquad [4.24]$$

$$V^s(p_0(0), \gamma_1, \gamma_2) = \sum_{i > n/2} \left(Y^i_{t_i} - p_0(0) - t_i\left[\gamma_1 - p_0(0)(\gamma_1 + \gamma_2)\right] \right)^2, \qquad [4.25]$$

where V^l does not depend on the initial value $p_0(0)$.

4.3.2. *Least-square estimators*

It should be noticed that [4.24] and [4.25] can be easily expressed in terms of $p_0(0)$ and

$$u_1 = \frac{\gamma_1}{\gamma_1 + \gamma_2}, \qquad [4.26]$$

$$u_2 = \gamma_1 - p_0(0)(\gamma_1 + \gamma_2). \qquad [4.27]$$

With this reparametrization, we have

$$V^l(u_1) = \sum_{i \leq n/2} \left(Y^i_{t_i} - u_1 \right)^2,$$

$$V^s(p_0(0), u_2) = \sum_{i > n/2} \left(Y^i_{t_i} - p_0(0) - t_i u_2 \right)^2.$$

The main advantage for these new equations is that the three variables have been decoupled from each other. It is obvious to see that $\hat{u}_1$ defined as the average value of the $Y^i_{t_i}$ in long time, i.e.

$$\hat{u}_1 = \frac{2}{n} \sum_{i \leq n/2} Y^i_{t_i},$$

minimizes V^l. In addition, $V^s(p_0(0), u_2)$ is the cost function associated with classical linear regression, with $p_0(0)$ as the intercept and u_2 as the slope. The well-known solution of this minimization problem is expressed in terms of empirical mean, variance and covariance of the measures in short time (Yan and Su 2009, Chapter 2),

$$\widehat{u}_2 = \frac{\sum_{i>n/2}\left(t_i - \frac{2}{n}\sum_{i>n/2} t_i\right)\left(Y^i_{t_i} - \frac{2}{n}\sum_{i>n/2} Y^i_{t_i}\right)}{\sum_{i>n/2}\left(t_i - \frac{2}{n}\sum_{i>n/2} t_i\right)^2},$$

$$\hat{p}_0(0) = \frac{2}{n}\sum_{i>n/2} Y^i_{t_i} - \widehat{u}_2 t_i. \qquad [4.28]$$

The reparametrization [4.26] and [4.27] can be inverted and we get the following estimators of γ_1 and γ_2:

$$\hat{\gamma}_1 = \frac{\hat{u}_1\hat{u}_2}{\hat{u}_1 - \hat{p}_0(0)}, \qquad [4.29]$$

$$\hat{\gamma}_2 = \frac{(1-\hat{u}_1)\hat{u}_2}{\hat{u}_1 - \hat{p}_0(0)}. \qquad [4.30]$$

4.3.3. *Numerical experiments*

To test the performance of the estimators [4.28], [4.29] and [4.30], we performed several numerical experiments by simulating sets of n trajectories according to equations [4.16] and [4.21], having chosen the atomic transition energy $\varepsilon_g = 10^{14}\,\text{rad}\cdot\text{s}^{-1}$, a value falling into the infrared spectrum, following the high interest arising nowadays on technologies based on infrared quantum dots Lu *et al.* (2019). The initial state of each trajectory is $|\psi_{\text{init}}\rangle = \frac{1}{\sqrt{2}}(|0\rangle + |1\rangle)$ and we fixed $\gamma_1 = 0.3\,\text{s}^{-1}$, $\gamma_2 = 0.1\,\text{s}^{-1}$.

As a first set of numerical experiments, we fixed the value $\beta = 0$, thus limiting ourselves to the case of direct photodetection, and simulated four groups of 10^2 sets of n trajectories, with $n = 50, 100, 500, 1000$. For each set of n trajectories, the procedure of sections 4.2.3 and 4.3.2 was applied to produce the estimation of the three parameters $p_0(0)$, γ_1 and γ_2. Figure 4.2 shows the distribution of the 10^2 estimator values for each value of n (top: $p_0(0)$; middle: γ_1; bottom: γ_2). As clearly seen, the performance of the estimators [4.28], [4.29] and [4.30] increases significantly with increasing n, while the number of outliers concurrently decreases.

The average value of each estimator is in very good agreement with the real value of the corresponding parameter (horizontal blue line in each panel).

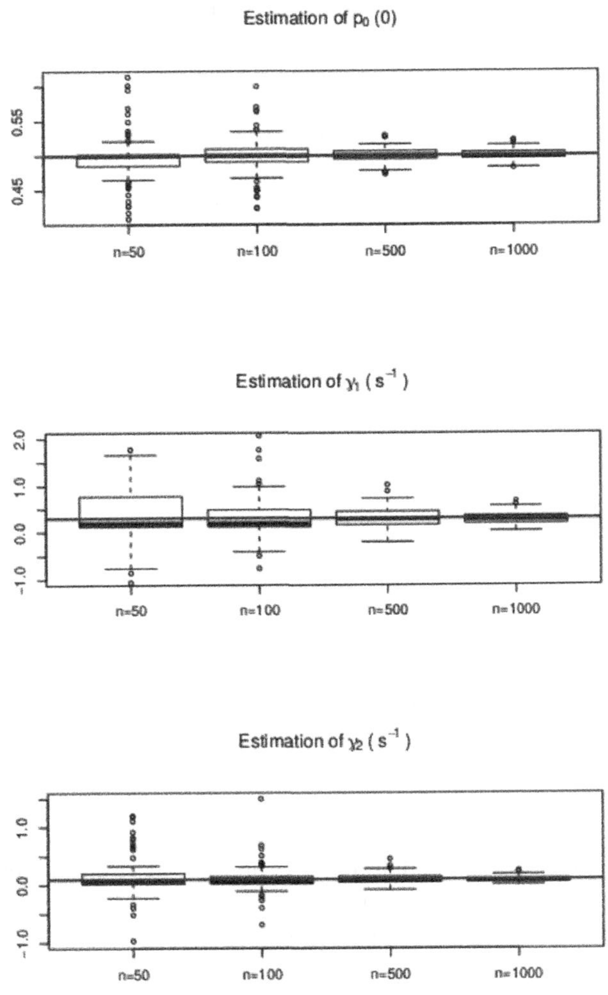

Figure 4.2. *Performance of estimators $\hat{p}_0(0)$ (equation [4.28], top panel), $\hat{\gamma}_1$ (equation [4.29], middle panel) and $\hat{\gamma}_2$ (equation [4.30], bottom panel), against the number n of simulated trajectories. Each boxplot gives the distribution of estimator values over 100 repetitions of n trajectories. All trajectories have the same initial state all having the same initial state $\frac{1}{\sqrt{2}}(|0\rangle + |1\rangle)$. Blue horizontal lines in each panel give the actual parameter value ($p_0(0) = 0.5$, $\gamma_1 = 0.3$, $\gamma_2 = 0.1$). Other system parameters are $\varepsilon_g = 10^{14}$ rad $\cdot$ s^{-1} and $\beta = 0$. For a color version of this figure, see www.iste.co.uk/barbu/data.zip*

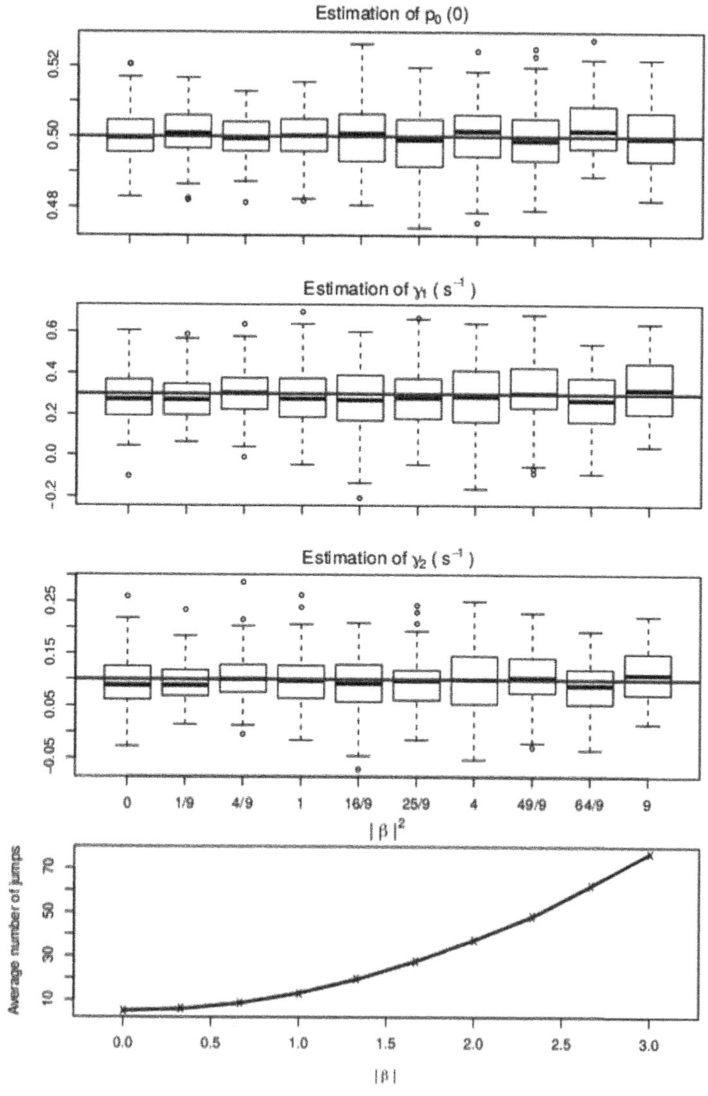

Figure 4.3. *Performance of estimators* $\hat{p}_0(0)$ *(equation [4.28], top panel),* $\hat{\gamma}_1$ *(equation [4.29], top-middle panel) and* $\hat{\gamma}_2$ *(equation [4.30], bottom-middle panel), against variations of* $|\beta|$*, which has a major effect on the average number of jumps per trajectory (bottom panel). Each boxplot gives the distribution of estimator values over 100 repetitions of* 10^3 *trajectories, all having the same initial state* $\frac{1}{\sqrt{2}}(|0\rangle + |1\rangle)$*. Blue horizontal lines in each panel give the actual parameter value (*$p_0(0) = 0.5$*,* $\gamma_1 = 0.3$*,* $\gamma_2 = 0.1$*). The other system parameter is* $\varepsilon_g = 10^{14}$ *rad·s*$^{-1}$*. For a color version of this figure, see www.iste.co.uk/barbu/data.zip*

As a second set of experiments, we investigated if and how different strategies for environmental measurements have an impact on the estimator performances. In particular, in the set of simulations shown in Figure 4.3, we fixed the number of trajectories for each estimation to $n = 10^3$ and tuned the value of $\beta \in \mathbb{R}$ in the range $[0,3]$. For each β-value, 10^2 groups of 10^3 trajectories were simulated, and estimators were calculated for each group. The boxplot distributions shown in the three uppermost panels of Figure 4.3, for respectively $p_0(0)$, γ_1 and γ_2, highlight the fact that the performance of the estimators does not seem to depend on the chosen measuring strategy. This is in spite of the fact that the structure of the simulated trajectories strongly depends on such strategy, as shown in the bottom panel of Figure 4.3 giving, for each simulated value of β, the average number of jumps in the corresponding set of trajectories, which increases quadratically with $|\beta|$.

4.4. Physical interpretation

The two parameters γ_1 and γ_2, whose value is estimated through the procedure presented in section 4.3.2, contain important information on physical properties of both the atom and the electromagnetic field. Their estimation can be thus used to access unknown physical properties of both systems. Recall that $\gamma_1 = \big(1 + n(\varepsilon_g, T)\big)\gamma_0$ and $\gamma_2 = n(\varepsilon_g, T)\gamma_0$, with the average photon number $n(\varepsilon_g, T)$ given by equation [4.14] and the atomic vacuum spontaneous emission rate

$$\gamma_0 = \frac{|\mathbf{d}|^2 \varepsilon_g^3}{3\hbar^4 \pi \epsilon_0 c^3}, \qquad [4.31]$$

where $\epsilon_0 \simeq 8.85 \times 10^{-12}\,\mathrm{F\,m^{-1}}$ is the vacuum permittivity, $c \simeq 2.998 \times 10^8\,\mathrm{m\,s^{-1}}$ is the speed of light in vacuum and $\mathbf{d}$ is the dipole moment vector of the atomic transition. This means, assuming the knowledge of the energy gap ε_g of the atomic transition, one has access to the temperature T of the electromagnetic field as

$$T = \frac{\varepsilon_g}{k_B \ln \frac{\gamma_1}{\gamma_2}}. \qquad [4.32]$$

Estimating γ_1 and γ_2 through atomic trajectories means that one can use the atom to probe the field temperature thus or, in other words, that the atom can be used as a field thermometer.

In addition, atomic properties can also be inferred. In particular, using equations [4.14], [4.31] and [4.32] together with the definition of γ_1 and γ_2 one obtains

$$|\mathbf{d}|^2 = (\gamma_1 - \gamma_2)\frac{3\hbar^4 \pi \epsilon_0 c^3}{\varepsilon_g^3}, \qquad [4.33]$$

which allows us to estimate the square norm of the atomic dipole through observations of the atomic PDP. This is exemplified in Figure 4.4, showing the estimation of both T and $|\mathbf{d}|$ through the estimators for γ_1 and γ_2.

For $\gamma_1 = 0.3\,\mathrm{s}^{-1}$, $\gamma_2 = 0.1\,\mathrm{s}^{-1}$ and $\varepsilon_g = 10^{14}\,\mathrm{rad}\cdot\mathrm{s}^{-1}$, one obtains $|\mathbf{d}| \sim 2\times 10^{-31}\,\mathrm{C}\cdot\mathrm{m}$ and $T \sim 700\,\mathrm{K}$, values which correspond to the blue horizontal lines in Figure 4.4. The good agreement shown in Figure 4.4 between the estimations and the real values of T and $|\mathbf{d}|$ shows that the procedure leading to the estimators [4.29] and [4.30] can indeed be used to probe unknown physical properties of both the atom and the electromagnetic field.

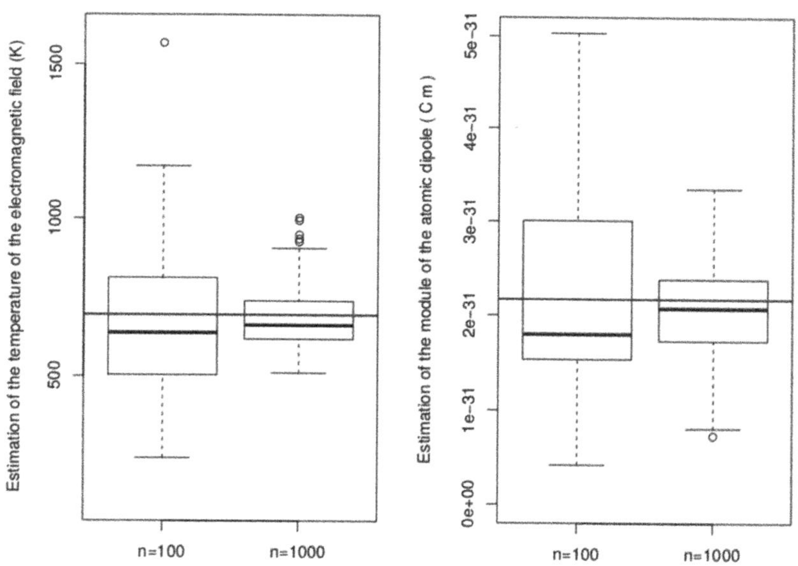

Figure 4.4. *Estimation of the environmental temperature T (equation [4.32], left panel) and the module of the atomic transition dipole (equation [4.33], right panel) for 100 repetitions of either 10^2 or 10^3 trajectories, for the same parameter values as in Figure 4.2. For comparison, the typical order of magnitude of the dipole moment of the hydrogen atom is 10^{-30} C · m. For a color version of this figure, see www.iste.co.uk/barbu/data.zip*

4.5. Concluding remarks

In this chapter, we have applied statistical parameter-estimation techniques to a specific class of Markovian PDPs encountered in quantum systems. When the dynamics of a quantum system is monitored continuously through measurements performed on a larger environment the system interacts with, its evolution becomes stochastic and can be described as a Markovian PDP. By applying standard least-square methods to simulated data of stochastic evolution of a two-level atom interacting with a thermal electromagnetic radiation, we have shown that important physical parameters such as the initial atomic state, the atomic dipole moment and the field temperature can be estimated.

4.6. References

Abdessalem, A.B., Azaïs, R., Touzet-Cortina, M., Gégout-Petit, A., and Puiggali, M. (2016). Stochastic modelling and prediction of fatigue crack propagation using piecewise-deterministic Markov processes. *Proceedings of the Institution of Mechanical Engineers, Part O: Journal of Risk and Reliability*, 230(4), 405–416.

Azaïs, R. (2014). A recursive nonparametric estimator for the transition kernel of a piecewise-deterministic Markov process. *ESAIM: Probability and Statistics*, 18, 726–749.

Azaïs, R., and Muller-Gueudin, A. (2016). Optimal choice among a class of nonparametric estimators of the jump rate for piecewise-deterministic Markov processes. *Electron. J. Statist.*, 10(2), 3648–3692.

Azaïs, R., Dufour, F., and Gégout-Petit, A. (2014). Non-parametric estimation of the conditional distribution of the interjumping times for piecewise-deterministic Markov processes. *Scandinavian Journal of Statistics*, 41(4), 950–969.

Bera, M.N., Acín, A., Ku, M., Mitchell, M.W., and Lewenstein, M. (2017). Randomness in quantum mechanics: Philosophy, physics and technology. *Reports on Progress in Physics*, 80(12), 124001.

Bertail, P., Clemençon, S., and Tressou, J. (2008). A storage model with random release rate for modelling exposure to food contaminants. *Mathematical Bioscience and Engineering*, 5(1), 35–60.

Breuer, H.-P., and Petruccione, F. (1995). Reduced system dynamics as a stochastic process in Hilbert space. *Phys. Rev. Lett.*, 74, 3788–3791.

Breuer, H.P., and Petruccione, F. (2002). *The Theory of Open Quantum Systems*. Oxford University Press, Oxford.

Brunner, N., Cavalcanti, D., Pironio, S., Scarani, V., and Wehner, S. (2014). Bell nonlocality. *Review of Modern Physics*, 86, 419–478.

Buckwar, E., and Riedler, M.G. (2011). An exact stochastic hybrid model of excitable membranes including spatio-temporal evolution. *J. Math. Biol.*, 63(6), 1051–1093.

Cohen-Tannoudji, C., Dupont-Roc, J., and Grynberg, G. (1989). *Photons and Atoms: Introduction to Quantum Electrodynamics*. John Wiley & Sons, New York.

Dalibard, J., Castin, Y., and Mølmer, K. (1992). Wave-function approach to dissipative processes in quantum optics. *Phys. Rev. Lett.*, 68, 580–583.

Doumic, M., Hoffmann, M., Krell, N., and Robert, L. (2015). Statistical estimation of a growth-fragmentation model observed on a genealogical tree. *Bernoulli*, 21(3), 1760–1799.

Doumic, M., Hoffmann, M., Reynaud-Bouret, P., and Rivoirard, V. (2011). Nonparametric estimation of the division rate of a size-structured population. *SIAM J. Numer. Anal.*, 50(2), 925–950.

Dum, R., Zoller, P., and Ritsch, H. (1992). Monte Carlo simulation of the atomic master equation for spontaneous emission. *Phys. Rev. A*, 45, 4879–4887.

Hekking, F. W.J., and Pekola, J.P. (2013). Quantum jump approach for work and dissipation in a two-level system. *Phys. Rev. Lett.*, 111, 093602. Available at: https://link.aps.org/doi/10.1103/ PhysRevLett.111.093602.

Hinds, E. and Blatt, R. (2012). Manipulating individual quantum systems. *Nature*, 492, 55.

Horodecki, R., Horodecki, P., Horodecki, M., and Horodecki, K. (2009). Quantum entanglement. *Rev. Mod. Phys.*, 81, 865–942.

Jaekel, M.-T., and Reynaud, S. (1997). Movement and fluctuations of the vacuum. *Reports on Progress in Physics*, 60(9), 863–887.

Kovacevic, R.M., and Pflug, G.C. (2011). Does insurance help to escape the poverty trap?—A ruin theoretic approach. *J. Risk Insurance*, 78(4), 1003–1027.

Leggio, B., Napoli, A., Messina, A., and Breuer, H.-P. (2013). Entropy production and information fluctuations along quantum trajectories. *Phys. Rev. A*, 88, 042111.

Lu, H., Carroll, G.M., Neale, N.R., and Beard, M.C. (2019). Infrared quantum dots: Progress, challenges, and opportunities. *ACS Nano*, 13(2), 939–953.

Lundeen, J.S., Sutherland, B.R., Patel, A., Stewart, C., and Bamber, C.J. (2011). Direct measurement of the quantum wavefunction. *Nature*, 474, 188–191.

Nielsen, M.A., and Chuang, I.L. (2011). *Quantum Computation and Quantum Information, 10th Anniversary Edition*. Cambridge University Press, Cambridge.

Plenio, M.B., and Knight, P.L. (1998). The quantum-jump approach to dissipative dynamics in quantum optics. *Rev. Mod. Phys.*, 70, 101–144.

Rotter, I., and Bird, J.P. (2015). A review of progress in the physics of open quantum systems: Theory and experiment. *Reports on Progress in Physics*, 78(11), 114001.

Von Neumann, J., and Wheeler, N.A. (2018). *Mathematical Foundations of Quantum Mechanics*. Princeton University Press, Princeton.

Wheeler, J.A., and Zurek, W.H. (1986). Quantum theory and measurement. *Synthese*, 67(3), 527–530.

Wiseman, H.M. (1994). Quantum theory of continuous feedback. *Phys. Rev. A*, 49, 2133–2150.

Wiseman, H.M. (1996). Quantum trajectories and quantum measurement theory. *Quantum and Semiclassical Optics: Journal of the European Optical Society Part B*, 8(1), 205–222.

Wiseman, H.M., and Milburn, G.J. (1993). Quantum theory of optical feedback via homodyne detection. *Phys. Rev. Lett.*, 70, 548–551.

Wiseman, H.M., and Milburn, G.J. (2009). *Quantum Measurement and Control*. Cambridge University Press, Cambridge.

Yan, X., and Su, X.G. (2009). *Linear Regression Analysis: Theory and Computing*. World Scientific Publishing Co., Singapore.

5

Identification of Patterns in a Semi-Markov Chain

Many models have been proposed aiming to identify a pattern in a sequence of symbols under the assumption that the sequence of symbols is described by a Markov chain. The sojourn time in a state in a Markov chain must be dictated by the geometric distribution in a discrete process. This is the main disadvantage of considering Markov chains in real applications. By contrast, semi-Markov processes (SMPs) provide a generalization of their Markov counterparts where the sojourn time in a state can be governed by any distribution function. The aim of this chapter is to provide the probability that a pattern occurs for the first time after n symbols in a semi-Markov sequence. Finally, in order to illustrate our model, we present an example concerning the DNA sequence of bacteriophage the concerned pattern, the SmaI enzyme.

5.1. Introduction

Finding a specific pattern through a sequence of symbols is useful in many areas, for instance in databases it could be the keyword for certain research; in biology it can make reference to a particular illness, in communication it could be a precise word from a conversation, etc. Seeking a particular pattern over a chain formed by a big quantity of symbols is a very cumbersome task and in most cases it cannot be achieved by simple inspection. There are a number of models and algorithms that propose an automatic treatment for the searching problem in a few seconds (see, for example, Aboluion 2011; Srivastava and Baptista 2016; Stefanov *et al.* 2011; Picard *et al.* 2011; Avena *et al.* 2013; Codish *et al.* 2017; Montemanni 2015). The problem of pattern matching has been studied using different approaches. To name a couple, we can consider combinatorial methods (see Guibas and Odlyzko 1981) and

Chapter written by Brenda Ivette GARCIA-MAYA and Nikolaos LIMNIOS.

probabilistic techniques. In probabilistic models, the efficiency and precision depend chiefly on the kind of dependency between symbols. For example, Stefanov *et al.* (1997) and Chadjiconstantinidis *et al.* (2000) treated the problem under the hypothesis that the sequence of symbols is derived from independently and identically distributed Bernoulli trials; Fu and Koutras (1994) and Robin and Daudin (1999) considered that the sequence of symbols is modeled by Bernoulli variables with same and different probability. Nevertheless, more complex hypotheses have been examined, for instance Antzoulakos (2001) and Fu and Chang (2002) proposed an algorithm based on an embedded Markov chain, Glaz *et al.* (2006) created a method that uses gambling teams and martingales in a two state Markov chain, Crochemore and Stefanov (2003) and Nuel (2008) treated the problem using string-matching algorithms based on automatic searching, Touyar *et al.* (2008) used Poisson approximations, etc. Even if Markov discrete processes better describe a discrete chain of symbols than Bernoulli trials, the main drawback in the Markov hypothesis is that it cannot take into account general distributions in the sojourn time in a state. The sojourn time in a state must be governed by the geometric distribution (in a discrete chain and by the exponential distribution in a continuous process); in contrast, discrete-time SMPs generalize the Markov hypothesis and they allow the distribution function in a state to be any of them.

In this chapter, we give a model to compute the probability that a pattern appears for the first time in a sequence after k symbols, in other words we compute the first hitting position of a pattern through a sequence. We also give some important properties of the pattern, like the number of times that it appears through the sequence, i.e. the occurrence rate of the pattern and its mean waiting position. As it has been said, the semi-Markov hypothesis fits better than the Markov one modeling a sequence of symbols, therefore this hypothesis will be taken into account. To obtain an efficient algorithm, we use the prefix set. The ensemble of prefixes is a group of symbols, which are part of the root of a pattern. To give an example, let us consider the pattern aba; if we construct the pattern step by step from its first symbol to its last one, the elements of this construction are called prefixes, i.e. the prefix set of aba is $\{a, ab, aba\}$ (more details about prefixes will be given in section 5.2). To provide an example of the problem tackled, consider the following sequence:

bbabbbabababaabababaaaaaabbbbbbbbaaaaabbbaababababababababab...

and suppose we want to find the pattern aba. For this string of letters, it is clear that the pattern appears for the first time at position 8 (starting from zero). This is what our algorithm performs, it gives an easy way to compute the probability that a pattern will be found for the first time at position k. Using the chain of the longest prefix of a pattern and all possible backwards times for each prefix (the time we have spent in the last letter of a prefix since the last state change, see section 5.2 for more details), it computes the distribution for the first hitting position to the pattern through the sequence. To show one of the applications of the proposed model, we test it in a bacteriophage DNA sequence where we search the pattern $CCCGGG$, which corresponds to the enzyme *SmaI*. We present the pattern occurrence rate, the distribution function, the expected value, the variance and the standard deviation of the random variable that represent the first hitting position to the pattern.

The chapter is organized as follows: in section 5.2, we introduce the prefixes chain. In section 5.3, we present the semi-Markov setting. In section 5.4, we introduce the (first) hitting position to the pattern. In section 5.5, we present a genomic application. In section 5.6, we give some concluding remarks.

5.2. The prefix chain

Suppose that the pattern and the sequence are taken from a finite set of symbols $\mathcal{A} = \{a_1, ..., a_l\}$, $2 \leq l < \infty$. To make things easier, the pattern will be denoted by the letter $\mathbf{w}$. Every $\mathbf{w}$ symbol can be represented in the following way:

$\mathbf{w} := w_1 w_2 \cdots w_h$ where $w_1, w_2, \cdots, w_h \in \mathcal{A}$ (see Lothaire 1983). The length of $\mathbf{w}$ is the number of symbols that forms the pattern and it is denoted by the symbol $|\mathbf{w}|$, observe that $h = |\mathbf{w}|$, for instance if the pattern is $\mathbf{w} = bbb$ and its length is $h = 3$. We consider that h takes values in the set $\mathbb{N}^* := \{1, 2, ...\}$. The prefixes set of $\mathbf{w}$ consist of writing $\mathbf{w}$ symbol by symbol from its first one until the pattern is completed. The prefixes set is denoted by E and its elements are $E = \{\varepsilon, w_1, w_1 w_2, ..., w_1 w_2 \cdots w_{h-1}, w\}$. The symbol ε expresses the empty prefix, which is used in case none of the symbols $\{w_1, w_1 w_2, ..., w_1 w_2 \cdots w_{h-1}, \mathbf{w}\}$ appear in the sequence. Observe that ε has no symbol, therefore $|\varepsilon| = 0$.

A sequence of prefixes that helps to find $\mathbf{w}$ through the sequence of symbols can be constructed (see, for example, Crochemore and Stefanov 2003; Nuel 2008). To establish this sequence, let us consider the following variables. Let $\delta : E \times \mathcal{A} \to E$ be a mapping such that for $p' \in E$ and $a \in \mathcal{A}$, $\delta(p', a)$ is the longest suffix of $p'a$ (concatenation of p' and a) in the prefix set E. To clarify this definition, observe the following example.

EXAMPLE 5.1.– If $\mathcal{A} = \{a, b\}$, $\mathbf{w} = ba$, the prefix set is $E = \{\varepsilon, b, ba\}$, then

$$\begin{aligned} &\delta(\varepsilon, a) = \varepsilon, \quad \delta(\varepsilon, b) = b, \\ &\delta(b, a) = ba, \ \delta(b, b) = b, \\ &\delta(ba, a) = \varepsilon, \ \delta(ba, b) = b. \end{aligned} \qquad [5.1]$$

The partial inverse of δ is defined by the function $\delta^{-1} : E \longrightarrow \mathcal{A}$ in the following way:

$$\delta^{-1}(p') = a \in \mathcal{A} \text{ where there exist a } p \in E \text{ such that } \delta(p, a) = p'. \qquad [5.2]$$

Roughly speaking, the partial inverse of prefix p' gives the last symbol of p', i.e. if $p' = ba$, $\delta^{-1}(p') = a$.

Before talking about the semi-Markov chain of symbols, let us introduce the essential background of semi-Markov discrete processes.

5.3. The semi-Markov setting

Let us consider a complete probability space $(\Omega, \mathcal{F}, \mathbb{P})$ on which we define all processes and random variables. Consider a stochastic chain $(Z_k)_{k\in\mathbb{N}}$ that takes values in the state space $\mathcal{A}$. Let process $(S_n)_{n\in\mathbb{N}}$ be the successive times points when a state changes in (Z_k) called renewal points or jump times points of process (Z_k). It is formally defined by $S_0 := 0$ and

$$S_{n+1} := \inf\{k > S_n : Z_k \neq Z_{S_n}\}, \quad n \geq 0, \qquad [5.3]$$

with the convention $\inf \emptyset = +\infty$. The embedded chain that records (Z_k) at renewal points (S_n) is denoted: $(J_n)_{n\in\mathbb{N}}$, i.e. $J_n = Z_{S_n}$. The sojourn time in a state is the time passed consecutively in a state since the last jump. It is denoted $(X_n)_{n\geq 0}$ where

$$X_0 := S_0 = 0 \text{ and } X_{n+1} := S_{n+1} - S_n, \quad n \in \mathbb{N}^*.$$

The random variables (X_n) are also called *waiting times* because they represent the time we have to wait until another renewal point. The random variable which counts the number of jumps (renewal points) in process (Z_k) until certain time k is $N(k)$,

$$N(k) := \max\{n \in \mathbb{N} : S_n \leq k\}.$$

Let i, j be two elements of $\mathcal{A}$, then if for any $k, n \in \mathbb{N}$, the following equation holds true:

$$\begin{aligned}&\mathbb{P}(J_{n+1} = j, S_{n+1} - S_n = k \mid J_0, ..., J_n = i; S_1, ..., S_n)\\&= \mathbb{P}(J_{n+1} = j, S_{n+1} - S_n = k \mid J_n = i)\end{aligned}$$

The stochastic chain (Z_k) is called SMP and (J_n, S_n) is its embedded Markov renewal chain (see, for example, Barbu *et al.* 2004). The semi-Markov kernel is defined as follows:

$$q_{ij}(k) := \mathbb{P}(J_{n+1} = j, S_{n+1} - S_n = k \mid J_n = i), \quad n \geq 0, k \in \mathbb{N}. \qquad [5.4]$$

It represents the probability that chain (Z_k) passes from state i at jump n to state j at jump $n + 1$ and it states k units of time i before changing. The cumulative semi-Markov kernel is defined by

$$Q_{ij}(k) := \mathbb{P}(J_{n+1}, X_{n+1} \leq k \mid J_n = i) = \sum_{l=0}^{k} q_{ij}(l), \quad i, j \in \mathcal{A}, k \in \mathbb{N}.$$

It is worth noting that the semi-Markov kernel considered here is independent of n, which means that the semi-Markov chain is homogeneous in time. Also observe that we use index $k \in \mathbb{N}$ for the calendar points, and index $n \in \mathbb{N}$ for the number of jumps of (Z_k). If the semi-Markov kernel is time homogeneous, we can easily see that (J_n) is a homogeneous Markov chain. We denote by p_{ij}, $i, j \in \mathcal{A}$ its transition probability, defined by

$$p_{ij} := \mathbb{P}(J_{n+1} = j \mid J_n = i), \quad i, j \in \mathcal{A}, \quad n \in \mathbb{N}.$$

Note also that p_{ij} can be expressed in terms of the semi-Markov kernel by $p_{ij} = \sum_{k=0}^{\infty} q_{ij}(k)$. The cumulative distribution function of the sojourn time in state $i \in \mathcal{A}$ is

$$H_i(k) := \sum_{l=0}^{k} \sum_{j \in \mathcal{A}} q_{ij}(l).$$

It is possible to define the distribution function of the waiting time in each state i given the next state j. This function is the conditional cumulative distribution of $X_{n+1}, n \in \mathbb{N}$

$$F_{ij}(k) := \mathbb{P}(X_{n+1} \leq k \mid J_n = i, J_{n+1} = j) = \begin{cases} \frac{Q_{ij}(k)}{p_{ij}}, & \text{if } p_{ij} \neq 0, \\ 1, & \text{if } p_{ij} = 0. \end{cases} \qquad [5.5]$$

The main difference between Markov and semi-Markov discrete time processes is the distribution function $F_{ij}(k)$. In a Markov chain, this function has to be geometric with parameter $1 - P(i, i)$, where P is the transition probability matrix; however, in the SMP the distribution function $F_{ij}(k)$ can be any type (see, for example, Limnios and Oprişan 2001; Barbu and Limnios 2008).

The maximum sojourn time in a state $i \in \mathcal{A}$ through process (Z_k) is the largest amount of consecutive time passed in i, and it is denoted as

$$r_i := \sup\{k \geq 0 : \sum_{j \in \mathcal{A}} q_{ij}(k) \neq 0\}. \qquad [5.6]$$

The backward process $(U_k)_{k \in \mathbb{N}}$ of a semi-Markov chain (Z_k) is the time passed in Z_k since the last jump $N(k)$. It is defined as follows:

$$U_k := k - \max\{S_m : S_m \leq k\} \text{ and } U_k := k \text{ if } S_1 > k. \qquad [5.7]$$

EXAMPLE 5.2.– If Z : *bbabaa...* is a sample of (Z_k), its backward recurrence time (U_k) is:

$$U_0 = 0, U_1 = 1, U_2 = 0, U_3 = 0, U_4 = 0, U_5 = 1, \ldots$$

Note that if k coincides with a renewal point, the value of U_k is zero. After a renewal point, U_k grows one by one until another renewal point, and so on. If r_i is the maximum sojourn time in state i, see equation [5.6], then the maximum value for the backward time in state $i \in \mathcal{A}$, is

$$l_i := r_i - 1. \qquad [5.8]$$

Even if it has been proven that semi-Markov chains can be transformed in a Markov discrete process using the backward times process (see, for example, Limnios and Oprişan 2001), the models made with these transformations can almost never be implemented because the resulting state space becomes huge (see, for example, Mode and Pickens 1988). For this reason, we shall use prefixes and its corresponding backward time for each prefix to compute the first hitting time to a pattern through a sequence. Once we have defined the essential background on the semi-Markov chain, we are ready to introduce the embedded prefixes sequence. The embedded prefixes sequence was introduced by Stefanov *et al.* (1997) to propose a methodology for deriving explicit expressions for occurrences of patterns in binary strings. This technique has been applied in lots of models (see, for example, Hopcroft *et al.* 2013; Stefanov *et al.* 2011, etc.). In this chapter, we are interested in finding a pattern w in the chain of symbols (Z_k), which is described by an SMP with state space $\mathcal{A}$. The embedded prefix chain of the pattern is defined as follows:

DEFINITION 5.1.– *The prefix chain of* $\mathbf{w} = w_1 \cdots w_h$ *defined in* E *is denoted by* $Y := (Y_k)_{k \in \mathbb{N}}$, *where*

$$Y_0 := \begin{cases} w_1 & \text{if } Z_0 = w_1, \\ \varepsilon & \text{if } Z_0 \neq w_1, \end{cases} \qquad [5.9]$$

and

$$Y_k := \delta(Y_{k-1}, Z_k), \quad k \geq 1. \qquad [5.10]$$

EXAMPLE 5.3.– If $\mathcal{A} = \{a, b, c\}$ and $\mathbf{w} = abc$, then the elements of E are listed as follows: $E = \{\varepsilon = 0, a = 1, ab = 2, abc = 3\}$. For the following sample of Z

Z : *bacababacabc*...

we have as result for Y

Y : 010121210123...

EXAMPLE 5.4.– If $\mathcal{A} = \{a, b, c\}$ and $\mathbf{w} = bbac$, then the elements of E are listed as follows: $E = \{\varepsilon = 0, b = 1, bb = 2, bba = 3, bbac = 4\}$. For the sample of Z

Z : *aabbbcbabbac*...

we have as result for Y

Y : 001220101234...

Observe that the number of positions that we have to wait for for an occurrence of w in (Z_k) corresponds to the number of positions we have to wait for for an occurrence of $\mathbf{w}$ in (Y_k) (see Nuel 2008). Therefore, the first hitting position (time) of $\mathbf{w}$ in sequence (Z_k) will be computed using the prefix sequence (Y_k). To achieve this goal, let us introduce the blocks of the pattern. If $\mathbf{w}$ is a pattern formed only by one symbol repeated h times, i.e. $\mathbf{w} = \underbrace{i \cdots i}_{h}$, $i \in \mathcal{A}$ and, $h \in \mathbb{N}^*$, we shall say that $\mathbf{w}$ is a block of $i's$ of length h and it shall be denoted by

$$\mathbf{w} = i^{(h)}, \quad i \in \mathcal{A}, \quad h \in \mathbb{N}, 1 \leq h < \infty. \qquad [5.11]$$

For instance, the sequence $\mathbf{w} := bbbb$ is a block of $b's$ with length $h = 4$. If $\mathbf{w}$ is not a block, it can be obtained by concatenating sequences that are blocks (see Karaliopoulou 2009). That is, $\mathbf{w}$ can be expressed as

$$\mathbf{w} = w(1)w(2) \cdots w(\eta), \qquad [5.12]$$

where $w(1) = i_1{}^{(n_1)}$, $w(2) = i_2{}^{(n_2)}$, ..., $w(\eta) = i_\eta{}^{(n_\eta)}$, for $i_1, i_2, ..., i_\eta \in \mathcal{A}$ and $n_1, n_2, ..., n_\eta \in \mathbb{N}^*$, such that $n_1 + n_2 + \cdots + n_\eta = |\mathbf{w}|$. For example, the pattern $\mathbf{w} := aabbbc$ is the concatenation of three blocks $w(1) = aa$, $w(2) = bbb$ and $w(3) = c$, where $n_1 = 2, n_2 = 3, n_3 = 1$, $i_1 = a$, $i_2 = b$ and $i_3 = c$. A prefix is also a pattern, therefore it can be represented by the concatenation of blocks.

We shall define the backward time of a prefix $p = w_1 w_2 \cdots w_{l-1} w_l$ for $1 \leq l \leq h$, as the size of its last block minus one. This definition comes from the time (the number of positions), we have stayed in the last symbol of prefix p since the last jump (renewal point). Note that in general the backward time of a prefix is different from the backward time U_k of symbols. The backward time of a prefix will be denoted by $U(p)$; for details, see the following example.

EXAMPLE 5.5.– Let $\mathbf{w} = aaabb$ be a pattern from the set $\mathcal{A} = \{a, b\}$, then the prefix set is

$\{a, aa, aaa, aaab, aaabb\}$. The backward time for each prefix is: $U(a) = 0$, $U(aa) = 1$, $U(aaa) = 2$, $U(aaab) = 0$, $U(aaabb) = 1$.

Observe that there are some prefixes where the chain (Y_k) can pass consecutively more than one unit of time; see example 5.4 where sequence (Y_k) passes two times in ε at the beginning of the sequence, and the values for these backward are 0 and 1. Under this consideration, we shall introduce the set of all possible backward times for each prefix through sequence (Z_k). Let l_p be the set of all possible values for the backward times of prefix p through process (Z_k). If $i \in \mathcal{A}$ is the partial inverse of p, i.e. $i = \delta^{-1}(p)$ (see equation [5.2]), l_i is the maximum value for the backward time in the letter i (see equation [5.8]) and n_1 is the size of the first block of $\mathbf{w}$ (see equation [5.12]), then

$$l_p^* = \begin{cases} [\![0, l_i]\!], & \text{if } |p| = 0, \\ \{U(p)\}, & \text{if } |p| \neq n_1 \text{ and } |p| \neq 0, \\ [\![U(p), l_i]\!], & \text{if } |p| = n_1. \end{cases} \quad [5.13]$$

Therefore, let

$$K_p := \{(p, n) : n \in l_p^*\} \quad [5.14]$$

be the set of binomials that represent the prefix p and the backward times for this prefix, by consequence the set

$$K := \bigcup_{p \in E} K_p \quad [5.15]$$

represents all prefixes and all their backward times. Clearly, $K_p \cap K_{p'} = \emptyset$, if $p \neq p'$. The set K is the state space of process (Y_k, U_k). The state space K is provided by Algorithm 1 in Garcia-Maya and Limnios (2019).

PROPOSITION 5.1.– The process $(Y_k, U_k)_{k \in \mathbb{N}}$ described above is a Markov chain with state space K. Let us consider a pattern $\mathbf{w} = w_1 w_2, ..., w_h$, where $w_1, w_2, ..., w_h \in \mathcal{A}$ are the symbols of $\mathbf{w}$, let p be a prefix of $\mathbf{w}$, consider the last letter of p, $i = \delta^{-1}(p)$ and the initial distribution of (Z_k), which is denoted by $\alpha(i) = \mathbb{P}(Z_0 = i)$, i.e. $\alpha(i)$ gives the probability that process (Z_k) starts at state $i \in \mathcal{A}$. Then the initial distribution of process (Y_k, U_k) is given by the expression

$$\mathbb{P}(Y_0 = p, U_0 = u) = \begin{cases} \alpha(w_1)\mathbb{1}_{\{u=0\}} & \text{if } p = w_1, \\ \alpha(i)\mathbb{1}_{\{u=0\}} & \text{if } p = \varepsilon, \quad i \in \mathcal{A} \setminus \{w_1\}, \\ 0 & \text{otherwise.} \end{cases} \quad [5.16]$$

The transition probability function is

$$\mathbb{P}(Y_{k+1} = p', U_{k+1} = v \mid Y_k = p, U_k = u)$$
$$= \begin{cases} \frac{p'_{ia}(u+1)}{1-H_i(u)}\mathbb{1}_{\{\delta(p,a)=p'\}} & \text{if } i \neq a, \quad v = 0, \\ \frac{1-H_i(u+1)}{1-H_i(u)}\mathbb{1}_{\{\delta(p,a)=p'\}} & \text{if } i = a, \quad v = u+1, \\ 0 & \text{otherwise.} \end{cases} \quad [5.17]$$

PROOF.– See Garcia-Maya and Limnios (2019). □

Let us denote by P be the transition probability matrix and by β the initial distribution of process (Y_k, U_k), i.e.

$$P((p, u), (p', v)) := \mathbb{P}(Y_{k+1} = p', U_{k+1} = v \mid Y_k = p, U_k = u) \quad [5.18]$$

and

$$\beta(p, u) := \mathbb{P}(Y_0 = p, U_0 = u). \quad [5.19]$$

In order to compute the transition probability matrix P, see Algorithm 2 in Garcia-Maya and Limnios (2019).

It is known that, if $(J_n, S_n)_{n\in\mathbb{N}}$ is an irreducible Markov chain, then process (Z_k, U_k) is also irreducible (see, for example, Chryssaphinou *et al.* 2008). In the next proposition, the same properties for (Y_k, U_k) are proven, i.e. the irreducibility and aperiodicity.

PROPOSITION 5.2.– Let $\mathcal{A}$ be an alphabet, such that $|\mathcal{A}| \geq 2$. If the Markov renewal process (J_n, S_n) with state space $\mathcal{A} \times \mathbb{N}$ is irreducible, then the process (Y_k, U_k) with state space K is also irreducible and aperiodic.

PROOF.– Let be $(p, u), (p', v) \in K$ and $i = \delta^{-1}(p)$. If $p' = w_1$ or $p' = \varepsilon$, the proof is a direct consequence of the irreducibility of (Z_k, U_k). If $p' = w_1 w_2 \cdots w_\ell$, with $1 < \ell \leq h$. Let $p'_1 = w_1, p'_2 = w_1 w_2, \cdots, p'_\ell = w_1 w_2 \cdots w_\ell$, be the consecutive prefixes from w_1 to $p' = w_1 w_2 \cdots w_\ell$ and $(p'_1, u_1), (p'_2, u_2), ..., (p', v)$ the consecutive couples in $E \times \mathbb{N}$, such that for $1 \leq i \leq \ell - 1$

$$u_{i+1} = \begin{cases} 0 & \text{if} w_i \neq w_{i+1}, \\ u_i + 1 & \text{elsewhere.} \end{cases}$$

Process (Z_k, U_k) is irreducible therefore, for $(i, u) \in \mathcal{A} \times \mathbb{N}$ there exist $n \in \mathbb{N}^*$ such that, for $m \in \mathbb{N}$

$$\mathbb{P}(Z_{m+n} = w_1, U_{m+n} = u_1 \mid Z_m = i, U_m = u) > 0,$$

by proposition 5.1

$$P((w_1, u_1), (w_1 w_2, u_2)) \times \cdots \times P((w_1 w_2 \cdots w_{\ell-1}, u_{\ell-1}), (w_1 w_2 \cdots w_\ell, u_\ell)) > 0.$$

Therefore, there exist $n_1 = n + \ell - 1$ such that

$$\mathbb{P}(Y_{m+n_1} = p', U_{m+n_1} = v \mid Y_m = p, U_m = u) > 0.$$

Now, to prove that (Y_k, U_k) is aperiodic, let p and p' be two elements of E such that $|p| = l$ and $|p'| = l + 1$, for $1 \leq l < h$. Let

$$n := \min\{k : P^k((p, u), (p, u)) > 0\}$$

and

$$m := \min\{k : P^k((p', v), (p', v)) > 0\}$$

be the minimum numbers of steps to pass from (p, u) to (p, u) and from (p', v) to (p', v), respectively. By equation [5.17], it is clear that $m = n + 1$. Let d be the period of (p, i) and (p', j), hence $d|n$ and $d|(n+1)$; thus, $d = 1$. □

5.4. The hitting time of the pattern

Note that the first hitting time of a pattern $\mathbf{w}$ through a random sequence of symbols (Z_k) is a random variable, and we observe the following examples:

EXAMPLE 5.6.– Let $\mathbf{w} = bab$ and consider the following sample of letters: Z : *aaaabbab....*

It is clear that $\mathbf{w}$ appears for first time in Z at position 7 (starting from zero), but considering the following sample of letters Z' : *babbbbbbaaaaa...*

We can note that in this case $\mathbf{w}$ appears for first time in the sequence at position 2, so the first hitting time is a random variable that depends on $\mathbf{w}$ and the sequence of letters.

The number of symbols in the chain (Z_k) before the first apparition of $\mathbf{w}$ will be denoted by $N_{\mathbf{w}}$; it is formally defined by

$$N_{\mathbf{w}} := \min\{k \geq 0 : (Y_k, U_k) = (\mathbf{w}, \cdot) \in K\}. \qquad [5.20]$$

As it has been noted in section 5.2, an occurrence of $\mathbf{w}$ in (Z_k) corresponds to an occurrence of $\mathbf{w}$ in (Y_k). The following proposition gives the probability law of $N_{\mathbf{w}}$.

PROPOSITION 5.3.– Let $\{\mathbf{w}^c, \mathbf{w}\}$ be a partition of the state space K such that $\mathbf{w} := \{(\mathbf{w}, \cdot) \in K\}$ and $\mathbf{w}^c := K \setminus \mathbf{w}$ is the complement of $\mathbf{w}$. Let $|\mathbf{w}| = h$ and $\mathbf{1}$ be a column vector of ones with size the cardinality of $\mathbf{w}^c$, i.e. $|\mathbf{w}^c|$. Let $P_{\mathbf{w}} = P|_{\mathbf{w}^c \times \mathbf{w}^c}$ and $\beta_{\mathbf{w}}$ be the restrictions, respectively, on $\mathbf{w}^c \times \mathbf{w}^c$ and $\mathbf{w}^c$ of the transition matrix P and the initial distribution β. Then

$$\mathbb{P}(N_{\mathbf{w}} = n) = \begin{cases} 0 & \text{if } n < h-1, \\ \beta_{\mathbf{w}}(P_{\mathbf{w}})^{h-1}\mathbf{1} & \text{if } n = h-1, \\ \beta_{\mathbf{w}} P_{\mathbf{w}}^{n-1}[I - P_{\mathbf{w}}]\mathbf{1} & \text{if } n \geq h. \end{cases} \qquad [5.21]$$

PROOF.– For $n < h - 1$, it is obvious. For $n = h - 1$, let $p_1 = w_1$, $p_2 = w_1 w_2, \cdots, p_{h-1} = w_1 w_1 \cdots w_{h-1}$, $w = w_1 w_1 \cdots w_{h-1} w_h$ be the consecutive prefixes from w_1 to w, then

$$\mathbb{P}(N_w = h-1) = \sum_{u_h \in l_w} \sum_{u_{h-1} \in l_{p_{h-1}}} \cdots \sum_{u_2 \in l_{p_2}} \sum_{u_1 \in l_{p_1}} P((p_{h-1}, u_{h-1}), (w, u_h))$$
$$\cdots P((p_1, u_1), (p_2, u_2)) \mathbb{P}(Y_0 = p_1, U_0 = u_1),$$

where the set l_p is the set of backward times that corresponds to prefix p (see equation [5.13]). For $n \geq h$

$$\begin{aligned}\mathbb{P}(N_{\mathbf{w}} > n) &= \mathbb{P}((Y_k, U_k) \in \mathbf{w}^c, k \in \{0, 1, ..., n\}) \\ &= \sum_{(q_n, u_n) \in \mathbf{w}^c} \sum_{(q_{n-1}, u_{n_1}) \in \mathbf{w}^c} \cdots \sum_{(q_0, u_0) \in \mathbf{w}^c} P((q_{n-1}, u_{n-1}), (q_n, u_n)) \\ &\cdots P((q_0, u_0), (q_1, u_1))\mathbb{P}(Y_0 = q_0, U_0 = u_0) \\ &= \sum_{(q_n, u_n) \in \mathbf{w}^c} \sum_{(q_0, u_0) \in \mathbf{w}^c} P^n((q_0, u_0), (q_n, u_n))\mathbb{P}(Y_0 = q_0, U_0 = u_0),\end{aligned}$$

and therefore

$$\begin{aligned}\mathbb{P}(N_{\mathbf{w}} = n) &= \beta_{\mathbf{w}}(P_{\mathbf{w}})^{n-1}\mathbf{1} - \beta_{\mathbf{w}}(P_{\mathbf{w}})^n\mathbf{1} \\ &= \beta_{\mathbf{w}}(P_{\mathbf{w}})^{n-1}[I - P_{\mathbf{w}}]\mathbf{1}. \ \square\end{aligned}$$

The previous preposition proved how to calculate the probability of the first hitting position of $\mathbf{w}$ in (Z_k). It is interesting to compute the expectation and the standard deviation of $N_{\mathbf{w}}$; to obtain this calculus, we shall use the generation function.

PROPOSITION 5.4.– Under the same notations as in proposition 5.3, where

$f(h-1) := \mathbb{P}(N_{\mathbf{w}} = h-1)$ and $\mathbb{1}$ is the indicator function. The generating function of $N_{\mathbf{w}}$, i.e. $G(s) := \mathbb{E}[s^{N_{\mathbf{w}}}]$, for $|s| \leq 1$ is

$$G(s) = s^{h-1}f(h-1)\mathbb{1}_{\{h \geq 1\}} + s\beta_w(sP_{\mathbf{w}})^{h-1}(I - sP_{\mathbf{w}})^{-1}(I - P_{\mathbf{w}})\mathbb{1}_{\{h \geq 1\}}\mathbf{1}.$$

PROOF.– By definition of the generating function and proposition 5.3,

$$\begin{aligned}G(s) &= \sum_{k \geq h-1} \mathbb{P}(N_{\mathbf{w}} = k)s^k \\ &= s^{h-1}f(h-1)\mathbb{1}_{\{h \geq 1\}} + \sum_{k \geq h} s^k \beta_{\mathbf{w}} P_{\mathbf{w}}^{k-1}[I - P_{\mathbf{w}}]\mathbf{1}.\end{aligned} \quad [5.22]$$

The set $\mathbf{w}^c$ is a proper subset of the state space K of an irreducible and aperiodic Markov chain, therefore the following expression holds true (see Neuts 1981)

$$\sum_{k \geq 0}(sP_{\mathbf{w}})^k = (I - sP_{\mathbf{w}})^{-1}. \quad [5.23]$$

By consequence

$$\begin{aligned}\sum_{k \geq h} s^k \beta_{\mathbf{w}} P_{\mathbf{w}}^{k-1}[I - P_{\mathbf{w}}]\mathbf{1} &= s^h \beta_{\mathbf{w}} P_{\mathbf{w}}^{h-1} \sum_{h \geq 0}(sP_{\mathbf{w}})^k(I - P_{\mathbf{w}})\mathbf{1} \\ &= s\beta_{\mathbf{w}}(sP_{\mathbf{w}})^{h-1}(I - sP_{\mathbf{w}})^{-1}(I - P_{\mathbf{w}})\mathbb{1}_{\{h \geq 1\}}\mathbf{1},\end{aligned}$$

hence

$$\begin{aligned} G(s) &= s^{h-1} f(h-1)\mathbb{1}_{\{h\geq 1\}} \\ &\quad + s\beta_{\mathrm{w}}(sP_{\mathrm{w}})^{h-1}(I - sP_{\mathrm{w}})^{-1}(I - P_{\mathrm{w}})\mathbb{1}_{\{h\geq 1\}}\mathbf{1}. \ \square \end{aligned}$$

We shall use the derivate of $G(s)$ to obtain the variance and the expected value of N_{w}; before introducing these results let us compute the n derivate of $(I - sP_{\mathrm{w}})$.

LEMMA 5.1.– The n derivate of $(I - sP_{\mathrm{w}})$ holds the following property:

$$\frac{d}{ds}(I - sP_{\mathrm{w}})^{-n} = n(I - sP_{\mathrm{w}})^{-(n+1)} P_{\mathrm{w}} = nP_{\mathrm{w}}(I - sP_{\mathrm{w}})^{-(n+1)}. \quad [5.24]$$

Obtaining the derivate of $G(s)$ is easy to compute the variance and the expected value of N_{w} using the expressions $\mathbb{E}[N_{\mathrm{w}}] = \left.\frac{dG(s)}{ds}\right|_{s=1}$ and $\mathbb{V}\mathrm{ar}[N_{\mathrm{w}}] = \left.\frac{d^2G(s)}{ds^2}\right|_{s=1} + \left.\frac{dG(s)}{ds}\right|_{s=1} - \left(\left.\frac{dG(s)}{ds}\right|_{s=1}\right)^2$. These results will be computed in the following preposition:

PROPOSITION 5.5.– The mean and variance of N_{w} are, respectively

$$\mathbb{E}[N_{\mathrm{w}}] = (h-1)f(h-1)\mathbb{1}_{\{h\geq 2\}} + \beta_{\mathrm{w}}[Q + hP_{\mathrm{w}}^{h-2}]P_{\mathrm{w}}\mathbf{1} \quad [5.25]$$

and for $h \geq 3$

$$\mathbb{V}\mathrm{ar}[N_{\mathrm{w}}] = 2(h-1)f(h-1) + a + b + c + d - (\mathbb{E}[N_{\mathrm{w}}])^2, \quad [5.26]$$

where:
$$\begin{aligned} a &= 2\beta_{\mathrm{w}} P_{\mathrm{w}}^2 Q^2 \mathbf{1} \\ b &= 2\beta_{\mathrm{w}} P_{\mathrm{w}} Q \mathbf{1} \\ c &= h\beta_{\mathrm{w}} P_{\mathrm{w}}^h Q \mathbf{1} \\ d &= h^2 P_{\mathrm{w}}^{h-1} \mathbf{1} \\ Q &= (I - P_{\mathrm{w}})^{-1}. \end{aligned}$$

PROOF.– See Garcia-Maya and Limnios (2019). □

5.5. A genomic application

The proposed model has many applications in different fields, for instance in communication, linguistics and data search. We propose an application of our model in a particular biological sequence. In this application, the DNA is considered like a semi-Markov discrete process (Z_k) with state space $\mathcal{A} = \{A, T, G, C\}$ where each letter represents the nucleotides: adenine, thymine, guanine and cytosine, respectively. In the sequel, we present the corresponding example.

Genome sequencing is often compared to a stochastic code ruled by four nitrogenous bases: adenine (A), thymine (T), guanine (G) and cytosine (C). Given a genome sequence, it is interesting to recognize a pattern, counting the number of, or determining the first position that this appears (see, for example, Abadi and Vergne 2008; Li *et al.* 2016, 2018; Sigwart and Garbett 2018; Hebert *et al.* 2003; Robin *et al.* 2007). This pattern could be, for example, responsible for the production of a particular protein, a genetic disorder, a specific malformation or could be useful to detect similar (or dissimilar) genes in different organisms. In this case, the genomic set of symbols is $\mathcal{A} = \{A, T, G, C\}$ and we consider the pattern $\mathbf{w} = CCCGGG$, which is the enzyme *SmaI*. The *SmaI* enzyme, widely used in genomic analysis, is a DNA cutter. That is, when it finds the "$CCCGGG$" fragment, it applies and cuts the DNA in the middle of this fragment, i.e. into "...CCC" and "GGG...". The probability that the *SmaI* enzyme appears after k nucleotides is observed in Figure 5.1. The random variable $N_{\mathbf{w}}$ counts the numbers of nucleotides before the apparition of the enzyme and it is defined according to equation [5.20]. The probability function of $N_{\mathbf{w}}$ is denoted as $f_{\mathbf{w}}(k) := \mathbb{P}(N_{\mathbf{w}} = k)$; this probability is computed using proposition 5.3.

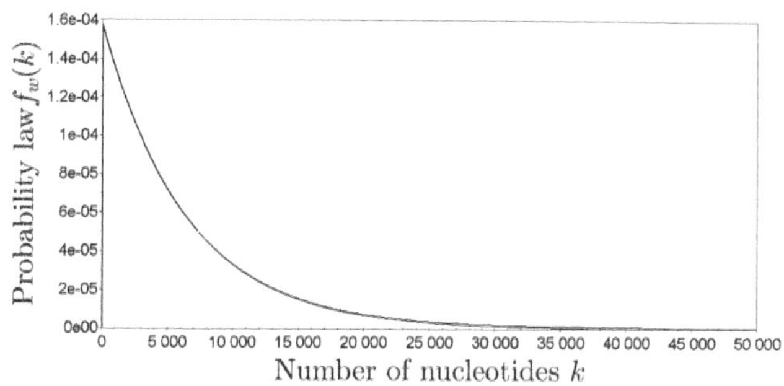

Figure 5.1. *Probability law of $N_{\mathbf{w}}$, i.e. $f_{\mathbf{w}}(k) := \mathbb{P}(N_{\mathbf{w}} = k)$*

In Figure 5.2, we can observe that the *SmaI* enzyme does not appear frequently in the DNA. The pattern occurrence rate is given by the rate function:

$$\lambda(k) = 1 - \frac{\bar{F}_{\mathbf{w}}(k)}{\bar{F}_{\mathbf{w}}(k-1)}, \qquad [5.27]$$

where $\bar{F}_{\mathbf{w}}(k) = 1 - F_{\mathbf{w}}(k)$ and $F_{\mathbf{w}}(k) := \sum_{l=0}^{k} f_{\mathbf{w}}(l)$ (see, for example, Barbu and Limnios 2008). Figure 5.3 shows the values for the distribution function $F_w(k)$. In the continuous time, the rate function takes values also greater than one; in the discrete-time case, it takes values only in the interval $[0, 1]$. For this reason, in another context, Roy and Gupta (1992) proposed another rate function as follows:

$$r(k) = -\ln(1 - \lambda(k)). \qquad [5.28]$$

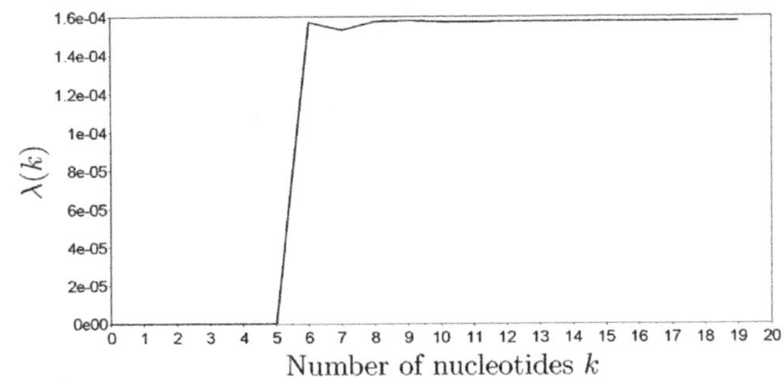

Figure 5.2. *Rate of occurrence of the word* $\mathbf{w}$ *at position* k *in the DNA sequence, i.e.* $\lambda(k)$

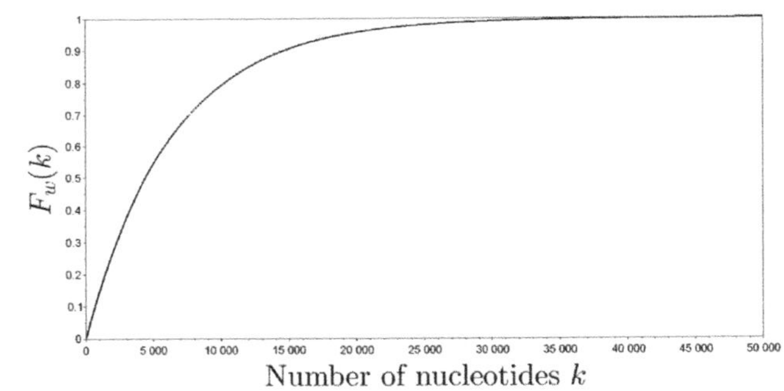

Figure 5.3. *Cumulative distribution function of* $N_{\mathbf{w}}$*, i.e.* $F_{\mathbf{w}}(k) := \mathbb{P}(N_{\mathbf{w}} \leq k)$

Nevertheless, when $\lambda(k)$ is close to 0, we have obviously

$$r(k) \cong \lambda(k). \qquad [5.29]$$

In the case where $\lambda(k)$ is large (close to one), $r(k)$ takes values larger than one that are very different from $\lambda(k)$. In the case of the present example, the value of $\lambda(k)$ are very small. So, Figure 5.2 represents also the function $r(k)$ in the present case.

Considering that DNA sequence has 48,502 nucleotides, i.e. it has size $M = 48,502$ and using the expressions:

$$\mathbb{E}[N_{\mathbf{w}} \cdot \mathbb{1}_{\{N_{\mathbf{w}} \leq M\}}] = \sum_{k=h-1}^{M} k f_{\mathbf{w}}(k), \qquad [5.30]$$

$$\mathbb{E}[T_{\mathbf{w}}^2 \cdot \mathbb{1}_{\{N_{\mathbf{w}} \leq M\}}] = \sum_{k=h-1}^{M} k^2 f_{\mathbf{w}}(k), \qquad [5.31]$$

where $f_{\mathbf{w}}(k) := \mathbb{P}(N_{\mathbf{w}} = k)$ is obtaining according with proposition 5.3; the expected value $\mathbb{E}[N_{\mathbf{w}} \cdot \mathbb{1}_{\{N_{\mathbf{w}} \leq M\}}]$ and the standard deviation $\sigma[N_{\mathbf{w}} \cdot \mathbb{1}_{\{N_{\mathbf{w}} \leq M\}}]$ are computed. Hence, we obtain:

$$\begin{aligned}\mathbb{E}[N_{\mathbf{w}} \cdot \mathbb{1}_{\{N_{\mathbf{w}} \leq M\}}] &= 6335.9,\\ \sigma[N_{\mathbf{w}} \cdot \mathbb{1}_{\{N_{\mathbf{w}} \leq M\}}] &= 6266.7.\end{aligned}$$

Considering different lengths for the DNA sequence, i.e. considering different values for M, the mean value of $N_{\mathbf{w}}$ for each M is observed in Figure 5.4. If we take into a count that the DNA sequence has infinity length, the mean value and the standard deviation of the random variable $N_{\mathbf{w}}$ are computed according with proposition 5.5; therefore, for the bacteriophage DNA sequence and the *SmaI* enzyme, we have the values

$$\begin{aligned}\mathbb{E}[N_{\mathbf{w}}] &= 6367.6,\\ \sigma[N_{\mathbf{w}}] &= 6354.9.\end{aligned}$$

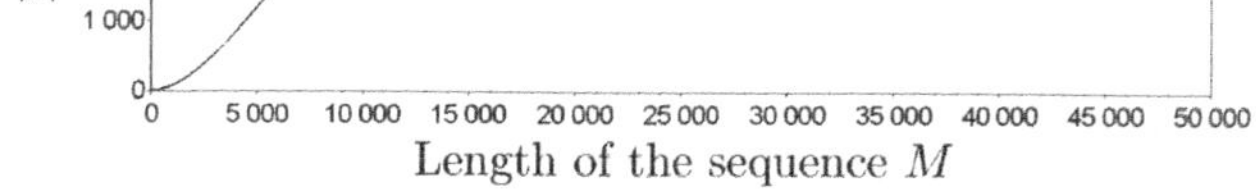

Figure 5.4. *Expected value for $N_{\mathbf{w}}$ for different values of M in the DNA*

Note in Figure 5.4, the value $\mathbb{E}[N_{\mathbf{w}} \cdot \mathbb{1}_{\{N_{\mathbf{w}} \leq M\}}]$ reaches $\mathbb{E}[N_{\mathbf{w}}]$ as M becomes large enough, as it was expected by the dominated converge theorem. It is worth noting that the standard deviation is high due to the rate of occurrence of the word being small. After position 9, it becomes geometric, as we can see in Figure 5.2. According to the geometric distribution, with success probability p, the variance of the random number of trials up to the first success is given by

$$Var(X) = \frac{1-p}{p^2}.$$

We observe that variance grows if p decreases. In our model, the rate of occurrence of the word is tiny; we can see in Figure 5.2 that after position 9, we have $p = \lambda(9) = 1.6 \times 10^{-4}$, which means that the probability to have a success is small, and this gives the big value for the standard deviation.

5.6. Concluding remarks

In this chapter, we proposed a model and algorithm based on the semi-Markov hypothesis to compute the probability distribution of the position of the first hitting time of a pattern in a semi-Markov chain at position k. We gave some important properties of the pattern like its occurrence rate. This model needs less memory space and it is more efficient than the one proposed by Chryssaphinou *et al.* (2008), the only model to our knowledge that computes the first hitting time of a pattern under the semi-Markov hypotheses. The algorithm proposed here depends only on the set of symbols $\mathcal{A}$, the pattern size, the prefixes of the pattern and the semi-Markov kernel. The process (Y_k, U_k) is a Markov chain where its transition matrix P is easily written. For a pattern $\mathbf{w}$ with $|\mathbf{w}| = h$ and $\gamma = \max\{l_p^* : p \in E\}$, we need a state space of cardinality $(h + |\mathcal{A}| - 1) \times \gamma$ at most that grows in a linear way depending on the size of the pattern h and the number of elements in $\mathcal{A}$. The small number of states facilitates the computational calculations. It is worth noting that for patterns of length one, i.e. a symbol, the proposed model recovers the Markov process (Z_k, U_k).

5.7. References

Abadi, M., and Vergne, N. (2008). Poisson approximation for search of rare words in DNA sequences. *Am. J. Prob. Math. Stat.*, 4, 233–244.

Aboluion, N.A. (2011). The construction of DNA codes using a computer algebra system. PhD Thesis, University of Glamorgan.

Antzoulakos, D.L. (2001). Waiting times for patterns in a sequence of multistate trials. *Journal of Applied Probability*, 38(2), 508–518.

Avena, L., da Costa, C., and den Hollander, F. (2013). Stochastic models for genetic evolution. Lecture Notes, Mathematical Institute, Leiden University.

Barbu, V.S., and Limnios, N. (2008). *Semi-Markov Chains and Hidden Semi-Markov Models Toward Applications–Their Use in Reliability and DNA Analysis*. Springer, New York.

Barbu, V.S., Boussemart, M., and Limnios, N. (2004). Discrete-time semi-Markov model for reliability and survival analysis. *Communications in Statistics–Theory and Methods*, 33(11), 2833–2868.

Chadjiconstantinidis, S., Antzoulakos, D., and Koutras, M. (2000). Joint distributions of successes, failures and patterns in enumeration problems. *Advances in Applied Probability*, 32(3), 866–884.

Chryssaphinou, O., Karaliopoulou, M., and Limnios, N. (2008). On discrete time semi-Markov chains and applications in words occurrences. *Communications in Statistics–Theory and Methods*, 37(8), 1306–1322.

Codish, M., Frank, M., and Lagoon, V. (2017). The DNA word design problem: A new constraint model and new results. *Proceedings of the Twenty-Sixth International Joint Conference on Artificial Intelligence (IJCAI-17)*, 585–591.

Crochemore, M., and Stefanov, V.T. (2003). Waiting time and complexity for matching patterns with automata. *Information Processing Letters*, 87(3), 119–125.

Fu, J.C., and Chang, Y. (2002). On probability generating functions for waiting time distributions of compound patterns in a sequence of multistate trials. *Journal of Applied Probability*, 39(1), 70–80.

Fu, J., and Koutras, M. (1994). Distribution theory of runs: A Markov chain approach. *Journal of the American Statistical Association*, 89(427), 1050–1058.

Garcia-Maya, B.I., and Limnios, N. (2019). Identification of words in biological sequences under the semi-Markov hypothesis. *Journal of Computational Biology*. Available at: http://doi.org/10.1089/cmb.2019.0253.

Glaz, J., Kulldorff, M., Pozdnyakov, V., and Steele, J.M. (2006). Gambling teams and waiting times for patterns in two-state Markov chains. *Journal of Applied Probability*, 43(1), 127–140.

Guibas, L.J., and Odlyzko, A.M. (1981). String overlaps, pattern matching and nontransitive games. *Journal of Combinatorial Theory, Series A*, 30(2), 183–208.

Hebert, P.D., Cywinska, A., Ball, S.L., and Dewaard, J.R. (2003). Biological identifications through DNA barcodes. *Proceedings of the Royal Society of London, Series B: Biological Sciences*, 270(1512), 313–321.

Hopcroft, J.E., Motwani, R., and Ullman, J.D. (2013). *Introduction to Automata Theory, Languages and Computation*, 3rd edition. Pearson, ACM, New York.

Karaliopoulou, M. (2009). On the number of word occurrences in a semi-Markov sequence of letters. *ESAIM: Probability and Statistics*, 13, 328–342.

Li, Z., Cao, H., Cui, Y., and Zhang, Y. (2016). Extracting DNA words based on the sequence features: Non-uniform distribution and integrity. *Theoretical Biology and Medical Modelling*, 13(1), 2.

Li, F., Zhang, M., Tian, B., Chen, B., Fu, G., and Ji, D. (2018). Recognizing irregular entities in biomedical text via deep neural networks. *Pattern Recognition Letters*, 105, 105–113.

Limnios, N., Oprişan, G. (2001). *Semi-Markov Processes and Reliability*. Statistics for Industry and Technology, Birkhäuser, Boston.

Lothaire, M. (1983). *Combinatorics on Words*. Addison Wesley, Massachusetts.

Mode, C.J., and Pickens, G.T. (1988). Computational methods for renewal theory and semi-Markov processes with illustrative examples. *The American Statistician*, 42(2), 143–152.

Montemanni, R. (2015). Combinatorial optimization algorithms for the design of codes: A survey. *Journal of Applied Operational Research*, 7(1), 37.

Neuts, M. (1981). *Matrix-geometric Solutions–An Algorithmic Approach*. The Johns Hopkins University Press, Baltimore.

Nuel, G. (2008). Pattern Markov chains: Optimal Markov chain embedding through deterministic finite automata. *Journal of Applied Probability*, 45(1), 226–243.

Picard, F., Schbath, S., Lebarbier, E., Neuvial, P., and Chiquet, J. (2011). Statistiques et génome. *La Gazette des Mathématiciens*, 130, 51–82.

Robin, S., and Daudin, J.-J. (1999). Exact distribution of word occurrences in a random sequence of letters. *Journal of Applied Probability*, 36(1), 179–193.

Robin, S., Schbath, S., and Vandewalle, V. (2007). Statistical tests to compare motif count exceptionalities. *BMC Bioinformatics*, 8(1), 84.

Roy, D., and Gupta, R. (1992). Classifications of discrete lives. *Microelectronics Reliability*, 32(10), 1459–1473.

Sigwart, J.D., and Garbett, A. (2018). Biodiversity assessment, DNA barcoding, and the minority majority. *Integrative and Comparative Biology*, 58(6), 1146–1156.

Srivastava, S., and Baptista, M.S. (2016). Markovian language model of the DNA and its information content. *Royal Society Open Science*, 3(1), 150527.

Stefanov, VT., and Pakes, A.G. (1997). Explicit distributional results in pattern formation. *The Annals of Applied Probability*, 7(3), 666–678.

Stefanov, V.T., Robin, S., and Schbath, S. (2011). Occurrence of structured motifs in random sequences: Arbitrary number of boxes. *Discrete Applied Mathematics*, 159(8), 826–831.

Touyar, N., Schbath, S., Cellier, D., and Dauchel, H. (2008). Poisson approximation for the number of repeats in a stationary Markov chain. *Journal of Applied Probability*, 45(2), 440–455.

PART 2

Autoregressive Processes

6

Time Changes and Stationarity Issues for Continuous Time Autoregressive Processes of Order p

Smooth time transforms of scalar continuous-time autoregressive (AR) processes of order one, say $\text{CAR}_1(p)$, with time-dependent coefficients are shown here to be $\text{CAR}_1(p)$ processes also, with explicit underlying stochastic differential equations (SDE). Necessary and sufficient conditions are given for a $\text{CAR}_1(p)$ process to be stationary or to become stationary through a pertinent time change. For that purpose, a new criterion for stationarity is given, which is proven to be equivalent with the classical notion of asymptotic stability. Many results are obtained thanks to a thorough study of the associated multivariate $\text{CAR}_p(1)$ process. Several examples are presented; in particular, multiplicative stationarity – involving logarithmic time change – is detailed. Illustration through simulation is provided.

6.1. Introduction

Transforming a stochastic process into another one through a deformation of the index set may lead to a full class of new properties for the resulting process. This concept was formalized in Perrin and Senoussi (1999), after having been introduced in geostatistics – statistics related to continuous-time random fields – in an application to environmental data in Sampson and Guttorp (1992). For time processes, classical time changes are considered in Gray and Zhang (1988), with a particular interest to the logarithmic time change, leading to the so-called notion of multiplicative stationarity; see also Woodward *et al.* (2012) and the references therein.

By transformation of time, a process often loses most of its original properties, but may acquire other new interesting ones. For example, a stationary process

---

Chapter written by Valérie GIRARDIN and Rachid SENOUSSI.

generally loses stationarity, while a non-stationary process may acquire this essential property after undergoing an adapted time change. Stationarity is thus structurally linked to the very definition of the time index. Weak stationarity essentially states that the covariance kernel of a random process is invariant under diagonal shifts. Hence the covariance function, depending on two variables for non-stationary processes, can be reduced to a function depending on one variable only. This reduction in dimension is well known to be invaluable for statistical applications, and, hence, identifying time changes, yielding crucial importance. Note that if some time change exists, inducing weak stationarity for a given process, this process will be said to be reducible to weak stationarity (RWS).

More precisely, this chapter focuses on RWS scalar continuous-time AR processes of order $p \geq 1$, that is, $\text{CAR}_1(p)$ processes with – possibly time varying – deterministic coefficients. The considered processes will mainly be Gaussian, so that weak stationarity – in the second-order sense – amounts to strict stationarity – in the distribution sense. In section 6.2, useful results related to time changes for diffusion processes are recalled. In section 6.3, the $\text{CAR}_1(p)$ processes are precisely defined, together with both classical and new properties. New necessary and sufficient conditions are established for these processes to be stationary, which is proven to be consistent with the classical concept of stability for differential systems. The covariance structure governing RWS AR processes and their derivatives is detailed. Many examples are given with simulated trajectories. Section 6.4 details how the differential rules for time changes apply. In particular, AR properties are shown to be preserved under time change, with closed-form formulas relating the two dynamics. The concept of reducibility to stationarity is then fully developed. When existing, the time change yielding stationarity is proven to be unique – up to affine changes. Simple necessary and sufficient conditions for a $\text{CAR}_1(p)$ to be RWS are given, with explicit time change. Many examples with time changes that are classical in the literature are presented; the case of multiplicative stationary (MS) AR processes is especially detailed. For the sake of completeness, the appendix gathers some technical results related to Faà di Bruno's formula, with links to Bell polynomials.

6.2. Basics

NOTATION.– With rare exceptions, scalar elements (either constants, functions, sequences or processes) are denoted by capital letters (e.g. X), vector elements by bold letters ($\mathbf{x}$) and matrix elements by capital bold ($\mathbf{X}$). Also, $\mathbf{I}_p$ stands for the $p \times p$ identity matrix, while $\mathbf{D}_{k,\tau} = \text{Diag}\left(1, \tau, \ldots, \tau^k\right)$ denotes the $(k+1) \times (k+1)$ diagonal matrix obtained from the k first powers of some real number τ. Of particular interest is the $p \times p$-matrix $\mathbf{E}_{p,p} = \mathbf{e}_p \mathbf{e}_p^T$, where $\mathbf{e}_p = (0, \ldots, 0, 1)$ is the pth vector of the standard basis of $\mathbb{R}^p$.

For any function f, the jth derivative is denoted by $f^{(j)}$ with the convention $f^{(0)} = f$, $f^{(1)} = f'$, while the column vector $\mathbf{f}^{(j_1,j_2)} = \left(f^{(j_1)}, \ldots, f^{(j_2)}\right)^T$ denotes the set of its j_1th to j_2th derivatives.

DEFINITION 6.1.– *Let (α, β) and (a, b) be two open intervals of $\mathbb{R}$ that may be not finite. A strictly increasing k-differentiable function $\psi : (\alpha, \beta) \to (a, b)$, with $\psi(\alpha) = a$ and $\psi(\beta) = b$, is called a k-smooth time change.*

Let X be a scalar continuous time process defined on the interval (a, b), and let $\psi : (\alpha, \beta) \to (a, b)$ be a k-smooth time change. The process defined on (α, β) by $X_\psi(t) = X(\psi(t))$ is called the ψ-time transform of X.

The derivative ψ' of any 1-smooth time change ψ is almost everywhere (a.e.) positive, and $\psi(t) = a + \int_\alpha^t \psi'(u)du$ for all $t \in (\alpha, \beta)$. The inverse function ψ^{-1} of ψ exists and is also a 1-smooth time change. Further, $(\psi^{-1})'(s)) = 1/\psi'(\psi^{-1}(s))$ is almost everywhere positive on (a, b).

Determining well-chosen time changes that transform a non-stationary process into a weakly stationary one requires the following notion of reducibility to weak stationarity.

DEFINITION 6.2.– *A square integrable continuous time process $\mathbf{x}$ defined for $s \in (a, b)$ is said to be (weakly) stationary if $\mathbf{m}(s) = \mathbf{E}[\mathbf{x}(s)]$ is a constant function and some matrix function $\mathbf{R} : \mathbb{R} \to \mathbb{R}^{d \times d}$ exists such that $\mathrm{Cov}[\mathbf{x}(s_1), \mathbf{x}(s_2)] = \mathbf{R}(s_2 - s_1)$ for all $s_1, s_2 \in (a, b)$.*

A square integrable continuous time process $\mathbf{y}$ defined for $t \in (\alpha, \beta)$ is said to be RWS if a time change $\psi : (\alpha, \beta) \to (a, b)$ exists such that $\mathbf{x}(s) = \mathbf{y}(\psi^{-1}(s))$ is stationary.

Random time changes could be considered as well, but, in general, random time transforms of CAR processes are no longer CAR processes, and hence fall beyond the scope or the present work. However, the general formula for random time transforms of SDEs driven by semi-martingales, proven in (Kobayashi 2011, theorem 4.2), applies as follows to smooth deterministic time changes.

PROPOSITION 6.1.– Let X be a diffusion process governed by

$$dX(s) = F(s)ds + B(s)dW(s), \quad s \in (a, b), \tag{6.1}$$

where $F(s)$ and $B(s)$ are adapted continuous processes. Let $\psi : (\alpha, \beta) \longrightarrow (a, b)$ be a 1-smooth deterministic time change.

Then the time transform X_ψ is also a diffusion process, satisfying

$$dX_\psi(t) = F_\psi(t)\psi^{(1)}(t)dt + B_\psi(t)dW_\psi(t), \quad t \in (\alpha, \beta).$$

In particular, the following characterization of the time transform W_ψ of a Brownian motion W holds true.

PROPOSITION 6.2.– Let W be a classical Brownian motion, and let ψ be a 1-smooth time change. The time transform W_ψ of W is a Gaussian process, distributed as the stochastic integral $\int_0^t \sqrt{\psi'(s)}dW(s)$. In other words, $dW_\psi(t) = \sqrt{\psi'(t)}d\widetilde{W}(t)$ for some Brownian motion $\widetilde{W}(t)$.

PROOF.– Both $W_\psi(t) = W(\psi(t))$ and $\int_0^t \sqrt{\psi'(s)}dW(s)$ are centered Gaussian processes. Since they share the same covariance function $\mathbf{E}[W_\psi(t_1)W_\psi(t_2))] = \psi(t_1) \wedge \psi(t_2)$, and the same diffusion equation, they also share the same probability distribution. □

Note that the time transform W_ψ does not coincide with – nor easily ensues from – the original Brownian motion W in the almost sure sense.

6.3. Stationary AR processes

A continuous-time scalar AR process $X = (X(s))_{s\in\mathbb{R}_+}$ of order p, say CAR$_1(p)$, is defined by the SDE

$$dX^{(p-1)}(s) = \left[A_p(s)X(s) + \cdots + A_1(s)X^{(p-1)}(s)\right] ds + B(s)dW(s), \qquad [6.2]$$

where B and A_j for $j = 1, \ldots, p$, are real valued functions, with A_p not the null function, and $W(s)$ is a scalar Brownian motion and $X^{(j)}(s)$ denotes the jth derivative of $X(s)$.

Setting $\mathbf{x}(s) = (X(s), X^{(1)}(s), \ldots, X^{(p-1)}(s)\,)^T$ and $\mathbf{w}(s) = \mathbf{e}_p W(s)$ makes it a continuous-time p–dimensional AR process of order 1, or CAR$_p(1)$, governed by

$$d\mathbf{x}(s) = \mathbf{A}(s)\mathbf{x}(s)ds + B(s)d\mathbf{w}(s), \qquad [6.3]$$

where $\mathbf{A}(s) = (A_{i,j}(s))$, the companion matrix of $\mathbf{a}(s) = (A_p(s), \ldots, A_1(s))$, satisfies $A_{i,j}(s) = \mathbb{1}_{\{j=i+1\}}$ for $i = 1, \ldots, p-1$, and $j = 1, \ldots, p$, and $A_{p,j}(s) = A_{p-j+1}(s)$, for $j = 1, \ldots, p$. Note that [6.2] can thus be written as

$$dX^{(p-1)}(s) = \mathbf{a}^T(s)\mathbf{x}(s)ds + B(s)dW(s). \qquad [6.4]$$

6.3.1. *Formulas for the two first-order moments*

The general solution of [6.3] reads

$$\mathbf{x}(s) = \boldsymbol{\mathcal{E}}(s)\left[\mathbf{x}(0) + \int_0^s \boldsymbol{\mathcal{E}}^{-1}(v)\mathbf{e}_p B(v)dW(v)\right], \qquad [6.5]$$

where $\boldsymbol{\mathcal{E}}(s)$ is the unique solution of the ordinary matrix differential equation

$$d\boldsymbol{\mathcal{E}}(s) = \mathbf{A}(s)\boldsymbol{\mathcal{E}}(s)ds, \quad s \in \mathbb{R}_+, \qquad [6.6]$$

with initial condition $\boldsymbol{\mathcal{E}}(0) = \mathbf{I}_d$, the identity matrix; see Jacod (1982), Protter (2005) and also Girardin and Senoussi (2019) for details.

If all A_j and B are continuous functions on $\mathbb{R}_+$, then $X(s)$ is almost surely continuously $(p-2)$-differentiable. Its $(p-1)$th derivative may only be continuous but still satisfies [6.2]. So, if $\mathbf{x}(0)$ is Gaussian, then $\mathbf{x}(s)$ is also Gaussian and the distribution of the process is entirely characterized by its mean and covariance functions. In this case, stationarity in the weak sense is equivalent to strong stationarity, that is, in the sense of distributions.

To go further into the characterization of the scalar CAR$_1(p)$ defined by [6.2], let $M(s) = \mathbf{E}[X(s)]$ denote the mean function and $C(s_1, s_2) = \mathrm{Cov}[X(s_1), X(s_2)]$ the covariance function of X.

PROPOSITION 6.3.– Suppose that $\mathbf{A}(s)$ and $B(s)$ in [6.3] are deterministic piecewise continuous functions. Then the vector mean function $\mathbf{m}(s) = \mathbf{E}[\mathbf{x}(s)] = (M_1(s), \cdots, M_p(s))^T$ of $\mathbf{x}(s)$ defined by [6.3] is given by $M_1(s) = M(s)$ and

$$M_{j+1}(s) = \mathbf{E}[X^{(j)}(s)] = M^{(j)}(s), \quad j = 1, \ldots, p-1. \qquad [6.7]$$

Moreover, if $\mathbf{x}(0)$ is independent of W, then the matrix covariance function $\mathbf{C}(s_1, s_2) = \mathbf{E}[(\mathbf{x}(s_1) - \mathbf{m}(s_1))\,(\mathbf{x}(s_2) - \mathbf{m}(s_2))^T]$ of $\mathbf{x}(s)$ is given by

$$\mathbf{C}(s_1, s_2) = \boldsymbol{\mathcal{E}}(s_1)\left[\mathbf{C}(0,0) + \int_0^{s_1 \wedge s_2} B^2(v)\boldsymbol{\mathcal{E}}^{-1}(v)\mathbf{E}_{p,p}(\boldsymbol{\mathcal{E}}^{-1}(v))^T dv\right]\boldsymbol{\mathcal{E}}^T(s_2). \qquad [6.8]$$

Further, setting $\mathbf{C}(s_1, s_2) = (C_{i,j}(s_1, s_2))_{1 \le i,j \le p}$, we get $C_{1,1}(s_1, s_2) = C(s_1, s_2)$ and

$$C_{i,j}(s_1, s_2) = \partial^{i+j-2}_{s_1^{i-1}, s_2^{j-1}} C_{1,1}(s_1, s_2), \quad 1 \le i, j \le p. \qquad [6.9]$$

PROOF.– By construction, $M(s) = M_1(s)$ and $C(s_1, s_2) = C_{1,1}(s_1, s_2)$. Further, $\mathbf{m}(s)$ satisfies $d\mathbf{m}(s) = \mathbf{A}(s)\mathbf{m}(s)ds$, whose solution is $\mathbf{m}(s) = \boldsymbol{\mathcal{E}}(s)\mathbf{m}(0)$. Taking expectation on both sides of [6.5] proves [6.7].

We can assume, without loss of generality, that $\mathbf{x}(s)$ is centered. Since $\mathbf{x}(0)$ and W are independent, elementary stochastic calculus yields [6.8].

Finally, classical mean square differentiability calculus (see Loève 1978) yields

$$\begin{aligned} C_{i,j}(s_1,s_2) &= \mathbf{E}[X^{(i-1)}(s_1)X^{(j-1)}(s_2)] = \mathbf{E}[\partial^{i-1}_{s_1^{i-1}}X(s_1)\partial^{j-1}_{s_2^{j-1}}X(s_2)] \\ &= \mathbf{E}[\partial^{i+j-2}_{s_1^{i-1},s_2^{j-1}}X(s_1)X(s_2)] = \partial^{i+j-2}_{s_1^{i-1},s_2^{j-1}}\mathbf{E}[X(s_1)X(s_2)] \\ &= \partial^{i+j-2}_{s_1^{i-1},s_2^{j-1}}C_{1,1}(s_1,s_2), \end{aligned}$$

which finishes the proof. □

More practically, $\mathbf{C}(s_1, s_2) = \mathbf{C}^T(s_2, s_1)$, and [6.8] reads

$$\mathbf{C}(s_1,s_2) = \begin{cases} \mathcal{E}(s_1)\mathcal{E}^{-1}(s_2)\mathbf{C}(s_2,s_2) & \text{if } s_1 \geq s_2, \\ \mathbf{C}(s_1,s_1)(\mathcal{E}^{-1}(s_1)^T\mathcal{E}^T(s_2) & \text{if } s_1 \leq s_2. \end{cases} \qquad [6.10]$$

In particular, for $\mathrm{CAR}_1(p)$ processes with constant coefficient $\mathbf{A}$, the solution $\mathcal{E}$ of [6.6] has an exponential matrix form, that is, $\mathcal{E}(s) = \mathrm{e}^{\mathbf{A}s}$. Then, $\mathcal{E}^{-1}(s) = \mathrm{e}^{-\mathbf{A}s}$, which simplifies the form of the solution [6.5].

The closed-form expression of $\mathbf{x}(s)$ involves complex-valued stochastic integrals

$$Z_{k,\lambda}(s) = \int_0^s v^k \mathrm{e}^{\lambda v} dW(v), \quad k \in \mathbb{N},\ \lambda \in \mathbf{C},$$

with bilateral covariance functions

$$\begin{aligned} E\left[Z_{k_1,\lambda_1}(s_1)\overline{Z}_{k_2,\lambda_2}(s_2)\right] &= \int_0^{s_1 \wedge s_2} v^{k_1+k_2}\mathrm{e}^{(\lambda_1+\overline{\lambda_2})v}dv \\ &= \frac{J_{k_1+k_2}\left((\lambda_1+\overline{\lambda_2})(s_1 \wedge s_2)\right)}{(\lambda_1+\overline{\lambda_2})^{k_1+k_2+1}}, \end{aligned}$$

where

$$J_k(s) = \int_0^s v^k \mathrm{e}^v dv = k!\left[\mathrm{e}^s \sum_{l=0}^{k} \frac{s^l(-1)^{k-l}}{l!} - (-1)^k\right].$$

6.3.2. *Examples*

Let us develop the particular cases of $\mathrm{CAR}_1(2)$ and $\mathrm{CAR}_1(3)$ models with constant regression coefficients.

First, let $X(s)$ be a $\text{CAR}_1(2)$ model, governed by [6.2] for $p = 2$, with constant coefficients B and $\mathbf{A} = \begin{pmatrix} 0 & 1 \\ A_2 & A_1 \end{pmatrix}$. Let $P(\lambda) = \lambda^2 - A_1\lambda - A_2$ be the characteristic polynomial of $\mathbf{A}$, with roots $\lambda_{1,2} = A_1/2 \pm \sqrt{\Delta}/2$, that may be either real or complex, distinct or equal, according to the value of the discriminant $\Delta = A_1^2 + 4A_2$.

For illustration, let us take $A_2 = -(A_1/2)^2$. Then $\Delta = 0$, and $\lambda = A_1/2$ is a double root, and

$$\mathrm{e}^{\mathbf{A}s} = \mathrm{e}^{\lambda s} \begin{pmatrix} 1 - \lambda s & s \\ -\lambda^2 s & 1 + \lambda s \end{pmatrix}.$$

In this case, $\mathbf{x}(s) = \left(X(s), X^{(1)}(s)\right)$ is given by

$$\begin{pmatrix} X(s) \\ X^{(1)}(s) \end{pmatrix} = \mathrm{e}^{\mathbf{A}s} \begin{pmatrix} X(0) \\ X^{(1)}(0) \end{pmatrix} + B\mathrm{e}^{\mathbf{A}s} \begin{pmatrix} -Z_{1,-\lambda}(s) \\ Z_{0,-\lambda}(s) - \lambda Z_{1,-\lambda}(s) \end{pmatrix}.$$

Figure 6.1 shows paths of the two components of the $\text{CAR}_1(2)$ process $\mathbf{x}(s)$ with $A_1 = -2, A_2 = -1$ and $B = 1$ that lead to $\Delta = 0$ and the double root $\lambda = -1$.

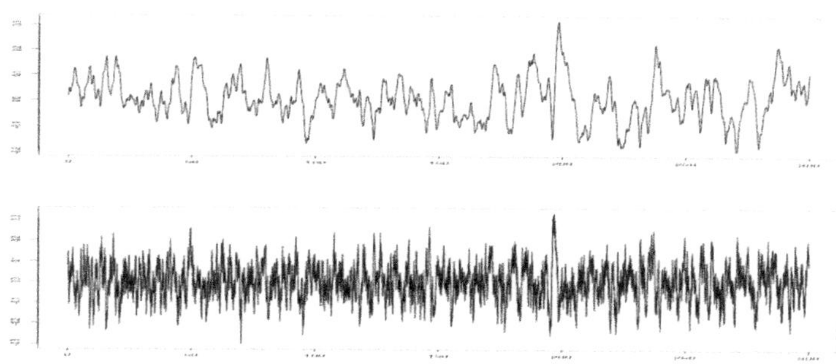

Figure 6.1. *Simulated paths of a stationary* $CAR_1(2)$ *model with* $A_1 = -2,\ A_2 = -1$ *and* $B = 1$. *Top: The process* $X(s), s \in [0, 300]$. *Bottom: Its derivative* $X^{(1)}(s)$. *Observe the smoothness of* $X(s)$ *compared to the roughness of its derivative*

Now let X be a $\text{CAR}_1(3)$ model, governed by [6.2] for $p = 3$, with constant coefficients B and

$$\mathbf{A} = \begin{pmatrix} 0 & 1 & 0 \\ 0 & 0 & 1 \\ A_3 & A_2 & A_1 \end{pmatrix},$$

with characteristic polynomial $P(\lambda) = \lambda^3 - A_1\lambda^2 - A_2\lambda - A_3) = (\lambda - \lambda_1)(\lambda - \lambda_2)(\lambda - \lambda_3)$. Considering the case of three distinct roots, for which the Jordan form of $\mathbf{A}$ is the diagonal matrix $\mathrm{Diag}(\lambda_1, \lambda_2, \lambda_3)$, leads to

$$\mathrm{e}^{\mathbf{A}s} = \begin{pmatrix} 1 & 1 & 1 \\ \lambda_1 & \lambda_2 & \lambda_3 \\ \lambda_1^2 & \lambda_2^2 & \lambda_3^2 \end{pmatrix} \begin{pmatrix} \mathrm{e}^{\lambda_1 s} & 0 & 0 \\ 0 & \mathrm{e}^{\lambda_2 s} & 0 \\ 0 & 0 & \mathrm{e}^{\lambda_3 s} \end{pmatrix} \begin{pmatrix} 1 & 1 & 1 \\ \lambda_1 & \lambda_2 & \lambda_3 \\ \lambda_1^2 & \lambda_2^2 & \lambda_3^2 \end{pmatrix}^{-1}. \qquad [6.11]$$

Similar formulas hold true for any $\mathrm{CAR}_1(p)$ model with distinct eigenvalues; the inverse matrix on the right-hand side of [6.11] is the inverse of the transpose of a Vandermonde matrix.

For illustration, let us take $A_3 = -4, A_2 = -6, A_1 = -4$, with the three distinct roots $\lambda_1 = -2, \lambda_2 = -1 + i$ and $\lambda_3 = -1 - i$.

Setting $f(s) = \mathrm{e}^{-2s} + 2\mathrm{e}^{-s}\sin(s)$, $g(s) = \mathrm{e}^{-2s} + \mathrm{e}^{-s}\left[2\sin(s) - \cos(s)\right]$, and $h(s) = (\mathrm{e}^{-2s} + \mathrm{e}^{-s}\left[\sin(s) - \cos(s)\right])/2$, yields

$$\mathrm{e}^{\mathbf{A}s} = \begin{pmatrix} f(s) & g(s) & h(s) \\ f^{(1)}(s) & g^{(1)}(s) & h^{(1)}(s) \\ f^{(2)}(s) & g^{(2)}(s) & h^{(2)}(s) \end{pmatrix}.$$

Therefore,

$$\mathbf{x}(s) = \begin{pmatrix} X(s) \\ X^{(1)}(s) \\ X^{(2)}(s) \end{pmatrix} = \mathrm{e}^{\mathbf{A}s} \begin{pmatrix} X(0) \\ X^{(1)}(0) \\ X^{(2)}(0) \end{pmatrix} + B\mathrm{e}^{\mathbf{A}s} \begin{pmatrix} \frac{1}{2}\left[\zeta_1(s) - \zeta_2(s) - \zeta_3(s)\right] \\ \zeta_1(s) + \zeta_3(s) \\ -2\zeta_1(s) + \zeta_2(s) - \zeta_3(s) \end{pmatrix},$$

where $\zeta_1(s) = \int_0^s \mathrm{e}^{2v} dW(v)$, $\zeta_2(s) = \int_0^s \mathrm{e}^v \sin(v) dW(v)$ and, finally, $\zeta_3(s) = \int_0^s \mathrm{e}^v \cos(v) dW(v)$. Figure 6.2 shows simulation of paths of the three components of $\mathbf{x}(s)$. Observe the loss of smoothness from $X(s)$ to $X^{(1)}(s)$ and then to $X^{(2)}(s)$.

6.3.3. *Conditions for stationarity of CAR$_1(p)$ processes*

Let X be a $\mathrm{CAR}_1(p)$ model, governed by [6.2], and let $\mathbf{x}$ be its vector form $\mathbf{x} = (X, X^{(1)}, \dots, X^{(p-1)})^T$ governed by [6.3]. The stationarity of $X$is equivalent to the stationarity of its $p-1$ derivatives.

PROPOSITION 6.4.– A scalar $\mathrm{CAR}_1(p)$ process X is stationary if and only if its vector form $\mathbf{x}$ is stationary.

PROOF.– If X is stationary, then $M_1(s) = M(s) \equiv M(0)$, and hence $M_j(s) = M^{(j-1)}(s) \equiv 0$ for $j \geq 2$, so $\mathbf{m}(s) = (M(0), 0, \cdots, 0)$ is a constant vector. We also have $C_{1,1}(s_1, s_2) = C(s_1, s_2) = R(s_2 - s_1)$, and then because of [6.9], for $1 \leq i, j \leq p$, we have $C_{i,j}(s_1, s_2) = \partial^{i+j-2}_{s_1^{i-1}, s_2^{j-1}} R(s_2 - s_1) = (-1)^{i-1} R^{(i+j-2)}(s_2 - s_1)$.

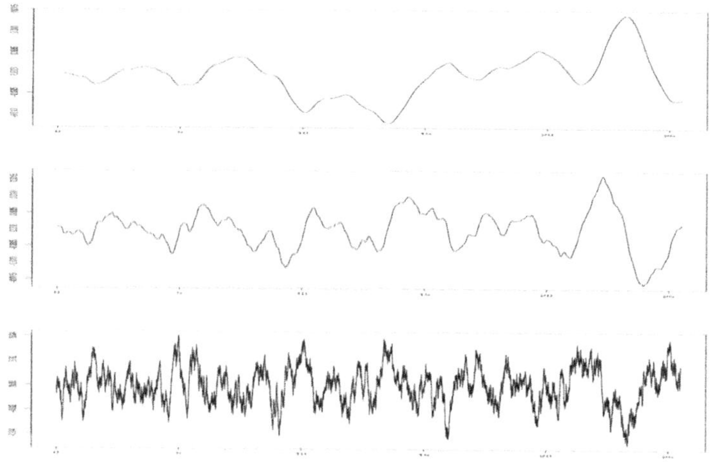

Figure 6.2. *Simulated paths of a stationary CAR*$_1(3)$ *with* $A_1 = -4,\ A_2 = -6, A_3 = -4$ *and* $B = 1$*. Top: The process* $X(s), s \in [0, 25]$*. Middle: Its first-order derivative* $X^{(1)}(s)$*. Bottom: Its second-order derivative* $X^{(2)}(s)$

The converse is trivial. □

THEOREM 6.1.– Let $\mathbf{A}$ and B in [6.4] be continuous functions. Let $\mathbf{x}(0)$ have mean $\mathbf{m}(0)$ and invertible covariance matrix Σ. Suppose also that $\mathbf{x}(0)$ is independent of $(W(s))_{s \geq 0}$.

Then $\mathbf{x}(s)$ is stationary if and only if the mean $\mathbf{m}(0)$ is null, $\mathbf{A}(s) = \mathbf{A}$ and $B(s) = B$ are constant, and Σ satisfies

$$\mathbf{A}\Sigma + \Sigma\mathbf{A}^T = -B^2\mathbf{E}_{p,p}. \qquad [6.12]$$

If this is the case, then $\mathbf{m}(s) \equiv 0$ and

$$\mathbf{C}(s_1, s_2) = \begin{cases} e^{\mathbf{A}(s_1 - s_2)}\,\Sigma, & \text{if } s_1 \geq s_2, \\ \Sigma\, e^{\mathbf{A}^T(s_2 - s_1)} & \text{if } s_1 \leq s_2. \end{cases} \qquad [6.13]$$

PROOF.– If $\mathbf{A}(s) = \mathbf{A}$ and $B(s) = B$, then $\boldsymbol{\mathcal{E}}(s) = e^{\mathbf{A}s}$, and hence $\mathbf{m}(s) = e^{\mathbf{A}s}\mathbf{m}(0) = \mathbf{0}$. For the covariance function, it is sufficient to consider [6.10] for $s_1 \geq s_2$. We compute $\mathbf{C}(s_1, s_2) = e^{\mathbf{A}(s_1 - s_2)}\mathbf{C}(s_2, s_2)$, and it only remains to prove that $\mathbf{C}(s, s)$ is constant. Since $\mathbf{A}$ commutes with $e^{-\mathbf{A}v}$, and $\mathbf{A}^T$ with $e^{-\mathbf{A}^T v}$, integrating by part yields $\int_0^s B^2 \mathrm{e}^{-\mathbf{A}v}\mathbf{E}_{pp}\mathrm{e}^{-\mathbf{A}^T v}dv = \mathrm{e}^{-\mathbf{A}s}\Sigma\mathrm{e}^{-\mathbf{A}^T s} - \Sigma$. Finally, because of [6.8], $\mathbf{C}(s,s) = \Sigma$ and $\mathbf{C}(s_1,s_2) = \exp^{\mathbf{A}(s_1-s_2)}\Sigma$.

Conversely, if $\mathbf{x}$ is stationary, then $\mathbf{m}(s) \equiv \mathbf{m}(0)$ and $\mathbf{C}(s_1,s_2) = \mathbf{R}(s_2-s_1)$ with $\mathbf{C}(s,s) = \mathbf{R}(0) = \Sigma$. First, $\mathbf{m}^{(1)}(s) = \mathbf{A}(s)\mathbf{m}(0) \equiv 0$, that implies $M_2 = \cdots = M_p = 0$ and $A_p(s)M_1 \equiv 0$ for all $s \geq 0$. Since A_p is not null, $M_1(0) = 0$, that is $\mathbf{m}(0) = 0$. Second, [6.10] becomes, for $s_1 \leq s_2$,

$$\mathbf{C}(s_1,s_2) = \mathbf{R}(s_2-s_1) = \mathcal{E}(s_1)\mathcal{E}^{-1}(s_2)\Sigma. \qquad [6.14]$$

Since $\mathcal{E}'(s) = \mathbf{A}(s)\mathcal{E}(s)$, we compute $\partial_s\mathcal{E}^{-1}(s) = -\mathcal{E}^{-1}(s)\partial_s\mathcal{E}(s)\mathcal{E}^{-1}(s) = -\mathcal{E}^{-1}(t)A(s)\mathcal{E}(s)\mathcal{E}^{-1}(s) = -\mathcal{E}^{-1}(s)A(s)$. So, differentiating both sides of [6.14] yields $\partial_{s_1}\mathbf{C}(s_1,s_2) = \mathbf{A}(s_1)\mathcal{E}(s_1)\mathcal{E}^{-1}(s_2)\Sigma$ and $\partial_{s_2}\mathbf{C}(s_1,s_2) = -\mathcal{E}(s_1)\mathcal{E}^{-1}(s_2)$ $\mathbf{A}(s_2)\Sigma$. Since $\partial_{s_2}\mathbf{C}(s_1,s_2) = \mathbf{R}'(s_2-s_1) = -\partial_{s_1}\mathbf{C}(s_1,s_2)$, we compute $\mathcal{E}(s_1)$ $\mathcal{E}^{-1}(s_2)\mathbf{A}(s_2)\Sigma = \mathbf{A}(s_1)\mathcal{E}(s_1)\mathcal{E}^{-1}(s_2)\Sigma$.

Since Σ is invertible, $\mathcal{E}^{-1}(s)\mathbf{A}(s)\mathcal{E}(s)$ is thus a constant matrix $\mathbf{L}$. In other words, $\mathcal{E}^{-1}(s)\mathbf{A}(s)\mathcal{E}(s) = \mathcal{E}^{-1}(s)\mathcal{E}'(s) = \mathbf{L}$, that is, $\mathcal{E}'(s) = \mathcal{E}(s)\mathbf{L}$. Therefore, $\mathcal{E}^T(s)$ is solution of $(\mathcal{E}^T)'(s) = \mathbf{L}^T\mathcal{E}^T(s)$, and hence $\mathcal{E}^T(s) = \mathrm{e}^{\mathbf{L}^T s}$. Since the latter matrix commutes with $\mathbf{L}^T$, we get $\mathcal{E}'(s) = \mathcal{E}(s)\mathbf{L} = \mathbf{L}\mathcal{E}(s)$. Recalling that $\mathcal{E}'(s) = \mathbf{A}(s)\mathcal{E}(s)$ and that $\mathcal{E}(s)$ is invertible yields $\mathbf{A}(s) = \mathbf{L}$, and thus $\mathbf{A}$ is constant.

Let us prove [6.12]. Using $\mathcal{E}(s) = \mathrm{e}^{\mathbf{A}s}$ and the stationarity for the variance, we get

$$\mathbf{C}(s,s) = \Sigma = \mathrm{e}^{\mathbf{A}s}\left[\Sigma + \int_0^s B^2(v)\mathrm{e}^{-\mathbf{A}v}\mathbf{E}_{pp}\mathrm{e}^{-\mathbf{A}^T v}dv\right]\mathrm{e}^{\mathbf{A}^T s}, \quad s \geq 0,$$

or $\mathrm{e}^{-\mathbf{A}s}\Sigma\mathrm{e}^{-\mathbf{A}^T s} = \Sigma + \int_0^s B^2(v)\mathrm{e}^{-\mathbf{A}v}\mathbf{E}_{pp}\mathrm{e}^{-\mathbf{A}^T v}dv$. Since $\mathbf{A}$ commutes with $\mathrm{e}^{-\mathbf{A}s}$, and $\mathbf{A}^T$ with $\mathrm{e}^{-\mathbf{A}^T s}$, differentiating both sides of the above equation yields [6.12]. In particular, $B^2(t)$ is the (p,p)-term of this equation, and hence B is constant. □

For making explicit the covariance matrix Σ of a stationary $\mathrm{CAR}_1(p)$ process in [6.12], let us define the companion matrix of second kind $\mathcal{A} = (\mathcal{A}_{j,r})_{1\leq j,r\leq p}$ associated with a vector $\mathbf{a} = (A_p,\ldots,A_1)^T$ by setting $A_0 = -1$ and

$$\begin{cases} \mathcal{A}_{j,r} = \mathbf{I}_{\{1\leq 2r-j\leq p+1\}}\,\tau_{2r-j,j}\,A_{p+j-2r+1}, & 1 \leq j \leq p-1, \\ \mathcal{A}_{p,r} = \mathbf{I}_{\{1\leq 2r-p\leq p\}}\,\tau_{2r-p,p}\,A_{2p-2r+1}, & j = p, \end{cases}$$

where $\tau_{i,j} = 0$ if $j-i$ is odd, and $\tau_{i,j} = (-1)^{(j-i)/2}$ if $j-i$ is even, for all $i,j \in \mathbb{N}$.

LEMMA 6.1.– If $\mathbf{x}$ is stationary with covariance matrix $\Sigma = (\sigma_{i,j})_{1\leq i,j\leq p}$, then the covariances are $\sigma_{i,j} = \tau_{i,j}\sigma^2_{(i+j)/2}$, for $1 \leq i,j \leq p$, and the variances $\sigma_{i,i} = \sigma_i^2$ are solution of the linear system

$$\mathcal{A}(\sigma_1^2, \ldots, \sigma_p^2)^T = (0, \ldots, 0, -B^2/2)^T. \qquad [6.15]$$

PROOF.– If $\mathbf{x}$ is stationary, then Σ is positive definite, and [6.12] becomes $\Sigma_{k=1}^p a_{i,k}\sigma_{k,j} + \Sigma_{k=1}^p a_{j,k}\sigma_{k,i} = -B^2\mathbb{1}_{\{(i,j)=(p,p)\}}$, for $1 \leq i,j \leq p$.

Since $a_{i,k} = \mathbb{1}_{\{k=i+1\}}$ for $1 \leq i \leq p-1$ and $a_{p,k} = A_{p-k+1}$, we get $\sigma_{i+1,j} = -\sigma_{j+1,i}$ for $1 \leq i,j \leq p-1$. Setting $j = i$ yields $\sigma_{i+1,i} = -\sigma_{i+1,i}$, that is $\sigma_{i+1,i} = 0$. For $i < j \leq p$, we recursively obtain $\sigma_{i,j} = \tau_{i,j}\sigma^2_{(j+i)/2}$.

The p remaining equations (when $i = p$) are $\Sigma_{k=1}^p a_{p,k}\sigma_{k,j} + \sigma_{j+1,p} = 0$ for $j = 1, \ldots, p-1$ and $\Sigma_{k=1}^p a_{p,k}\sigma_{k,p} = -B^2/2$, or equivalently,

$$\begin{cases} \Sigma_{k=1}^p A_{p-k+1}\, \tau_{k,j}\, \sigma^2_{(k+j)/2} + \tau_{j+1,p}\, \sigma^2_{(j+1+p)/2} = 0, & j = 1, \ldots, p-1, \\ \Sigma_{k=1}^p A_{p-k+1}\, \tau_{k,p}\, \sigma^2_{(k+p)/2} = -B^2/2. & \end{cases}$$

Since $\tau_{i,j}$ is defined for all $i,j \geq 0$ and satisfies $\tau_{j+1,p} = -\tau_{p+1,j}$, [6.15] holds. □

More precisely, the $p \times p$ covariance matrix Σ of a stationary $\mathrm{CAR}_1(p)$ process has the form

$$\Sigma = \begin{pmatrix} \sigma_1^2 & 0 & -\sigma_2^2 & 0 & \ddots \\ 0 & \sigma_2^2 & 0 & -\sigma_3^2 & \ddots \\ -\sigma_2^2 & 0 & \sigma_3^2 & 0 & \ddots \\ 0 & -\sigma_3^2 & 0 & \sigma_4^2 & \ddots \\ \ddots & \ddots & \ddots & \ddots & \ddots \end{pmatrix},$$

where the variances $\sigma_1^2, \ldots, \sigma_p^2$ satisfy [6.15]. Note that, for instance,

$$\mathcal{A} = \begin{pmatrix} A_2 & 1 \\ 0 & A_1 \end{pmatrix} \text{ for } p = 2, \quad \text{and} \quad \mathcal{A} = \begin{pmatrix} A_3 & -A_1 & 0 \\ 0 & A_2 & 1 \\ 0 & -A_3 & A_1 \end{pmatrix} \text{ for } p = 3.$$

COROLLARY 6.1.– A $\mathrm{CAR}_1(1)$ process X is stationary if and only if $A(s) = A$, $B(s) = B$, $\mathbf{E}(X(0)) = 0$ and $\mathrm{Var}(X(0)) - B^2/2A = \sigma^2$. Then its covariance function is given by $C(s_1, s_2) = \sigma^2 \mathrm{e}^{A|s_2 - s_1|}$.

A $\mathrm{CAR}_1(2)$ process X is stationary if and only if $A_2(s) = A_2$, $A_1(s) = A_1$, $B(s) = B$, $\mathbf{E}[\mathbf{x}(0)] = 0$ and $\mathrm{Cov}(\mathbf{x}(0)) = \Sigma = (\sigma_{i,j})_{1\leq i,j\leq 2}$ with $\sigma_{1,2} = \sigma_{2,1} = 0$, $\sigma_1^2 = B^2/2A_1A_2$ and $\sigma_2^2 = -B^2/2A_1$. Then the covariance function of $\mathbf{x}(s)$ is given by $\mathbf{C}(s_1, s_2) = \Sigma e^{\mathbf{A}^T(s_2-s_1)}$ if $s_1 \leq s_2$, and $e^{\mathbf{A}(s_1-s_2)}\Sigma$ if $s_1 \geq s_2$.

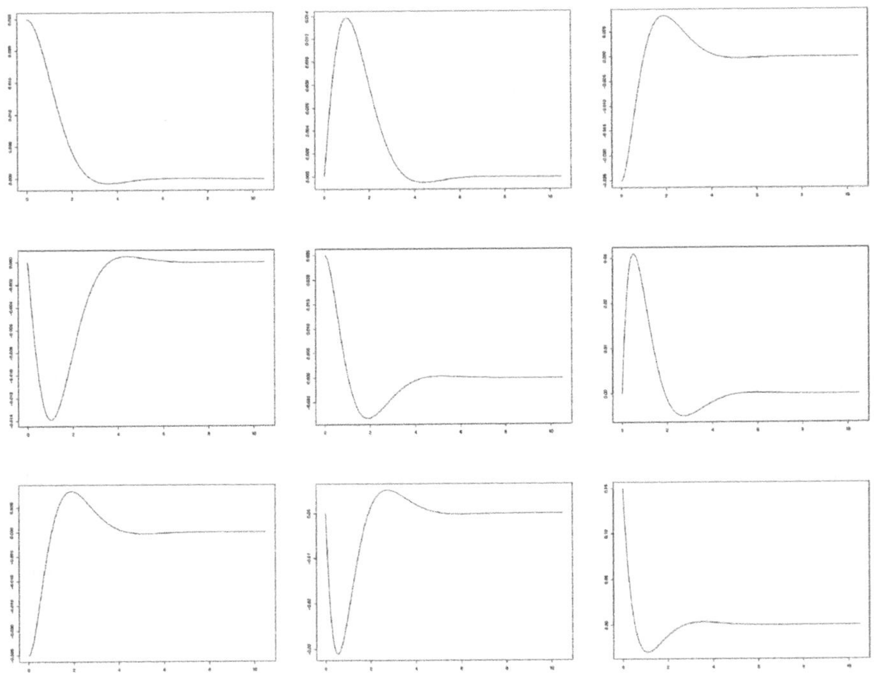

Figure 6.3. *Covariance structure of the stationary $\mathrm{CAR}_1(3)$ process with constants coefficients $A_1 = -4$, $A_2 = -6$, $A_3 = -4$ and $B = 1$. The graph of the function $C_{i,j}(h) = \mathbf{E}[X^{(i-1)}(s+h)X^{(j-1)}(s)]$, for $h \in [0, 10]$ and all s is at raw i and column j*

A $\mathrm{CAR}_1(3)$ process X is stationary if and only if $A_3(s) = A_3, A_2(s) = A_2$, $A_1(s) = A_1$, $B(s) = B$, $\mathbf{E}[\mathbf{x}(0)] = 0$, and

$$\mathrm{Cov}(\mathbf{x}(0)) = \Sigma = \frac{B^2}{2(A_3 + A_1A_2)} \begin{pmatrix} \frac{A_1}{A_3} & 0 & -1 \\ 0 & 1 & 0 \\ -1 & 0 & -A_2 \end{pmatrix}. \quad [6.16]$$

Then the covariance function of $\mathbf{x}(s)$ is given by $\mathbf{C}(s_1, s_2) = \Sigma e^{\mathbf{A}(s_2-s_1)}$ if $s_1 \leq s_2$ and $e^{\mathbf{A}|s_1-s_2|}\Sigma$ otherwise.

Figure 6.3 shows an example of a stationary covariance structure, namely the matrix coefficients $C_{i,j}(h) = \mathbf{E}[X^{(i-1)}(s+h)X^{(j-1)}(s)]$, for $1 \leq i, j \leq 3$, and $h \geq 0$. Note that the $C_{i,j}(h)$ are independent of s but are not symmetric, since, in general, $\mathbf{E}[X^{(i-1)}(s_1)X^{(j-1)}(s_2)] \neq \mathbf{E}[X^{(j-1)}(s_1)X^{(i-1)}(s_2)]$ for $i \neq j$.

Theorem 6.1 assumes for $\text{CAR}_1(p)$ stationary processes that the covariance matrix $\Sigma = \mathbf{E}[\mathbf{x}(0)\mathbf{x}^T(0)] = \mathbf{E}[\mathbf{x}(s)\mathbf{x}^T(s)]$ is invertible. This always holds true except in the following degenerate case.

PROPOSITION 6.5.– If $\mathbf{x}$ is stationary with singular covariance matrix Σ, then both coefficients A_p and B are null.

PROOF.– Since $\mathbf{x}$ is stationary, $\mathbf{E}[\mathbf{x}(s)] = \mathbf{0}$ and $\Sigma = \mathbf{E}[\mathbf{x}(s)\mathbf{x}^T(s)]$. So, if Σ is not invertible, then a non-zero vector $\boldsymbol{\nu} = (\nu_1, \ldots, \nu_p)^T$ exists such that $\Sigma\, \boldsymbol{\nu} = \mathbf{0}$. The variance of the centered scalar process $\mathbf{n}(s) = \boldsymbol{\nu}^T\mathbf{x}(s)$ is $\mathbf{E}[\mathbf{n}^2(s)] = \boldsymbol{\nu}^T\Sigma\boldsymbol{\nu} = 0$, and hence $\mathbf{n}(s) = 0$ for all $s \geq 0$, or, almost surely, $\nu_1 X(s) + \nu_2 X^{(1)}(s) + \cdots + \nu_p X^{(p-1)}(s) = 0$.

Since $\boldsymbol{\nu} \neq 0$, some index k, $1 \leq k \leq p$, exists such that $\nu_k \neq 0$ and $\nu_{k+1} = \cdots = \nu_p = 0$. If $k = 1$, then $X(s) = 0$ almost surely and nothing is to be proven. If $k > 1$,

$$X^{(k-1)}(s) = -\left[\frac{\nu_1}{\nu_k}X(s) + \frac{\nu_2}{\nu_k}X^{(1)}(s) + \cdots + \frac{\nu_{k-1}}{\nu_k}X^{(k-2)}(s)\right].$$

Differentiating $(p-k)$ times both sides of the above equation, we get

$$dX^{(p-1)}(s) = -\left[\frac{\nu_1}{\nu_k}X^{(p-k)}(s) + \frac{\nu_2}{\nu_k}X^{(p-k+1)}(s) + \cdots + \frac{\nu_{k-1}}{\nu_k}X^{(p-2)}(s)\right] ds.$$

In other words, whatever the value of k, for $2 \leq k \leq p$, the process X is governed by an AR dynamics with $A_p = B = 0$, whose solution $\mathbf{x}(s) = \exp^{\mathbf{A}s}\mathbf{x}(0)$ is de facto independent of the white noise W. □

Let us now link stationarity with the classical Lyapounov stability criterion. The following definition is given in Gantmacher (1998).

DEFINITION 6.3.– *Let* $d\mathbf{y}(t) = \mathbf{A}\mathbf{y}(t)dt$ *be a linear differential system. Let* P *be the characteristic polynomial associated with the matrix* $\mathbf{A}$*, defined by*

$$P(\lambda) = \det(\mathbf{A} - \lambda\mathbf{I}_p) = (\lambda - \lambda_1)^{n_1} \ldots (\lambda - \lambda_r)^{n_r}, \qquad [6.17]$$

with $n_1 + \cdots + n_r = p$. *When* $\mathcal{R}e(\lambda_j) < 0$ *for all* $j = 1, \ldots, r$, *the differential system is said to be asymptotically stable.*

Clearly, for a centered $\text{CAR}_1(1)$ process governed by $dX(s) = AX(s) + BdW(s)$, the classical condition $A < 0$ for stationarity is equivalent to the above definition of asymptotic stability. Actually, for any order p, theorem 6.1 is consistent with the classical condition for weak stationarity, equivalent to the asymptotic stability property (see also Brockwell 2000).

THEOREM 6.2.– A centered $\mathrm{CAR}_1(p)$ process X, governed by [6.2], with constant coefficients $\mathbf{A}$ and B, is weakly stationary if and only if $dy(t) = \mathbf{A}y(t)dt$ is an asymptotically stable system.

PROOF.– Suppose the characteristic polynomial [6.17] has $r \leq p$ distinct, roots λ_j, possibly complex. Let $\mathbf{v}_j = (V_{1,j}, \ldots, V_{p,j})$, for $j = 1, \ldots, r$, be eigenvectors associated with the λ_j, possibly complex.

First, $V_{p,j} \neq 0$ for all $j = 1, \ldots, r$. Indeed, since $A_p \neq 0$, the matrix $\mathbf{A}$ is regular and all λ_j are non-null. The very structure of $\mathbf{A}$, together with $\mathbf{A}^T\mathbf{v}_j = \lambda_j\mathbf{v}_j$, implies that $V_{p,j} \neq 0$, since otherwise $\mathbf{v}_j = 0$.

Next, because of theorem 6.1, if X is stationary, then a positive definite matrix Σ exists such that [6.12] holds true. Then, for any $(\lambda_j, \mathbf{v}_j)$, we have $\overline{\mathbf{v}}_j^T\mathbf{A}\Sigma\mathbf{v}_j + \overline{\mathbf{v}}_j^T\Sigma\mathbf{A}^T\mathbf{v}_j = -B^2\overline{\mathbf{v}}_j^T\mathbf{E}_{p,p}\mathbf{v}_j = -B^2\left\|\overline{\mathbf{v}}_j^T\mathbf{e}_p\right\|^2$, that is, $\overline{\lambda}_j\overline{\mathbf{v}}_j^T\Sigma\mathbf{v}_j + \lambda_j\overline{\mathbf{v}}_j^T\Sigma\mathbf{v}_j = -B^2\left\|\overline{\mathbf{v}}_j^T\mathbf{e}_p\right\|^2$. In other words, $(\overline{\lambda}_j+\lambda_j)(\overline{\mathbf{v}}_j^T\Sigma\mathbf{v}_j) = -B^2\left\|\overline{\mathbf{v}}_j^T\mathbf{e}_p\right\|^2 = -B^2\left\|V_{p,j}\right\|^2$. Since $V_{p,j} \neq 0$, and $\overline{\mathbf{v}}_j^T\Sigma\mathbf{v}_j > 0$ by the positive definiteness property, $\mathcal{R}e(\lambda) = (\overline{\lambda}_j + \lambda_j)/2 < 0$. The system is asymptotically stable.

Conversely, let us assume that the system is asymptotically stable and prove that a positive definite matrix Σ satisfying [6.12] exists. The matrix equation $\mathbf{U}\Sigma - \Sigma\mathbf{V} = \mathbf{W}$ has a unique solution Σ if and only if the eigenvalues of the matrices $\mathbf{U}$ and $\mathbf{V}$ are all distinct (see Gantmacher 1998, Volume 1, Chapter 8, section 3). Setting $\mathbf{U} = \mathbf{A}$, $\mathbf{V} = -\mathbf{A}^T$ and $\mathbf{W} = -B^2\mathbf{E}_{p,p}$, all the eigenvalues λ_j of $\mathbf{A}$ have negative real parts, whereas all the eigenvalues $-\lambda_j$ of $-\mathbf{A}^T$ have positive real parts. Consequently, [6.12] has a unique solution.

It remains to prove that Σ is positive definite. Since Σ^T is also a solution, due to uniqueness, $\Sigma = \Sigma^T$ is symmetric. Let us prove that Σ is regular. Let $\mathbf{u}^T = (U_1, \ldots, U_p)$ be such that $\Sigma\mathbf{u} = \mathbf{0}$. Then $\mathbf{u}^T\Sigma = 0$ and by [6.12], we get $0 = \mathbf{u}^T\mathbf{A}\Sigma\mathbf{u} + \mathbf{u}^T\Sigma\mathbf{A}^T\mathbf{u} = -B^2\left(\mathbf{u}^T\mathbf{e}_p\right)^2 = -B^2\left(U_p\right)^2$, that is, $U_p = 0$. Therefore, $\mathbf{A}\Sigma\mathbf{u} + \Sigma\mathbf{A}^T\mathbf{u} = \Sigma\mathbf{A}^T\mathbf{u} = -B^2\mathbf{e}_p.\mathbf{e}_p^T\mathbf{u} = -B^2U_p\mathbf{e}_p = \mathbf{0}$, proving that $\Sigma\mathbf{u}^{[1]} = 0$, where $\mathbf{u}^{[1]} = \mathbf{A}^T\mathbf{u} = (U_1^{[1]}, \ldots, U_p^{[1]})$ is such that $U_p^{[1]} = 0$. Iterating the process and setting $\mathbf{A}^T\mathbf{u}^{[l]} = \mathbf{u}^{[l+1]} = (U_1^{[l+1]}, \ldots, U_p^{[l+1]})$ yields $U_p^{[l+1]} = 0$ for $l \geq 1$. Setting $U_0^{[l]} = 0$ for $l \geq 1$ for convenience, the recursive formula becomes $U_j^{[l+1]} = U_{j-1}^{[l]} + A_{p-j+1}U_p^{[l]}$, for $j = 1, \ldots, p$ and $l \geq 1$, which yields recursively $U_{p-1} = 0, \ldots, U_1 = 0$, and hence Σ is regular.

Finally, since Σ is symmetric, an orthogonal matrix $\mathbf{O}$ exists such that $\Delta = \mathbf{O}\Sigma\mathbf{O}^T$ is a diagonal matrix, say, Δ =Diag$(d_1, \ldots, d_p)$, with real eigenvalues d_j and eigenvectors $\mathbf{v}_j$ satisfying $\Sigma\mathbf{v}_j = d_j\mathbf{v}_j$. Let us prove that all the d_j are positive. First, they are not null, since Σ is regular. Next, using [6.12], we get, for any $(d_j, \mathbf{v}_j)$,

$$\begin{aligned}\mathbf{v}_j^T\mathbf{A}\Sigma\mathbf{u}_j + \mathbf{v}_j^T\Sigma\mathbf{A}^T\mathbf{v}_j &= d_j\mathbf{v}_j^T\mathbf{A}\mathbf{v}_j + d_j\mathbf{v}_j^T\mathbf{A}^T\mathbf{v}_j = d_j\mathbf{v}_j^T(\mathbf{A} + \mathbf{A}^T)\mathbf{v}_j \\ &= -B^2(\mathbf{v}_j^T\mathbf{e}_p)^2 \leq 0.\end{aligned}$$

The asymptotic stability of the system with coefficient $\mathbf{A}$ implies that the symmetric matrix $-(\mathbf{A} + \mathbf{A}^T)$ has positive eigenvalues, and hence is definite positive, so that $\mathbf{v}_j^T(\mathbf{A} + \mathbf{A}^T)\mathbf{v}_j > 0$. Therefore, all d_j are positive and hence Σ is positive definite. □

6.4. Time transforms

Here we will study some properties of smooth time transforms of CAR$_1(p)$ processes. In section 6.4.1, we show, first, that the class of CAR$_1(p)$ processes with time-varying coefficients is stable under smooth time changes through exhibiting the equations of the corresponding SDEs driving the time transforms. Then, since weakly stationary CAR$_1(p)$ processes necessarily have constant coefficients, we exhibit necessary and sufficient conditions for a CAR$_1(p)$ process to be RWS, together with the unique (up to affine transformations) corresponding time change. Section 6.4.2 focuses on the logarithmic time change that led, in Gray and Zhang (1988), to the definition of multiplicative stationarity.

6.4.1. *Properties of time transforms*

A necessary step in the investigation of RWS CAR$_1(p)$ processes is showing how the SDE [6.2] is modified under a p-smooth time change. The first statement heavily relies on Faà di Bruno's formula and Bell polynomials and related results are given in the appendix; in particular, the notation is to be found therein.

THEOREM 6.3.– Let X be a CAR$_1(p)$ process satisfying [6.2], and let $Y(t) = X(\psi(t))$ define the time transform of X by a p-smooth time change $s = \psi(t)$. Then Y is a CAR$_1(p)$ process satisfying

$$dY^{(p-1)}(t) = \widetilde{\mathbf{a}}^T(t)\mathbf{y}(t)dt + \widetilde{B}(t)d\widetilde{W}(t), \qquad [6.18]$$

where $\mathbf{y}(t) = (Y(t), Y^{(1)}(t), \ldots, Y^{(p-1)}(t))$ satisfies

$$\mathbf{y}(t) = \mathbf{B}_{p-1,\psi}(t)\ \mathbf{x}(\psi(t)), \qquad [6.19]$$

$\widetilde{W}(t)$ is a standard Brownian motion, and

$$\left(\widetilde{\mathbf{a}}^T(t), -1\right) = [\psi^{(1)}(t)]^p \left(\mathbf{a}^T(\psi(t)), -1\right) \mathbf{B}_{p,\psi}^{-1}(t), \qquad [6.20]$$

$$\widetilde{B}(t) = [\psi^{(1)}(t)]^{p-1/2} B(\psi(t)). \qquad [6.21]$$

PROOF.– Any X satisfying [6.2] is $(p-1)$-times differentiable and $X^{(p-1)}$ may be only continuous, but nevertheless satisfies [6.4]. Since ψ is p-times differentiable, [6.29] applies to the first $(p-1)$ derivatives of $Y(t) = X(\psi(t))$ to yield $Y^{(j)}(t) = \sum_{l=1}^{j} X^{(l)}(\psi(t)) B_{j,l} \circ \boldsymbol{\psi}^{(1,j)}(t)$, for $1 \le j \le p-1$, or $\mathbf{y}(t) = \mathbf{B}_{p-1,\psi}(t)\ \mathbf{x}(\psi(t))$, where the matrix $\mathbf{B}_{p-1,\psi}(t)$ is given by [6.30]. Equivalently, [6.32] says that

$$\mathbf{x}(\psi(t)) = \mathbf{B}_{p-1,\phi}(\psi(t))\mathbf{y}(t). \qquad [6.22]$$

Setting $\beta_l(t) = B_{p-1,l} \circ \boldsymbol{\phi}^{(1,p-1)}(\psi(t))$, the pth component in the latter equation can be written as

$$X^{(p-1)}(\psi(t)) = Y^{(1)}(t)\beta_1(t) + \cdots + Y^{(p-2)}(t)\beta_{p-2}(t) + Y^{(p-1)}(t)\beta_{p-1}(t),$$

and hence satisfies

$$dX^{(p-1)}(\psi(t)) = \Big[\beta_1'(t)Y^{(1)}(t) + [\beta_1(t) + \beta_2'(t)]\, Y^{(2)}(t) + \cdots$$
$$+ \left[\beta_{p-2}(t) + \beta_{p-1}'(t)\right] Y^{(p-1)}(t)\Big]\, dt + \beta_{p-1}(t) dY^{(p-1)}(t). \qquad [6.23]$$

Substituting ϕ for ψ and setting $j = p-1$ in [6.31] yields

$$B_{p,l} \circ \boldsymbol{\phi}^{(1,p)}(s) = \phi^{(1)}(s) B_{p-1,l-1} \circ \boldsymbol{\phi}^{(1,p-1)}(s) + \left(B_{p-1,l} \circ \boldsymbol{\phi}^{(1,p-1)}\right)'(s).$$

Since, $\beta_l'(t) = \psi^{(1)}(t)(B_{p-1,l} \circ \boldsymbol{\phi}^{(1,p-1)})'(\psi(t))$ and $\psi^{(1)}(t)\phi^{(1)}(\psi(t)) = 1$, we get

$$\psi^{(1)}(t) B_{p,l} \circ \boldsymbol{\phi}^{(1,p)}(\psi(t)) = B_{p-1,l-1} \circ \boldsymbol{\phi}^{(1,p-1)}(\psi(t))$$
$$+\psi^{(1)}(t)\left(B_{p-1,l} \circ \boldsymbol{\phi}^{(1,p-1)}\right)'(\psi(t)),$$

meaning that $\beta_{l-1}(t) + \beta_l'(t) = \psi^{(1)}(t) B_{p,l} \circ \boldsymbol{\phi}^{(1,p)}(\psi(t))$ holds not only for $l = 1, \ldots, p-1$ but also for $l = p$. Since

$$\left(0, B_{p,1} \circ \boldsymbol{\phi}^{(1,p)}, \ldots, B_{p,p-1} \circ \boldsymbol{\phi}^{(1,p)}, B_{p,p} \circ \boldsymbol{\phi}^{(1,p)}\right)(\psi(t)) = \mathbf{e}_{p+1}^T \mathbf{B}_{p,\phi}(\psi(t)),$$

and $B_{k,k} \circ \phi^{(1,k)}(\psi(t)) = (\phi^{(1)}(\psi(t)))^k = (\psi^{(1)}(t))^{-k}$ for $k \leq p$, [6.23] becomes, by writing $0 = 0 \times \eta^{(0)}$,

$$dX^{(p-1)}(\psi(t)) = \qquad [6.24]$$
$$\psi^{(1)}(t)\left[\mathbf{e}_{p+1}^T \mathbf{B}_{p,\phi}(\psi(t)) \begin{pmatrix} \mathbf{y}(t) \\ 0 \end{pmatrix} dt + B_{p,p} \circ \phi^{(1,p)}(\psi(t)) dY^{(p-1)}(t)\right].$$

Since $X^{(p-1)}(s)$ satisfies [6.1] with $F(s) = \mathbf{a}^T(s)\mathbf{x}(s)$, and according to [6.22], the time transform $\rho(t) = X^{(p-1)}(\psi(t))$ satisfies

$$\begin{aligned} d\rho(t) &= \psi'(t)\mathbf{a}^T(\psi(t))\mathbf{x}(\psi(t))dt + B(\psi(t))dW_\psi(t) \\ &= \psi^{(1)}(t)\mathbf{a}^T(\psi(t))\mathbf{B}_{p-1,\phi}(\psi(t))\mathbf{y}(t)dt + B(\psi(t))dW_\psi(t), \\ &= \psi^{(1)}(t)\left[(\mathbf{a}^T(\psi(t)), 0)\ \mathbf{B}_{p,\phi}(\psi(t))\begin{pmatrix} \mathbf{y}(t) \\ 0 \end{pmatrix} dt + \frac{B(\psi(t))}{\psi^{(1)}(t)} dW_\psi(t)\right] \end{aligned} \qquad [6.25]$$

Then [6.24] and [6.25] jointly yield

$$\begin{aligned} dY^{(p-1)}(t) = {} & [\psi^{(1)}(t)]^p \left(\mathbf{a}^T(\psi(t)), -1\right) \mathbf{B}_{p,\phi}(\psi(t)) \begin{pmatrix} \mathbf{y}(t) \\ 0 \end{pmatrix} dt \\ & + B(\psi(t))[\psi^{(1)}(t)]^{p-1} dW_\psi(t). \end{aligned}$$

Taking into account both the properties of $\mathbf{B}_{p,\psi}$ given in lemma 6.2, below, and of the Brownian time transform given in proposition 6.2, this ends the proof. □

Theorem 6.3 reads as follows for some simple examples of time transforms.

COROLLARY 6.2.– The time transform $Y(t) = X(\psi(t))$ of a $\text{CAR}_1(p)$ process X with constant coefficients $\mathbf{a}(s) = \mathbf{a}$ and $B(s) = B$ is a $\text{CAR}_1(p)$ process with time-varying coefficients $\widetilde{\mathbf{a}}(t)$ and $\widetilde{B}(t)$ given by $(\widetilde{\mathbf{a}}^T(t), -1) = [\psi^{(1)}(t)]^p(\mathbf{a}^T, -1)$ $\mathbf{B}_{p,\psi}^{-1}(t)$ and $\widetilde{B}(t) = [\psi^{(1)}(t)]^{p-1/2}B$.

If, moreover, $\mathbf{x}$ is weakly stationary with covariance matrix function given by [6.13] and Σ satisfying [6.12], then the $\text{CAR}_p(1)$ process $\mathbf{y}$ associated with Y is centered, with covariance matrices $\widetilde{\mathbf{C}}(t_1, t_2) = \mathbf{B}_{p-1,\psi}(t_1)e^{\mathbf{A}(t_1-t_2)}\ \Sigma\ \mathbf{B}_{p-1,\psi}^T(t_2)$ for $t_1 \geq t_2$.

COROLLARY 6.3.– The linear time transform $Y(t) = X(\alpha t + s^*)$ of a $\text{CAR}_1(p)$ process $X($ with coefficients $\mathbf{a}(s)$ and $B(s)$ is a $\text{CAR}_1(p)$ process with coefficients $\widetilde{\mathbf{a}}(t) = (\widetilde{A}_p(t), \ldots, \widetilde{A}_1(t)$ and $\widetilde{B}(t)$ given by $\widetilde{A}_j(t) = \alpha^j A_j(\alpha t + s^*)$, $j = 1, \ldots, p$, and $\widetilde{B}(t) = \alpha^{p-1/2}B(\alpha t + s^*)$.

PROOF.– According to lemma 6.3, the linear time change $\psi(t) = \alpha t + s^*$ yields the diagonal matrix $\mathbf{B}_{p,\psi}(t) = \mathbf{D}_{p,\alpha}$ with inverse $\mathbf{B}_{p,\phi}(\psi(t)) = \mathbf{D}_{p,\frac{1}{\alpha}}$. □

For the specific case of a stationary AR process and an affine time change, the formulas simplify as follows.

COROLLARY 6.4.– Let X be a stationary $\mathrm{CAR}_1(p)$ process, with constant coefficients $(A_p, \ldots, A_1)$ and B. Let $\mathbf{x}$ be the associated $\mathrm{CAR}_p(1)$ process, with covariance matrix function given by [6.13] where Σ satisfies [6.12].

Then, any linear time transform $Y(t) = X(\alpha t + s^*)$ is a stationary $\mathrm{CAR}_1(p)$ process too, with coefficients $\widetilde{A}_j = \alpha^j A_j$ for $j = 1, \ldots, p$, and $\widetilde{B} = \alpha^{p-1/2} B$. The associated $\mathrm{CAR}_p(1)$ process $\mathbf{y}(t)$ is centered with covariance $\widetilde{\mathbf{C}}(t_1, t_2) = \mathrm{e}^{\widetilde{\mathbf{A}}(t_1 - t_2)} \widetilde{\Sigma}$ for $t_1 \geq t_2$, where $\mathrm{e}^{\widetilde{\mathbf{A}} u} = \mathbf{D}_{p-1,\alpha} \mathrm{e}^{\alpha \mathbf{A} u} \mathbf{D}_{p-1,\frac{1}{\alpha}}$, and $\widetilde{\Sigma} = \mathbf{D}_{p-1,\alpha} \Sigma \mathbf{D}_{p-1,\alpha}$ satisfies $\widetilde{\mathbf{A}} \widetilde{\Sigma} + \widetilde{\Sigma} \widetilde{\mathbf{A}}^T = -\widetilde{B}^2 \mathbf{E}_{p,p}$.

PROOF.– The coefficients of $\mathbf{y}(t)$ in corollaries 6.2 and 6.3 can be written in matrix form $\widetilde{\mathbf{A}} = \alpha \mathbf{D}_{p-1,\alpha} \mathbf{A} \mathbf{D}^{-1}_{p-1,\alpha}$, with $\mathbf{D}^{-1}_{p-1,\alpha} = \mathbf{D}_{p-1,\frac{1}{\alpha}}$. Therefore, $\alpha B^2 \mathbf{D}_{p-1,\alpha} \mathbf{E}_{p,p} \mathbf{D}_{\alpha} = \widetilde{B}^2 \mathbf{E}_{p,p}$. Easy matrix computation shows that [6.12] becomes $\widetilde{\mathbf{A}} \widetilde{\Sigma} + \widetilde{\Sigma} \widetilde{\mathbf{A}}^T = -\widetilde{B}^2 \mathbf{E}_{p,p}$ with $\widetilde{\Sigma} = \mathbf{D}_{p-1,\alpha} \Sigma \mathbf{D}_{p-1,\alpha}$. For the covariance function, it is sufficient to note that $\mathrm{e}^{\widetilde{\mathbf{A}} t} = \mathbf{D}_{p-1,\alpha} \mathrm{e}^{\alpha \mathbf{A} t} \mathbf{D}^{-1}_{p-1,\alpha}$. □

The following result is a uniqueness theorem.

THEOREM 6.4.– If a $\mathrm{CAR}_1(p)$ process Y is RWS, then the time change ϕ is unique, up to an arbitrary ante affine time change – in other words, $\phi(s)$ and any $\widetilde{\phi}(s) = \phi(\alpha s + s^*)$, $\alpha > 0, s^* \in \mathbb{R}$ are equivalent time changes.

PROOF.– Suppose that ϕ_1 and ϕ_2 are both p-smooth time changes such that $X_1(s) = Y(\phi_1(s))$ and $X_2(s) = Y(\phi_2(s))$ both define stationary $AR_1(p)$ processes with respective constant coefficients $\mathbf{a}_1, B_1$ and $\mathbf{a}_2, B_2$. Considering the respective inverse time changes $\psi_1(t), \psi_2(t)$, we get $Y(t) = X_1(\psi_1(t)) = X_2(\psi_2(t))$, and thus, because of theorem 6.3, for $k = 1, 2$,

$$\left(\widetilde{\mathbf{a}}^T(t), -1\right) = (\psi_k^{(1)}(t))^p \left(\mathbf{a}_k^T, -1\right) \mathbf{B}^{-1}_{p,\psi_k}(t) \text{ and } \widetilde{B}(t) = (\psi_k^{(1)}(t))^{p-1/2} B_k. \quad [6.26]$$

First, setting $\alpha = (B_2/B_1)^{2/(2p-1)}$ gives $\psi_1^{(1)}(t) = \alpha \psi_2^{(1)}(t)$, that is $\psi_1(t) = \alpha \psi_2(t) + s^*$ for some real s^*, or equivalently $\phi_2(s) = \phi_1(\alpha s + s^*)$. In other words, ϕ_2 necessarily equals ϕ_1 up to an affine time change. Second, lemma 6.3 gives $\mathbf{B}^{-1}_{p,\psi_1}(t) = \mathbf{D}_{p,\frac{1}{\alpha}} \mathbf{B}^{-1}_{p,\psi_2}(t)$, so that [6.26] implies $\mathbf{a}_1 = \mathrm{Diag}(\alpha^p, \ldots, \alpha) \mathbf{a}_2$. □

We can now state a necessary and sufficient condition for a $\mathrm{CAR}_1(p)$ process with time-varying coefficients to be RWS.

THEOREM 6.5.– Let Y be a $\mathrm{CAR}_1(p)$ process satisfying [6.18], where $\widetilde{B}$ is a.e. non-zero and $(p-1)$-differentiable and the companion matrix $\widetilde{\mathbf{A}}(t)$ of $\widetilde{\mathbf{a}}(t)$ is a.e. non-singular. Let $\psi(t)$ be a primitive of $\widetilde{B}(t)^{2/(2p-1)}$. Then $Y(t)$ is RWS if and only if

$$[\psi^{(1)}(t)]^{-p}\left(\widetilde{\mathbf{a}}^T(t),-1\right)\mathbf{B}_{p,\psi}(t)=(\mathbf{a}^T,-1), \qquad [6.27]$$

is a time constant vector, and if some $t_0\in\mathbb{R}$ exists such that $\mathbb{E}[\mathbf{y}(t_0)]=\mathbf{0}$, and $\widetilde{\Sigma}=\mathbb{E}[\mathbf{y}(t_0)\mathbf{y}^T(t_0)]$ satisfies

$$\mathbf{A}\left[\mathbf{B}_{p-1,\psi}^{-1}(t_0)\widetilde{\Sigma}\mathbf{B}_{p-1,\psi}^{-1,T}(t_0)\right]+\left[\mathbf{B}_{p-1,\psi}^{-1}(t_0)\widetilde{\Sigma}\mathbf{B}_{p-1,\psi}^{-1,T}(t_0)\right]\mathbf{A}^T=-\mathbf{E}_{p,p}, \qquad [6.28]$$

where $\mathbf{A}$ is the companion matrix of $\mathbf{a}$.

If both [6.27] and [6.28] are satisfied, then the stationary $\mathrm{CAR}_1(p)$ process $X(s)=Y(\phi(s))$ is obtained by a unique, up to an ante affine time change, $\phi(s)$, that is the inverse of $\psi(t)$.

PROOF.– Suppose that $Y(t)=X(\widehat{\psi}(t))$ is the time transform of a stationary $\mathrm{CAR}_1(p)$ process X with time constant coefficients $\mathbf{a}$, B, and variance matrix Σ. By [6.20] and [6.21], we have $\psi^{(1)}(t)=\alpha\widehat{\psi}^{(1)}(t)$ with $\alpha=B^{2/(2p-1)}$, that is $\psi(t)=\alpha\widehat{\psi}(t)+s^*$, for some $s^*\in\mathbb{R}$. By the uniqueness theorem 6.4, we can thus take $\widehat{\psi}=\psi$, that is set $B^2=1$. Again using [6.20] in the reverse way, we get [6.27], which is indeed constant. Since X is stationary, theorem 6.1 yields that $\mathbb{E}[\mathbf{x}(s)]=0$ for all s, and that the covariance matrix is a constant function, precisely $\mathrm{Cov}(\mathbf{x}(s))=\Sigma$. In particular, for $s=\psi(t_0)$, we get $\mathbb{E}[\mathbf{y}(t_0)]=0$. Moreover, since $B^2=1$, Σ satisfies [6.12]. Therefore, [6.28] is proven, because [6.19] induces that $\widetilde{\Sigma}=\mathrm{Cov}(\mathbf{y}(t_0)=\mathbf{B}_{p-1,\psi}(t_0)\Sigma\mathbf{B}_{p-1,\psi}^T(t_0)$, or $\Sigma=\mathbf{B}_{p-1,\psi}^{-1}(t_0)\widetilde{\Sigma}\mathbf{B}_{p-1,\psi}^{-1,T}(t_0)$.

Let us now assume that both [6.27] and [6.28] hold. Since $\widetilde{B}(t)\neq 0$ a.e., ψ is strictly increasing. Let ϕ denote the inverse of ψ. Note that the derivative $\phi^{(1)}(s)$ satisfies the implicit equation $\phi^{(1)}(s)=\widetilde{B}^2(\phi(s)^{-1/(2p-1)}$. Because of theorem 6.3, the process defined by $X(s)=Y(\phi(s))$ is a $\mathrm{CAR}_1(p)$ process with coefficients $\mathbf{a}(s)$ and $B(s)$ satisfying

$$\begin{aligned}B^2(s)&=[\phi^{(1)}(s)]^{2p-1}\widetilde{B}^2(\phi(s))=\widetilde{B}^{-2}(\phi(s))\widetilde{B}^2(\phi(s))=1,\\(\mathbf{a}^T(s),-1)&=[\phi^{(1)}(s)]^p\left(\widetilde{\mathbf{a}}^T(\phi(s)),-1\right)\mathbf{B}_{p,\phi}^{-1}(s)\\&=[\psi^{(1)}(\phi(s))]^{-p}\left(\widetilde{\mathbf{a}}^T(\phi(s)),-1\right)\mathbf{B}_{p,\psi}(\phi(s)).\end{aligned}$$

Taking $t=\phi(s)$, under condition [6.27], we get that $(\mathbf{a}^T(s),-1)=[\psi^{(1)}(t)]^{-p}$ $(\widetilde{\mathbf{a}}^T(t),-1)\mathbf{B}_{p,\psi}(t)$ is a constant vector $(\mathbf{a}^T,-1)$. In other words, the time transform X is an $\mathrm{CAR}_1(p)$ process with constant coefficients $\mathbf{a}$ and $B=1$.

Finally, theorem 6.1 applies to prove that $X(s)$ is stationary. Since ψ can be any primitive, let us take $\psi(t_0) = 0$. The condition $\mathbb{E}[\mathbf{y}(t_0)] = 0$ thus becomes $\mathbb{E}[\mathbf{x}(0)] = 0$. Again using [6.19], we get $\Sigma = \mathbf{B}_{p-1,\psi}^{-1}(t_0)\widetilde{\Sigma}\mathbf{B}_{p-1,\psi}^{-1,T}(t_0)$. In other words, [6.28] says that Σ satisfies [6.12], which proves that X is stationary. □

For $p = 1$ and $p = 2$, theorem 6.5 reads as follows.

COROLLARY 6.5.– Let $Y(t)$ be a CAR$_1(1)$ process satisfying $dY(t) = \widetilde{A}(t)Y(t)dt + \widetilde{B}(t)d\widetilde{W}(t)$, with a.e. non-null coefficients $\widetilde{A}$ and $\widetilde{B}$. Then Y is RWS if and only if $\mathbf{E}[\mathbf{y}(t_0)] = 0$ for some t_0 and $\widetilde{A}(t)\widetilde{B}^{-2}(t)$ is a negative constant.

Let $Y(x)$ be an CAR$_1(2)$ process satisfying $dY^{(1)}(t) = [\widetilde{A}_2(t)Y(t) + \widetilde{A}_1(t)$ $Y^{(1)}(t)]dt + \widetilde{B}(t)d\widetilde{W}(t)$, with a.e. non-null coefficients $\widetilde{A}_2(t)$ and $\widetilde{B}(t)$. Then Y is RWS if and only if $\mathbf{E}[\mathbf{y}(t_0)] = 0$ for some t_0, and both $\widetilde{B}^{-\frac{4}{3}}(t)\widetilde{A}_2(t)$ and $\widetilde{B}^{-\frac{2}{3}}(t)[\widetilde{A}_1(t) - 2\widetilde{B}^{(1)}(t)/3\widetilde{B}(t)]$ are both negative and constant.

PROOF.– For $p = 1$, the inverse of ψ is given by $\psi^{(1)}(t) = \widetilde{B}^2(t)$, and the Bell polynomials are $\mathbf{B}_{0,0}(t) \equiv 1$ and $\mathbf{B}_{1,\psi}(t) = \text{Diag}(1, \psi^{(1)}(t))$ (see appendix). Therefore, condition [6.27] amounts to say that $\widetilde{B}^{-2}(t)\widetilde{A}(t)$ is a constant number A. Condition [6.28] thus becomes $2A\widetilde{\Sigma} = -1$. This holds true with a positive $\widetilde{\Sigma}$ if and only if $A < 0$.

For $p = 2$, we have $\widetilde{\mathbf{a}}^T(t) = (\widetilde{A}_2(t), \widetilde{A}_1(t))$, $\psi^{(1)}(t) = (\widetilde{B}^2(t))^{1/3}$, $\mathbf{B}_{1,\psi}(t) = \text{Diag}(1, \psi^{(1)}(t))$ and

$$\mathbf{B}_{2,\psi}(t) = \begin{pmatrix} 1 & 0 & 0 \\ 0 & \psi^{(1)}(t) & 0 \\ 0 & \psi^{(2)}(t) & \left(\psi^{(1)}(t)\right)^2 \end{pmatrix}.$$

Since $\psi^{(2)}(t) = 2/3\widetilde{B}^{-1/3}(t)\widetilde{B}^{(1)}(t)$, [6.27] asserts that $\widetilde{B}^{-4/3}(t)\widetilde{A}_2(t)$ and $\widetilde{B}^{-2/3}(t)\widetilde{A}_1(t) - \frac{2}{3}\widetilde{B}^{-1}(t)\widetilde{B}^{(1)}(t)$ are both constant, say A_2 and A_1. Condition [6.28] says that the covariance matrix $\widetilde{\Sigma} = (\widetilde{\sigma}_{i,j})_{1\leq i,j\leq 2}$ of $\mathbf{y}(t_0)$, such that

$$\Delta = \begin{pmatrix} 0 & 1 \\ A_2 & A_1 \end{pmatrix} \begin{pmatrix} 1 & 0 \\ 0 & (\psi^{(1)}(t_0))^{-1} \end{pmatrix} \begin{pmatrix} \widetilde{\sigma}_{1,1} & \widetilde{\sigma}_{1,2} \\ \widetilde{\sigma}_{1,2} & \widetilde{\sigma}_{2,2} \end{pmatrix} \begin{pmatrix} 1 & 0 \\ 0 & (\psi^{(1)}(t_0))^{-1} \end{pmatrix},$$

satisfies $\Delta + \Delta^T = -\mathbf{E}_{2,2}$. This holds if and only if $\widetilde{\sigma}_{1,2} = 0$, $\widetilde{\sigma}_{2,2} = -(2A_1)^{-1}$ $\psi^{(1)}(t_0)^2 > 0$ and $\widetilde{\sigma}_{1,1} = (2A_1A_2)^{-1} > 0$, that is if and only if both A_2 and A_1 are negative numbers. □

For instance, the CAR$_1(2)$ process satisfying $dY^{(1)}(t) = [A_2(1+t^2)^{-2}Y(t) + (A_1 - 2t)(1+t^2)^{-1}Y^{(1)}(t)]dt + B(1+t^2)^{-\frac{3}{2}}d\widetilde{W}(t)$, with $A_2 \neq 0$ and $B \neq 0$, is RWS if and only if $A_2 < 0$ and $A_1 < 0$. Then, $\psi^{(1)}(t) = B^{2/3}(1+t^2)^{-1}$. In that case, ψ is the arctangent function and ϕ the tangent time change.

6.4.2. *MS processes*

Here we will focus on the logarithm time change related to the notion of multiplicative stationarity. A centered process X is said to be MS if it is RWS via the logarithm function. Equivalently, the covariance function is ratio invariant, that is, $C(s_1, s_2) = R(s_1/s_2)$ (see Gray and Zhang 1988; Woodward *et al.* 2012).

Gray and Zhang (1988) defined and studied the MS transforms of stationary continuous-time AR processes, calling them continuous Euler processes. Later on, Vijverberg (2002) defined Euler sequences as the time series obtained by discretizing Euler processes at integer times (see also Vijverberg and Gray 2003).

Because of theorem 6.5, considering the form of the Bell matrix [6.35], more precisely the lower triangular $(p+1) \times (p+1)$-matrix of Stirling numbers of the second kind $\mathbf{S}_p = (B_{j,l}(1,\ldots,1))_{1\le j,l\le p}$ in [6.34] leads to the following necessary and sufficient condition for a $\text{CAR}_1(p)$ process to be MS.

COROLLARY 6.6.– Let Y be a $\text{CAR}_1(p)$ process defined for $t \geq 1$ by

$$dY^{(p-1)}(t) = \left[\widetilde{A}_p(t)Y(t) + \ldots \widetilde{A}_1(t)Y^{(p-1)}(t)\right] dt + \widetilde{B}(t)d\widetilde{W}(t), \quad t \geq 1,$$

with $\mathbf{E}[\mathbf{y}(1)] = \mathbf{0}$, and $\mathbf{E}[\mathbf{y}(1)\mathbf{y}^T(1)] = \widetilde{\Sigma}$.

Then Y is MS if and only if constants B and $\mathbf{a}^T = (A_p, \ldots, A_1)$, with companion matrix $\mathbf{A}$, exist such that

$$\widetilde{A}_j(t) = t^{-j}\sum_{k=0}^{j} A_{j-k}\left\{\begin{matrix} p-j+k \\ p-j \end{matrix}\right\}, \; j = 1,\ldots,p, \quad \widetilde{B}(t) = Bt^{\frac{1}{2}-p},$$

and the definite positive matrix $\Sigma = \mathbf{S}_{p-1}\mathbf{D}_{p-1,t}\widetilde{\Sigma}\mathbf{D}_{p-1,t}\mathbf{S}_{p-1}^T$ satisfies [6.12].

PROOF.– Theorem 6.5 and [6.35] together yield constant coefficients $\mathbf{a}$ and B such that $\widetilde{B}(t) = Bt^{-(p-1/2)}$ and $(\widetilde{\mathbf{a}}^T, -1) = (\mathbf{a}^T, -1)\mathbf{S}_p\text{Diag}(t^{-p}, \ldots, t^{-1}, 1)$. Elementary matrix calculation shows that Σ satisfies [6.28], and hence [6.12]. □

For illustration, let us study the cases $p = 2$ and $p = 3$. First, because of corollary 6.6, an MS $\text{CAR}_1(2)$ process satisfies

$$dY^{(1)}(t) = \left[A_2t^{-2}Y(t) + (A_1 - 1)t^{-1}Y^{(1)}(t)\right] dt + Bt^{-\frac{3}{2}}d\widetilde{W}(t), \quad t \geq 1,$$

with $A_2 < 0$ and $A_1 < 0$. If this is the case, then $\psi^{(1)}(t) = B^{2/3}t^{-1}$, and hence the time change is indeed the logarithm function.

The time transform of a general $\text{CAR}_3(1)$ process $\mathbf{x}(s)$, via the time change $\Psi(t) = \log t$ for $t \geq 1$, is $\mathbf{y}(t) = \mathbf{B}_{2,\log}(t)\mathbf{x}((\log t)$, that is

$$\begin{pmatrix} Y \\ Y^{(1)} \\ Y^{(2)} \end{pmatrix}(t) = \begin{pmatrix} 1 & 0 & 0 \\ 0 & \frac{1}{t} & 0 \\ 0 & -\frac{1}{t^2} & \frac{1}{t^2} \end{pmatrix} \begin{pmatrix} X \\ X^{(1)} \\ X^{(2)} \end{pmatrix} (\log t).$$

Since $dX^{(2)}(s) = \left[A_3X(s) + A_2X^{(1)}(s) + A_1X(s)\right] ds + BdW(s)$ for $s \geq 0$, we get, for all $t \geq 1$,

$$dY^{(2)}(t) = \left[\frac{A_3}{t^3}Y(t) + \frac{A_2 + A_1 - 1}{t^2}Y^{(1)}(t) + \frac{A_1 - 3}{t}Y(t)\right] dt + \frac{B}{t^{5/2}}d\widetilde{W}(t).$$

The covariance Σ of the $\text{CAR}_3(1)$ process $\mathbf{x}$ associated with a stationary $\text{CAR}_1(3)$ process X is given by [6.16]. Therefore, the covariance function of the MS $\text{CAR}_1(3)$ process $\mathbf{y}$ associated with an MS $\text{CAR}_1(3)$ process Y is $\mathbf{E}[\mathbf{y}(t_1)\mathbf{y}^T(t_2)] = \mathbf{B}_{2,\log}(t_1)\mathrm{e}^{\mathbf{A}\log(t_1/t_2)}\Sigma\mathbf{B}^T_{2,\log}(t_2)$ for $t_1 \geq t_2$.

Figure 6.4 shows, on the left, simulated trajectories of the components $\left(X(s), X^{(1)}(s), X^{(2)}(s)\right)$ of a stationary $\text{CAR}_1(3)$ process $\mathbf{x}(t)$ for $0 \leq s \leq 20$. On the right, the three components $\left(Y(t), Y^{(1)}(t), Y^{(2)}(t)\right)$ associated with the MS process $Y(t) = X(\log t)$ are shown; due to the large values of $t = \exp(s)$, only the restricted time interval $[1, 3.125]$ is shown.

6.5. Conclusion

Among classical time changes in the literature are the power functions $\psi(t) = t^\beta$ with $\beta > 0$, exponential functions $\psi(t) = \beta^t$ with $\beta > 1$, periodically increasing functions such as $\psi(t) = (t + \sin(t)\cos(t))/2$, and inverse hyperbolic sine functions $\psi(t) = \frac{1}{\sqrt{a}} \operatorname{arcsinh}(2at + b/\sqrt{-\Delta})$, with discriminant $\Delta = b^2 - 4ac < 0$ for $a > 0$. Other examples are $\psi(t) = (t^\lambda - 1)/\lambda$ in Jiang and Woodward (2006) or polynomials of order 2 in Roberton and Woodward (2010); see also Woodward *et al.* (2012) and the references therein. All of them can be dealt with by using theorem 6.5 and expressing the associated $\mathbf{B}_{p,\psi}$ matrix of Bell polynomials defined in the appendix.

This chapter is concerned with expliciting the rules governing time changes for scalar AR models driven by a Brownian innovation. Most of the results can be extended to AR processes driven by other types of orthogonal innovations, for instance, martingales obtained as compensated Lévy processes. Further, (Girardin and Senoussi 2019) studies rules governing back and forth discretization and embedding between discrete and continuous-time first-order multivariate AR processes, driven by innovations of the semimartingale type, including Brownian motion and jump processes.

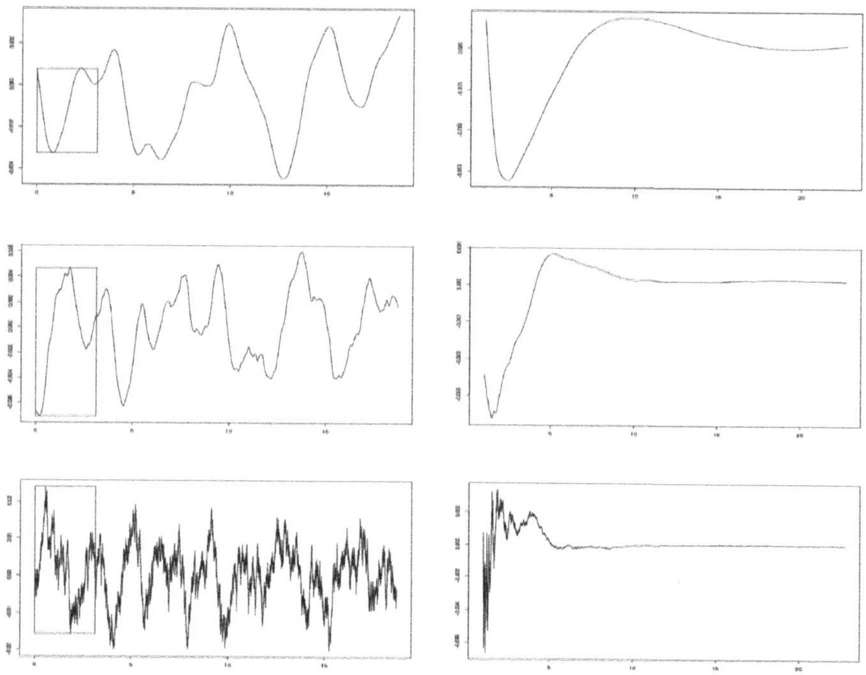

Figure 6.4. *Left: Paths of* $(X(s), X^{(1)}(s), X^{(2)}(s)), 0 \leq s \leq 20$ *for a stationary CAR*$_1(3)$ *process* X *with coefficients* $A_1 = -4$, $A_2 = -6, A_3 = -4$ *and* $B = 1$. *Right: Its time transform* $\mathbf{y}(t) = (Y(t) = X(\log t), Y^{(1)}(t), Y^{(2)}(t))$ *in [1,3.125] (rectangles within graphs on the left)*

6.6. Appendix

Here, some results on Faà di Bruno's formula with links to Bell polynomials are discussed. In the large and various literature on this topic, these results take different forms; see Aldrovandi (2001) and Wang and Wang (2007) and the references therein. We present them under the appropriate form to match their use in the main text, and consequently give the proofs for sake of completeness and clarity of reading.

Let f and ψ be k-times differentiable real functions defined on domains of $\mathbb{R}$ such that the composition $h(t) = f \circ \psi(t) = f(\psi(t))$ is well defined. Faà di Bruno's well-known formula gives the form of the derivatives $h^{(j)}(t)$ of h, namely

$$h^{(j)}(t) = \sum_{l=1}^{j} f^{(l)}(\psi(t))B_{j,l}\left(\psi^{(1)}(t), \ldots, \psi^{(j-l+1)}(t)\right), \ 1 \leq j \leq k, \quad [6.29]$$

where $B_{j,l}(u_1, \ldots, u_{j-l+1}), 1 \leq l \leq j$, are the partial or incomplete exponential Bell polynomials. The set of Bell polynomials forms a triangular array classically completed into a square array $\mathbf{B}$ of polynomials by the convention $B_{0,0} \equiv 1$, $B_{j,0} \equiv 0$ for all $j \geq 1$, and $B_{j,l} \equiv 0$ for all for $l > j$.

Even if $B_{j,l}$ depends only on the $j-l+1$ first variables, for the sake of presentation and in accordance with the notation $\mathbf{g}^{(l,j)}(t) = (g^{(l)}(t), \ldots, g^{(j)}(t))^T$ with $l \leq j$, we also set $B_{j,l}(\boldsymbol{\psi}^{(1,j-l+1)}(t)) = B_{j,l}(\boldsymbol{\psi}^{(1,j)}(t)) = B_{j,l} \circ \boldsymbol{\psi}^{(1,j)}(t)$. Then [6.29] can be written under the matrix form $\mathbf{h}^{(0,k)}(t) = \mathbf{B}_{k,\psi}(t)\mathbf{f}^{(0,k)}(\psi(t))$, where the $(k+1) \times (k+1)$ lower triangular matrix $\mathbf{B}_{k,\psi}(t) = (B_{j-1,l-1} \circ \boldsymbol{\psi}^{(1,j-1)}(t))$, with $1 \leq j, l \leq k+1$, is given by

$$\mathbf{B}_{k,\psi} = \begin{pmatrix} 1 & 0 & \cdots\cdots & 0 \\ 0 & B_{1,1} \circ \boldsymbol{\psi}^{(1)} & \ddots & \vdots \\ \vdots & \vdots & \ddots \quad \ddots & \vdots \\ \vdots & \vdots & \ddots & 0 \\ 0 & B_{k,1} \circ \boldsymbol{\psi}^{(1,k)} & \cdots\cdots & B_{k,k} \circ \boldsymbol{\psi}^{(1,k)} \end{pmatrix}. \qquad [6.30]$$

We are also interested in the form taken by [6.29] when ψ is a k-smooth time change.

LEMMA 6.2.– Let ψ be k-smooth time change with inverse function ϕ. Then, the coefficients of $\mathbf{B}$ given in [6.30] satisfy, for all $1 \leq l \leq j < k$,

$$B_{j+1,l} \circ \boldsymbol{\psi}^{(1,j+1)}(t) = \psi^{(1)}(t) B_{j,l-1} \circ \boldsymbol{\psi}^{(1,j)}(t) + \left(B_{j,l} \circ \boldsymbol{\psi}^{(1,j)}\right)'(t). \qquad [6.31]$$

Moreover, if $\psi^{(1)}(t) \neq 0$, then $\mathbf{B}_{k,\psi}(t)$ is also invertible and

$$\mathbf{B}_{k,\psi}^{-1}(t) = \mathbf{B}_{k,\phi}(\psi(t)). \qquad [6.32]$$

PROOF.– Using [6.29] for $h = f \circ \psi$, we have $h^{(j+1)}(t) = \sum_{l=1}^{j+1} f^{(l)}(\psi(t)) B_{j+1,l} \circ \boldsymbol{\psi}^{(1,j+1)}(t)$ for all $0 \leq j < k$. Differentiating yields

$$\begin{aligned} h^{(j+1)}(t) = & \sum_{l=1}^{j} \psi^{(1)}(t) f^{(l+1)}(\psi(t)) B_{j,l} \circ \boldsymbol{\psi}^{(1,j)}(t) \\ & + \sum_{l=1}^{j} f^{(l)}(\psi(t)) \left(B_{j,l} \circ \boldsymbol{\psi}^{(1,j)}\right)'(t). \end{aligned}$$

Since the function f is arbitrary, the identification of the terms associated with each $f^{(l)}$ gives [6.31].

Since the polynomials $B_{j,j}$ reduce to monomials $B_{j,j}(u) = u^j$ for $j \geq 1$, we get $\det(\mathbf{B}_{k,\psi}(t)) = \Pi_{j=0}^{k} B_{j,j}(\psi^{(1)}(t)) = [\psi^{(1)}(t)]^{k(k+1)/2} \neq 0$, proving that $\mathbf{B}_{k,\psi}(t)$ is invertible. If $\psi(t)$ is k-times differentiable and invertible at t, so is ϕ at $s = \psi(t)$. Therefore, [6.29] applies to $f(s) = h(\phi(s))$ to yield $\mathbf{f}^{(0,k)}(s) = \mathbf{B}_{k,\phi}(s)\mathbf{h}^{(0,k)}(\phi(s))$. Setting $s = \psi(t)$, we get $\mathbf{f}^{(0,k)}(\psi(t)) = \mathbf{B}_{k,\phi}(\psi(t))\mathbf{h}^{(0,k)}(t)$, which means exactly that $\mathbf{B}_{k,\psi}^{-1}(t) = \mathbf{B}_{k,\phi}(\psi(t))$. □

Two particular cases are of special interest.

Let $\psi(t)$ be a k-smooth time change with inverse $\phi(s)$. Let $\widetilde{\psi}(t) = \gamma\psi(\alpha t + s^*) + s^{**}$, with $\gamma \neq 0, \alpha \neq 0$, be a time change equivalent to ϕ up to affine changes. Clearly, the inverse $\widetilde{\phi}$ of $\widetilde{\psi}$ is of the same form, namely $\widetilde{\phi}(s) = \widetilde{\gamma}\phi(\widetilde{\alpha}s + \widetilde{s}^*)) + \widetilde{s}^{**}$, where $\widetilde{\gamma} = yy\alpha^{-1}$, $\widetilde{\alpha} = \gamma^{-1}$, $\widetilde{s}^* = -s^{**}\gamma^{-1}$ and $\widetilde{s}^{**} = -s^*\alpha^{-1}$.

LEMMA 6.3.– If $\widetilde{\psi}(t) = \gamma\psi(\alpha t + s^*) + s^{**}$, then

$$\mathbf{B}_{k,\widetilde{\psi}}(t) = \mathbf{D}_{k,\alpha}\mathbf{B}_{k,\psi}(\alpha t + s^*)\mathbf{D}_{k,\gamma}\,, \qquad [6.33]$$

where $\mathbf{D}_{k,\gamma} = Diag\left(1, \gamma, \gamma^2, \ldots, \gamma^k\right)$.

In particular, for $\widetilde{\psi}(t) = \alpha t + s^*$, [6.33] becomes $\mathbf{B}_{k,\widetilde{\psi}}(t) = \mathbf{D}_{k,\alpha}$.

PROOF.– We compute $\widetilde{\psi}^{(j)}(t) = \gamma\alpha^j\psi^{(j)}(\alpha t + s^*)$ and $B_{j,l}(\gamma\alpha u_1, \gamma\alpha^2 u_2, \ldots, \gamma\alpha^{j-l+1}u_{j-l+1}) = \gamma^l\alpha^j B_{j,l}(u_1, u_2, \ldots, u_{j-l+1})$. Therefore, we readily get $B_{j,l}(\widetilde{\psi}^{(1,j)}(t)) = \gamma^l\alpha^j B_{j,l}\left(\psi^{(1,j)}(\alpha t + s^*)\right)$ for $1 < l \leq j \leq k$, which in matrix form is [6.33]. For the particular case, just observe that $\psi(t) = t$ yields the constant matrix $\mathbf{B}_{k,\psi}(t) = \mathbf{I}_{k+1}$. □

The second particular case concerns the time changes $\psi(t) = \log t$, and $\phi(s) = \mathrm{e}^s$. It makes use of the classical symbols [ ] and { } for Stirling numbers of first and second kind, respectively, defined by

$$\begin{bmatrix} j \\ l \end{bmatrix} = B_{j,l}\left(0!, 1!, \ldots, (j-l)!\right) \quad \text{and} \quad \begin{Bmatrix} j \\ l \end{Bmatrix} = B_{j,l}\left(1, \ldots, 1\right). \qquad [6.34]$$

Since $\psi^{(j)}(t) = -(j-1)!(-t)^{-j}$ and $\phi^{(j)}(s) = \mathrm{e}^s$, for all $j \geq 1$, and noting that $B_{j,l}(\rho\alpha^1 u_1, \ldots, \rho\alpha^{j-l+1}u_{j-l+1}) = \rho^l\alpha^j B_{j,l}(u_1, \ldots, u_{j-l+1})$, we get

$$B_{j,l}\left(-0!(-t)^{-1}, -1!(-t)^{-2}, \ldots, -(j-l)!(-t)^{-(j-l+1)}\right) = (-1)^l(-t)^{-j}\begin{bmatrix} j \\ l \end{bmatrix}$$

and $B_{j,l}\left(\mathrm{e}^s, \ldots, \mathrm{e}^s\right) = \mathrm{e}^{ls}\begin{Bmatrix} j \\ l \end{Bmatrix}$.

In other words,

$$\mathbf{B}_{k,\log}(t) = \mathbf{D}_{k,(-t)^{-1}} \left(\begin{bmatrix} j \\ l \end{bmatrix} \right)_{0 \le l,j \le k} \mathbf{D}_{k,-1},$$

and

$$\mathbf{B}_{k,\exp}(s) = \left(\begin{Bmatrix} j \\ l \end{Bmatrix} \right)_{0 \le l,j \le k} \mathbf{D}_{k,e^s},$$

leading, as a direct consequence of lemma 6.2, to the inverse formula

$$\mathbf{B}_{k,\log}^{-1}(t) = \mathbf{B}_{k,\exp}(\log t) = \left(\begin{Bmatrix} j \\ l \end{Bmatrix} \right)_{0 \le l,j \le k} \mathbf{D}_{k,t}. \qquad [6.35]$$

6.7. References

Aldrovandi, R. (2001). *Special Matrices of Mathematical Physics: Stochastic, Circulant and Bell Matrices*. World Scientific Publishing, Singapore.

Brockwell, P.J. (2000). Heavy-tailed and non-linear continuous-time ARMA models for financial time series. In *Proceedings of the Hong Kong International Workshop on Statistics and Finance: An Interface: Centre of Financial Time Series, The University of Hong Kong, July 4–8, 1999*, Chan, W.-S., Li, W.K., Tong, H. (eds). Imperial College Press, London.

Gantmacher, F.R. (1998). *The Theory of Matrices*. Chelsea Publishing Company, New York.

Girardin, V., and Senoussi, R. (2019). Filling the gap between continuous and discrete time dynamics of autoregressive processes. *Journal of Time Series Analysis*. Available at: https://doi.org/10.1111/jtsa.12507.

Gray, H.L., and Zhang, N.F. (1988). On a class of nonstationary processes. *Journal of Time Series Analysis*, 9, 133–154.

Jacod, J. (1982). Équations différentielles stochastiques linéaires : la méthode de variation des constantes. *Séminaire de Probabilité (Strasbourg)*, 16(6), 442–446.

Jiang, H., Gray, H.L., and Woodward, W.A. (2006). Time-frequency analysis g(λ) stationary processes. *Computational Statistics & Data Analysis*, 51, 1997–2028.

Kobayashi, K. (2011). Stochastic calculus for a time-changed semimartingale and the associated stochastic differential equations. *Journal of Theoretical Probalility*, 24, 789–820.

Loève, M. (1978). *Probability Theory II*, 4th edition. Springer, New York.

Perrin, O., and Senoussi, R. (1999). Reducing non-stationary stochastic processes to stationarity by a time deformation. *Statistics & Probability Letters*, 43, 393–397.

Protter, P.E. (2005). *Stochastic Integration and Differential Equations: A New Approach*. Springer, Berlin.

Roberton, S.D., Gray, H.L., and Woodward, W. (2010). The generalized linear chirp process. *Journal of Statistical Planning and Inference*, 140, 3676–3687.

Sampson, D., and Guttorp, P. (1992). Nonparametric estimation of nonstationary spatial covariance structure. *JASA*, 87, 108–119.

Vijverberg, C. (2002). Discrete multiplicative stationary processes. Ph.D. thesis, Department of Statistical Science, Southern Methodist University.

Vijverberg, C.P., and Gray, H.L. (2003). Time deformation and the m-spectrum—A new tool for cyclical analysis. Technical Report SMU-T-314, Department of Statistical Science, Southern Methodist University.

Wang, W., and Wang, T. (2007). Matrices related to the bell polynomials. *Linear Algebra and its Applications*, 422, 139–154.

Woodward, W., Gray, H.L., and Elliot, A. (2012). *Applied Time Series Analysis*. CRC Press, Boca Raton.

7

Sequential Estimation for Non-parametric Autoregressive Models

In this chapter, we present sequential estimation methods for non-parametric autoregression models. First, we consider the pointwise estimation problem and then we study the estimation problem for the quadratic risk. As to the pointwise estimation, we develop truncated sequential kernel estimation procedures for which we provide the minimax properties in an adaptive setting. It should be emphasized that for autoregressive models the adaptive minimax estimation is possible only in the sequential analysis framework. Moreover, using the sequential kernel estimators we develop new model selection estimation procedures for which we show non-asymptotic sharp oracle inequalities for the robust quadratic risk. This means that the constructed procedures are optimal in the sharp oracle non-asymptotic inequalities sense.

7.1. Introduction

The tradition of considering the problem of statistical estimation as that of estimating a finite number of parameters goes back to Fisher. Statistical models that explain the data more deeply are usually more complex: the unknowns of these models are, in general, some functions with certain properties of regularity. The problem of non-parametric estimation consists of estimating, from the observations, an unknown function belonging to a certain rather large functional class.

The theory of non-parametric estimation has been considerably developed in the last three decades, focusing on some key themes, in particular the study of optimality of estimators and adaptive estimation. We are interested in particular in the optimality of the estimators when the sample size tends to infinity, such estimators

Chapter written by Ouerdia ARKOUN, Jean-Yves BRUA and Serguei PERGAMENCHTCHIKOV.

are called asymptotically efficient. Numerous problems of asymptotic efficiency have been studied over the past 40 years, both in parametric and non-parametric framework and for different models.

We focus here on showing the asymptotic efficiency of some estimators for the following autoregressive model:

$$y_k = S(x_k) y_{k-1} + \xi_k \,, \quad x_k = a + \frac{k}{n}(b-a) \,, \quad 1 \le k \le n \,, \qquad [7.1]$$

where $S(\cdot)$ is an unknown function, the initial value y_0 is a known fixed constant and the noise $(\xi_k)_{k\geq 1}$ is an i.i.d. sequence of unobservable random variables with $\mathbb{E}\xi_k = 0$ and $\mathbb{E}\xi_k^2 = 1$.

Model [7.1] is a generalization of an autoregressive process of the first order. It should be noted that the varying coefficient principle is well known in the regression analysis. It permits the use of more complex forms for regression coefficients and, therefore, the models constructed via this method are more adequate for applications (see, for example, Fan and Zhang 2008; Luo *et al.* 2009). There is a number of papers that study these varying coefficient autoregressive models. In Dahlhaus (1996b), the process [7.1] is considered with the function S having a parametric form. Moreover, Dahlhaus (1996a) studied spectral properties of the stationary process [7.1] with the non-parametric function S. Belitser (2000) considered this model with Lipschitz conditions and proposed a recursive estimator (the author established the convergence rate for quadratic risk). In all these papers, the authors propose some asymptotic (as $n \to \infty$) methods for different identification studies without considering optimal issues (in the minimax sense).

To our knowledge, for the first time, the minimax estimation problem for model [7.1] has been treated in Moulines *et al.* (2005) in the non-adaptive case, i.e. when the regularity of S is known. More specifically, it was shown that the convergence rate is optimal for the quadratic risk by using a recursive method for the autoregressive model of order d.

This chapter discusses the development of an adaptive sequential procedure for the estimation of the unknown functional coefficient in model [7.1]. The minimax non-adaptive and adaptive convergence rates are established for a Hölderian functional class, and an asymptotically efficient estimator for the absolute error risk is constructed. Then, in a framework of quadratic risk optimization, for non-parametric autoregressive processes, we propose new sequential model selection procedures that provide sharp oracle non-asymptotic inequalities for both robust and usual quadratic risks.

7.2. Main conditions

As in Arkoun and Pergamenchtchikov (2016), we assume that in the model [7.1] the i.i.d. random variables $(\xi_k)_{k\geq 1}$ have a density p (with respect to Lebesgue measure) from the functional class $\mathcal{P}_\varsigma$ defined as

$$\mathcal{P}_\varsigma := \left\{ p \geq 0 : \int_{-\infty}^{+\infty} p(x)\mathrm{d}x = 1\,, \int_{-\infty}^{+\infty} x\,p(x)\,\mathrm{d}x = 0\,, \right.$$

$$\left. \int_{-\infty}^{+\infty} x^2\,p(x)\,\mathrm{d}x = 1\,, \quad \sup_{k\geq 1} \frac{\int_{-\infty}^{+\infty} x^{2k}\,p(x)\,\mathrm{d}x}{\varsigma^k (2k-1)!!} \leq 1 \right\}\,, \qquad [7.2]$$

where $\varsigma \geq 1$ is some parameter, which may be a function of the number observation n, i.e. $\varsigma = \varsigma(n)$, such that for any $\delta > 0$

$$\lim_{n\to\infty} \frac{\varsigma(n)}{n^\delta} = 0\,. \qquad [7.3]$$

Note that the $(0,1)$-Gaussian density belongs to $\mathcal{P}_\varsigma$. In the sequel, we denote this density by p_0. It is clear that for any $q > 0$

$$\mathbf{m}_q^* = \sup_{p\in\mathcal{P}} \mathbf{E}_p\,|\xi_1|^q < \infty\,, \qquad [7.4]$$

where $\mathbf{E}_p$ is the expectation with respect to the density p from $\mathcal{P}_\varsigma$.

To obtain the stability (uniformly with respect to the function S) of model [7.1], we assume that for some fixed $0 < \varepsilon < 1$ and $L > 0$ the unknown function S belongs to the ε - *stability set* introduced in Arkoun and Pergamenchtchikov (2016) as

$$\Theta_{\varepsilon,L} = \left\{ S \in \mathbf{C}_1([a,b],\mathbb{R}) : |S|_* \leq 1-\varepsilon \quad \text{and} \quad |\dot{S}|_* \leq L \right\}\,,$$

where $\mathbf{C}_1([a,b],\mathbb{R})$ is the Banach space of continuously differentiable $[a,b] \to \mathbb{R}$ functions and $|S|_* = \sup_{a\leq x\leq b}|S(x)|$. Now, using this set we introduce the corresponding Hölder classes, which we will use to study the model [7.1].

DEFINITION 7.1.– *Let* $L > 0$, $0 < \varepsilon < 1$ *and* $1 < \beta < 2$. *Then the stable Hölder class at the point* $a < z_0 < b$ *is defined by*

$$\mathcal{H}^{(\beta)}(z_0,\varepsilon,L,K) = \left\{ S \in \Theta_{\varepsilon,L} : \sup_{a\leq x\leq b} \frac{|\dot{S}(x) - \dot{S}(z_0)|}{|x-z_0|^{\beta-1}} \leq K \right\}\,, \qquad [7.5]$$

with a known regularity $1 < \beta < 2$.

Moreover, to provide efficient pointwise estimation we will use the weak local Hölder classes, which were introduced for the first time in Galtchouk and Pergamenshchikov (2006b) for diffusion processes and then modified in Arkoun and Pergamenchtchikov (2008) for the models [7.1].

DEFINITION 7.2.– *For any* $0 < \delta, \varepsilon, h < 1$ *and* $1 < \beta < 2$, *the weak stable local Hölder class at the point* z_0 *is defined by*

$$\mathcal{U}^{(\beta)}(z_0, \varepsilon, \delta, h) = \left\{ S \in \Theta_{\varepsilon, L_\delta} \, : \, |\Omega_h(z_0, S)| \leq \delta h^\beta \right\}, \qquad [7.6]$$

where $L_\delta = \delta^{-1}$ *and* $\Omega_h(z_0, S) = \int_{-1}^{1} (S(z_0 + uh) - S(z_0))\, du$.

One can show (see, for example, Galtchouk and Pergamenshchikov 2006b) that for any $0 < \varepsilon, \delta, h < 1$ and $1 < \beta < 2$

$$\mathcal{H}^{(\beta)}(z_0, \varepsilon, \delta^{-1}, \delta) \subset \mathcal{U}^{(\beta)}(z_0, \varepsilon, \delta, h). \qquad [7.7]$$

This means that the weak Hölder condition is weaker than the usual Hölder one.

7.3. Pointwise estimation with absolute error risk

This section is devoted to the non-parametric estimation for autoregressive models. We consider the problem of estimating an unknown function at a fixed point using data governed by autoregressive models defined in [7.1]. To define the risk associated with the use of an estimator and thus to measure its quality, we use the loss function related to the absolute error. This work follows the minimax approach for which the goal is to find a lower bound of the asymptotic minimax risk and then to construct an estimator, said to be asymptotically efficient, for which the maximum risk asymptotically reaches this bound.

For a non-parametric autoregressive model where the autoregressive function is supposed to belong to the weak Hölder class [7.6] with known regularity β, we show that a kernel estimator is asymptotically efficient. When the regularity of the autoregressive function is unknown, we get the minimax adaptive convergence rate of estimators for the Hölderian classes [7.5]. In terms of asymptotic efficiency, we are now able to define the maximum risk of an estimator and to describe the minimax approach used for this purpose.

7.3.1. *Minimax approach*

Let us begin by giving precisely the definition of a estimator for the considered model [7.1].

DEFINITION 7.3.– *For the autoregressive model* [7.1], *an estimator of* S *at point* z_0 *is a random variable* $\omega \mapsto \widetilde{S}_n = \widetilde{S}_n(z_0, y_1, \ldots, y_n)$ *that is measurable with respect to the* σ*-algebra generated by* $y_1, \ldots, y_n$.

In this chapter, we consider the estimation problem in the case when the noise distribution, i.e. the distribution of $(\xi_k)_{k\geq 1}$, is unknown. Therefore, it is natural, similarly to Galtchouk and Pergamenshchikov (2006a), to measure the estimation quality of an estimator $\widetilde{S}_n$ by the robust risk defined as

$$\mathcal{R}_n(\widetilde{S}_n, S) := \sup_{p\in\mathcal{P}_\varsigma} \mathbf{E}_{S,p}|\widetilde{S}_n - S(z_0)| \,, \qquad [7.8]$$

where $\mathbf{E}_{S,p}$ means the expectation with respect to the distribution $\mathbf{P}_{S,p}$ of the process [7.1] corresponding to the function S and the noise distribution p.

We are therefore led to consider the asymptotic behavior of the minimax problem defined as

$$\inf_{\widetilde{S}_n} \sup_{S\in\Sigma} \mathcal{R}_n(\widetilde{S}_n, S) \,, \qquad [7.9]$$

where the infimum is considered for all the estimators and Σ is some fixed functional class. The primary goal of the minimax approach is to find an estimator $\widehat{S}_n$ whose maximal value of the risk [7.8] over a chosen functional class Σ is minimal on all possible estimators, i.e.

$$\inf_{\widetilde{S}_n} \sup_{S\in\Sigma} \upsilon(S)\,\mathcal{R}_n(\widetilde{S}_n, S) = \sup_{S\in\Sigma} \upsilon(S)\,\mathcal{R}_n(\widehat{S}_n, S) \,, \qquad [7.10]$$

where $\upsilon(S)$ is some normalizing coefficient such that

$$0 < \inf_{S\in\Sigma} \upsilon(S) \leq \sup_{S\in\Sigma} \upsilon(S) < \infty \,. \qquad [7.11]$$

Such estimators are called minimax. Unfortunately, it is not possible to find these estimators for any fixed $n \geq 1$ in the case of dependent observations. For this reason, we consider the minimax problem [7.10] in the asymptotic setting.

The goal of the approach is to find an estimator S_n^*, a sequence $(\varphi_n)_{n\in\mathbb{N}^*}$ of real positive numbers such that φ_n goes to $+\infty$, as n tends to $+\infty$, and some constants $0 < c \leq C < \infty$ satisfying

$$\liminf_{n\to\infty} \varphi_n \inf_{\widetilde{S}_n} \sup_{S\in\Sigma} \mathcal{R}_n(\widetilde{S}_n, S) \geq c \qquad [7.12]$$

and

$$\limsup_{n\to\infty} \varphi_n \sup_{S\in\Sigma} \mathcal{R}_n(S_n^*, S) \leq C \,. \qquad [7.13]$$

Now we can define the asymptotic minimax properties.

DEFINITION 7.4.– *If the properties* [7.12] *and* [7.13] *hold true, then the family* $(\varphi_n)_{n\in\mathbb{N}^*}$ *is called minimax convergence rate over the functional class* Σ *and the estimator* S_n^* *is called asymptotically minimax.*

REMARK 7.1.– It should be noted that, usually, to establish the lower bound [7.12], one uses the local asymptotic normality (LAN) method developed by Ibragimov and Hasminskii (1979) for the minimax non-parametric pointwise estimation problems.

DEFINITION 7.5.– *An estimator* $\widehat{S}_n$ *is said to be asymptotically efficient if*

$$\lim_{n\to\infty} \frac{\inf_{\widetilde{S}_n} \sup_{S\in\Sigma} \upsilon(S)\mathcal{R}_n(\widetilde{S}_n, S)}{\sup_{S\in\Sigma} \upsilon(S)\mathcal{R}_n(\widehat{S}_n, S)} = 1\,, \qquad [7.14]$$

where $\upsilon(\cdot)$ *is some coefficient satisfying the property* [7.11].

REMARK 7.2.– Note that the use of the normalizing coefficient υ in the criteria [7.10] is due to the fact that, generally, the minimal risk value over all possible estimators for the statistical models with dependent observations depends on the unknown function. This coefficient is well known for efficient non-parametric estimation problems. For the first time, such coefficient was derived by Pinsker (1981) for the filtration problem in a continuous time Gaussian model. Then for continuous time semi-martingale regression models, such coefficients were obtained by Konev and Pergamenshchikov (2009b, 2012, 2015). Then for diffusion models, such coefficients were found by Galtchouk and Pergamenshchikov (2011, 2015, 2020). For regression models in discrete time, such constants were obtained by Nussbaum (1985) and Galtchouk and Pergamenshchikov (2009b). For this reason, to take into account the influence of the model efficiency property we need to include this coefficient in the minimization problem [7.10].

7.3.2. *Adaptive minimax approach*

The minimax approach is called adaptive when one of the parameters of the functional class Σ in the problem [7.8] is assumed to be unknown, for example, the regularity of the autoregressive function S in the model [7.1]. In an adaptive setting, we choose Σ as the Hölderian class defined by [7.5], in which the regularity parameter β is unknown and $\beta \in \mathcal{B}$, where $\mathcal{B}$ is an interval. In this case, for any estimator $\widetilde{S}_n$ of $S(z_0)$, we introduce the adaptive version of the risk [7.9]

$$\mathcal{R}_{n,\beta}(\widetilde{S}_n) = \sup_{S\in\mathcal{H}^{(\beta)}} \mathcal{R}_n(\widetilde{S}_n, S)\,, \qquad [7.15]$$

where the functional class $\mathcal{H}^{(\beta)} = \mathcal{H}^{(\beta)}(z_0, \varepsilon, L, K)$ is defined in [7.5]. Then, as for the non-adaptive minimax approach, we look for an adaptive (i.e. independent of

$\beta \in \mathcal{B}$) and an asymptotically efficient estimator, as well as the exact asymptotic bound of the adaptive minimax risk

$$\inf_{\widetilde{S}_n} \sup_{\beta \in \mathcal{B}} \mathcal{R}_{n,\beta}(\widetilde{S}_n) \,.$$

The question we are also asking is the existence of an optimal adaptive estimator in convergence rate, i.e. an estimator, independent of $\beta \in \mathcal{B}$, which converges at this rate on each class $\mathcal{H}^{(\beta)}$ defined in [7.5]. More precisely:

DEFINITION 7.6.– *The family* $(\phi_n)_{n\in\mathbb{N}^*}$ *is called the adaptive minimax rate of convergence of estimators on the family of classes* $\{\mathcal{H}^{(\beta)}\}_{\beta\in\mathcal{B}}$ *if*

– *there exists a constant* $0 < c < \infty$ *such that:*

$$\liminf_{n\to\infty} \inf_{\widetilde{S}_n} \sup_{\beta\in\mathcal{B}} \phi_n \mathcal{R}_{n,\beta}(\widetilde{S}_n) \geq c \,,$$

where the inf is taken over all possible estimators;

– *and for a certain estimator* S^*_n *and a constant* $0 < C < \infty$, *we have:*

$$\limsup_{n\to\infty} \sup_{\beta\in\mathcal{B}} \phi_n \mathcal{R}_{n,\beta}(S^*_n) \leq C \,. \qquad [7.16]$$

The estimator S^*_n *is called the adaptive optimal in the minimax sense and* ϕ_n *is the adaptive minimax convergence rate for the family* $\{\mathcal{H}^{(\beta)}\}_{\beta\in\mathcal{B}}$.

REMARK 7.3.– It should be noted that, generally speaking, the minimax convergence rate from definition 7.4 is not the same as for the adaptive estimation given in definition 7.6. Indeed, as it is shown by Lepskii (1990) for the pointwise non-parametric estimation of Hölderian functions observed with Gaussian white noises, there are no estimators satisfying the property [7.16] for the minimax convergence rate $(\phi_n)_{n\in\mathbb{N}^*}$ for the set $\mathcal{B}$ containing at least two elements.

7.3.3. *Non-adaptive procedure*

This section deals with the non-parametric estimation of the autoregression coefficient function S in the model [7.1] at a given point $a < z_0 < b$, when the smoothness of S is known. For this problem, we make use of the following modified kernel estimator:

$$\widehat{S}_n(z_0) = \frac{1}{A_n} \sum_{k=1}^{n} Q(u_k)\, y_{k-1}\, y_k\, \mathbf{1}_{(A_n \geq d)}, \qquad [7.17]$$

where $Q(\cdot)$ is a kernel function,

$$A_n = \sum_{k=1}^{n} Q(u_k) y_{k-1}^2 \quad \text{with} \quad u_k = \frac{x_k - z_0}{h}\,;$$

d and h are some positive parameters.

Our first goal is to find the convergence rate of the minimax risk taken over the stable local Hölder class at the point z_0 defined in [7.5]. We assume that the regularity $\underline{\beta} \leq \beta \leq \overline{\beta}$, where $\underline{\beta} = 1 + \underline{\alpha}$ and $\overline{\beta} = 1 + \overline{\alpha}$ for some fixed parameters $0 < \underline{\alpha} < \overline{\alpha} < 1$.

We will find an asymptotical (as $n \to \infty$) positive lower bound for the minimax risk with the normalizing coefficient

$$\varphi_n = n^{\frac{\beta}{2\beta+1}}\,. \qquad [7.18]$$

To obtain this convergence rate, we set in [7.17]

$$h = n^{-\frac{1}{2\beta+1}} \quad \text{and} \quad d = \kappa_n\, nh\,, \qquad [7.19]$$

where $\kappa_n \geq 0$,

$$\lim_{n\to\infty} \kappa_n = 0 \quad \text{and} \quad \lim_{n\to\infty} \frac{h}{\kappa_n^2} = 0\,. \qquad [7.20]$$

We choose a kernel function Q such that

$$\int_{-1}^{1} Q(z)\,\mathrm{d}z > 0 \quad \text{and} \quad \int_{-1}^{1} z\,Q(z)\,\mathrm{d}z = 0\,. \qquad [7.21]$$

The two following theorems assert that the estimator [7.17] with the parameters [7.19]–[7.21] is asymptotically optimal with respect to the convergence rate [7.18] over the Hölder class [7.5].

THEOREM 7.1.– For any $L > 0$, $K > 0$ and $0 < \varepsilon < 1$, the following lower bound holds true

$$\liminf_{n\to\infty} \inf_{\widetilde{S}_n} \sup_{S\in\mathcal{H}^{(\beta)}} \varphi_n\, \mathcal{R}_n(\widetilde{S}_n, S) > 0,$$

where the infimum is taken over all estimators and $\mathcal{H}^{(\beta)} = \mathcal{H}^{(\beta)}(z_0, \varepsilon, L, K)$ is the stable the Hölder class defined in [7.5].

Now we obtain an upper bound for the kernel estimator [7.17].

THEOREM 7.2.– For any $L > 0$, $K > 0$ and $0 < \varepsilon < 1$, the kernel estimator [7.17] with the parameters [7.19]–[7.21] satisfies the following inequality:

$$\limsup_{n\to\infty} \sup_{S\in\mathcal{H}^{(\beta)}} \varphi_n \mathcal{R}_n(\widehat{S}_n, S) < \infty \,,$$

where $\mathcal{H}^{(\beta)} = \mathcal{H}^{(\beta)}(z_0, \varepsilon, L, K)$.

At the next step, we study sharp asymptotic properties for the minimax estimators [7.17]. In this case, we find in theorem 7.3 a positive constant giving the exact asymptotic lower bound for the minimax risk with the normalizing coefficient [7.18]. Moreover, theorem 7.4 shows that, for the estimator [7.17] with the parameters [7.19]–[7.20] and the indicator kernel $Q = \mathbf{1}_{[-1,1]}$, the asymptotic upper bound of the minimax risk coincides with this constant, i.e. this estimator is asymptotically efficient.

THEOREM 7.3.– For any $\delta > 0$ and $0 < \varepsilon < 1$

$$\liminf_{n\to\infty} \varphi_n \inf_{\widetilde{S}_n} \sup_{S\in\mathcal{U}^{(\beta)}} \upsilon(S)\mathcal{R}_n(\widetilde{S}_n, S) \geq \mathbf{E}|\eta| \,,$$

where the infimum is taken over all possible estimators, the class $\mathcal{U}_n^{(\beta)} = \mathcal{U}^{(\beta)}(z_0, \varepsilon, \delta, n^{-\frac{1}{2\beta+1}})$ is defined in [7.6], $\upsilon(S) = \gamma^{-1/2}(S)$, $\gamma(S) = 1 - S^2(z_0)$ and η is a Gaussian random variable with the parameters $(0, 1/2)$.

THEOREM 7.4.– The estimator [7.17] with the parameters [7.19] and [7.20] and $Q(z) = 1_{[-1,1]}$ satisfies the following inequality:

$$\limsup_{\delta\to 0} \limsup_{n\to\infty} \varphi_n \sup_{S\in\mathcal{U}_n^{(\beta)}} \upsilon(S)\, \mathcal{R}_n(\widehat{S}_n, S) \leq \mathbf{E}|\eta| \,,$$

where the class $\mathcal{U}_n^{(\beta)}$, the function $\upsilon(S)$ and the random variable η are defined in theorem 7.3.

Theorems 7.3 and 7.4 directly imply the following result.

COROLLARY 7.1.– The estimator [7.17] with the parameters [7.19] and [7.20] and $Q(z) = 1_{[-1,1]}$ is efficient, i.e.

$$\lim_{n\to\infty} \frac{\inf_{\widetilde{S}_n} \sup_{S\in\mathcal{U}^{(\beta)}} \upsilon(S)\mathcal{R}_n(\widetilde{S}_n, S)}{\sup_{S\in\mathcal{U}_n^{(\beta)}} \upsilon(S)\, \mathcal{R}_n(\widehat{S}_n, S)} = 1 \,.$$

REMARK 7.4.– It should be noted that the inclusion [7.7] means that the "natural" normalizing coefficient for the functional class [7.6] is the sequence [7.18]. Theorems 7.3 and 7.4 extend the usual Hölder approach for pointwise estimation by keeping the minimax convergence rate [7.18].

7.3.4. *Sequential kernel estimator*

In the perspective of obtaining an optimal estimator in an adaptive framework (when the regularity β is unknown), similarly to Galtchouk and Pergamenshchikov (2001), we will apply in the next section the adaptive procedure proposed by Lepskii (1990) to the model [7.1] based on the sequential kernel estimates.

Here, we construct the sequential kernel estimator using the method proposed in Borisov and Konev (1977) for the parametric case. It should be noted that to apply the Lepskii procedure, the kernel estimators must have the tail distribution of the Gaussian type. To obtain this property, one needs to use the sequential approach.

First of all, we need the first ν observations for the auxiliary estimation of $S(z_0)$:

$$\widehat{S}_\nu = \frac{1}{A_\nu} \sum_{j=1}^{\nu} Q(u_j)\, y_{j-1}\, y_j \,, \quad A_\nu = \sum_{j=1}^{\nu} Q(u_j)\, y_{j-1}^2 \,, \qquad [7.22]$$

where the kernel $Q(\cdot)$ is the indicator function of the interval $[-1;1]$, $u_j = (x_j - z_0)/h$ and h is some positive bandwidth, which will be specified later. In the sequel for any $0 \le k < m \le n$, we set

$$A_{k,m} = \sum_{j=k+1}^{m} Q(u_j)\, y_{j-1}^2 \,,$$

i.e. $A_\nu = A_{0,\nu}$. It is clear that to estimate $S(z_0)$ on the basis of the kernel estimate with the kernel Q, we can use the observations $(y_j)_{k_* \le j \le k^*}$, where

$$k_* = [n\overline{z}_0 - n\overline{h}] + 1 \quad \text{and} \quad k^* = [n\overline{z}_0 + n\overline{h}] \,, \qquad [7.23]$$

where $\overline{z}_0 = (z_0 - a)/(b-a)$ and $\overline{h} = h/(b-a)$. Here, $[a]$ is the integral part of a number a. In this definition, we assume that the bandwidth $0 < \overline{h} < \overline{z}_0$ and $n \ge \overline{h}^{-1}$. So for the first estimation, we choose ν as

$$\nu = \nu(h, \alpha) = k_* + \iota \,, \qquad [7.24]$$

where

$$\iota = \iota(h,\alpha) = [\widetilde{\epsilon} n h] + 1 \quad \text{and} \quad \widetilde{\epsilon} = \widetilde{\epsilon}(h,\alpha) = h^\alpha / \ln n \,.$$

REMARK 7.5.– In order to construct an adaptive procedure, we will use the sequential estimator in which the number of observations is given by a stopping time, which unfortunately depends on the function $S(z_0)$. It should also be noted that to construct an efficient procedure one needs to use all observations. To provide this property, we need to construct an auxiliary estimator for $S(z_0)$. For this, we will use the kernel estimator based on the number observations given by [7.24]. As we will see later, this approach allows us to obtain asymptotically efficient property.

Next, similarly to Arkoun (2011), we use a kernel sequential procedure based on the observations $(y_j)_{\nu \le j \le n}$. The key idea from the sequential procedure proposed by Borisov and Konev (1977) is to replace the random denominator by some threshold $H > 0$ in the kernel estimator [7.17], which transforms this estimator to a linear function of observations replacing the number of observations n by the following stopping time:

$$\tau_H = \inf\{k \geq \nu + 1 : A_{\nu,k} \geq H\},$$

where $\inf\{\emptyset\} = n$ and the positive threshold H will be chosen as a positive random variable measurable with respect to the σ - field $\{y_1, \ldots, y_\nu\}$. Therefore, we define on the set $\{\tau_H < n\}$

$$S_h^* = \frac{1}{H} \left(\sum_{j=\nu+1}^{\tau_H - 1} Q(u_j)\, y_{j-1}\, y_j \,+\, \varkappa_H\, Q(u_{\tau_H})\, y_{\tau_H - 1}\, y_{\tau_H} \right), \qquad [7.25]$$

where the correcting coefficient $\varkappa_H$ is defined on the set $\{A_{\nu,\tau_H} \geq H\}$ as

$$A_{\nu,\tau_H - 1} + \varkappa_H\, Q(u_{\tau_H})\, y^2_{\tau_H - 1} = H$$

and $\varkappa_H = 1$ on the set $\{A_{\nu,\tau_H} < H\}$. If $\tau_H = \nu + 1$, then we set $A_{\nu,\tau_H - 1} = 0$, i.e. we use the convention that $\sum_{j=l}^{m} = 0$ for $l > m$.

REMARK 7.6.– Note that we calculate the estimator [7.25] only for the case when $n > \overline{h}^{-1}$. Otherwise, we cannot calculate the kernel estimator, since there are no points x_j in the neighborhood $[z_0 - h\,,\, z_0 + h]$ of z_0.

Now, to obtain an efficient estimator we need to use all n observations, i.e. asymptotically for sufficiently large n the stopping time $\tau_H \approx n$. Similarly to Konev and Pergamenshchikov (1984), one can show that $\tau_H \approx \gamma(S)\, H$ as $H \to \infty$, where $\gamma(S) = 1 - S^2(z_0)$. Therefore, to use all observations we have to choose H as the number of observations divided by $\gamma(S)$. But in our case we use $k^* - k_*$ observations to estimate $S(z_0)$. Therefore, to obtain optimal estimation, we need to define H as $(k^* - k_*)/\gamma(S)$. Taking into account that $k^* - k_* \approx 2nh$ and that $\gamma(S)$ is unknown, we define the threshold H as

$$H = H(h, \alpha) = \rho n h\,, \quad \rho = \rho(h, \alpha) = \frac{2(1 - \widetilde{\epsilon})}{\gamma(\widetilde{S}_\nu)}\,, \qquad [7.26]$$

where $\widetilde{S}_\nu$ is the projection of the estimator $\widehat{S}_\nu$ on the interval $]-1+\varepsilon, 1-\varepsilon[$, i.e.

$$\widetilde{S}_\nu = \min(\max(\widehat{S}_\nu, -1+\varepsilon), 1-\varepsilon)\,.$$

In this section, we choose the bandwidth h in the following form:

$$h = h(\beta) = (\kappa_n)^{\frac{1}{2\beta+1}}\,, \qquad [7.27]$$

where the sequence $\kappa_n > 0$ satisfies

$$\kappa_* = \liminf_{n\to\infty} n\kappa_n > 0 \quad \text{and} \quad \lim_{n\to\infty} n^{\delta}\,\kappa_n = 0 \qquad [7.28]$$

for any $0 < \delta < 1$.

7.3.4.1. *Property of the stopping time τ_H*

PROPOSITION 7.1.– Assume that the threshold H is chosen in the form [7.26] and the bandwidth h satisfies the conditions [7.27] and [7.28]. Then for any $m \geq 1$ and for any $1 < \underline{\beta} \leq \overline{\beta} < 2$

$$\limsup_{n\to\infty} \sup_{\underline{\beta}\leq\beta\leq\overline{\beta}} h^{-m} \sup_{S\in\Theta_{\varepsilon,L}} \sup_{p\in\mathcal{P}_{\varsigma}} \mathbf{P}_{S,p}(\tau_H = n) < \infty\,,$$

where the bandwidth h is given in [7.27].

7.3.4.2. *Decomposition of the estimate error*

In the sequel, we present the following decomposition of the estimate error on which the proof of the optimality of the sequential kernel estimator is based.

First, we set

$$\check{\varkappa}_j = \mathbf{1}_{\{\tau_H \neq j\}} + \varkappa_H\,\mathbf{1}_{\{\tau_H = j\}}\,.$$

Then, taking this into account, we can represent the estimate error as

$$S^*_h - S(z_0) = -S(z_0)\,\mathbf{1}_{(\tau_H = n)} + \left(B_n(h) + \frac{1}{\sqrt{H}}\,\zeta_n(h)\right)\,\mathbf{1}_{(\tau_H < n)}\,,$$

with the approximative terms

$$B_n(h) = \frac{\sum_{j=\nu+1}^{\tau_H} \check{\varkappa}_j\,Q(u_j)\,(S(x_j) - S(z_0))\,y^2_{j-1}}{H}$$

and

$$\zeta_n(h) = \frac{\sum_{j=\nu+1}^{\tau_H} \check{\varkappa}_j\,Q(u_j)\,y_{j-1}\,\xi_j}{\sqrt{H}}\,.$$

Now we study the properties of the stochastic term. In the following lemma, we obtain that the sequential kernel estimator has the same tail of distribution form as a Gaussian random variable.

LEMMA 7.1.– The sequence $\zeta_n(h)$ satisfies the following properties:

– the sequence $\zeta_n(h)\mathbf{1}_{(\tau_H<n)}$ converges in law to $\mathcal{N}(0,1)$ as $n \to \infty$ uniformly in $p \in \mathcal{P}_\varsigma$ and $S \in \Theta_{\varepsilon,L}$;

– and for all $z \geq 2$

$$\sup_{n\geq 1}\sup_{h>0}\sup_{S\in\mathbf{C}[0,1]} \mathbf{P}_{S,p}\left(\zeta_n(h)\,\mathbf{1}_{(\tau_H<n)} \geq z\right) \leq 2e^{-z^2/8}\,. \qquad [7.29]$$

REMARK 7.7.– It is well known that to provide the adaptive pointwise estimation, the stochastic term has to have the Gaussian distribution tail, i.e. satisfy the inequality [7.29]. It should be noted that the non-sequential kernel estimator [7.17] does not have the above properties in the case of the model [7.1]. Therefore, in this case, the adaptive pointwise estimation is possible only in the sequential framework.

7.3.5. *Adaptive sequential procedure*

We construct a sequential adaptive procedure for estimating the autoregressive function at a given point in non-parametric autoregression models with Gaussian noise. We make use of the sequential kernel estimators. The optimal adaptive convergence rate is given as well as the upper bound for the minimax risk.

We will construct an adaptive minimax sequential estimation for the function S belonging to the functional class [7.5] with unknown regularity β. To this end, we will use the modification of the Lepskii method proposed in Arkoun (2011) based on the sequential estimators [7.25]. We set

$$d_n = \frac{n}{\ln n} \quad \text{and} \quad \phi_n = (d_n)^{\frac{\beta}{2\beta+1}}\,. \qquad [7.30]$$

Moreover, we choose the bandwidth h in the form [7.27] with $\kappa_n = 1/d_n$, i.e. we set

$$\check{h} = \check{h}(\beta) = \left(\frac{1}{d_n}\right)^{\frac{1}{2\beta+1}}\,.$$

Now, for some fixed $0 < \underline{\alpha} < \overline{\alpha} < 1$ and for $\underline{\beta} = 1 + \underline{\alpha}$, $\overline{\beta} = 1 + \overline{\alpha}$, we define the grids on the intervals $[\underline{\beta}\,,\,\overline{\beta}]$ and $[\underline{\alpha}\,,\,\overline{\alpha}]$ as

$$\beta_k = \underline{\beta} + \frac{k}{m}(\overline{\beta} - \underline{\beta}) \quad \text{and} \quad \alpha_k = \underline{\alpha} + \frac{k}{m}(\overline{\alpha} - \underline{\alpha})$$

for $0 \leq k \leq m$ with $m = [\ln d_n] + 1$, and we set

$$\phi_{n,k} = \phi_n(\beta_k) \quad \text{and} \quad \check{h}_k = \check{h}(\beta_k)\,.$$

Replacing in [7.24] and [7.26] the parameters h and α, we define

$$\check{\nu}_k = \nu(\check{h}_k, \alpha_k) \quad \text{and} \quad \check{H}_k = H(\check{h}_k, \alpha_k)\,.$$

Now using these parameters in the estimators [7.22] and [7.25],

$$\check{S}_k = S^*_{\check{h}_k}(z_0) \quad \text{and} \quad \check{\omega}_k = \max_{0\le j\le k} \left(|\check{S}_j - \check{S}_k| - \frac{\check{\lambda}}{N_j} \right), \qquad [7.31]$$

where

$$\check{\lambda} > \check{\lambda}_* = 4\sqrt{2} \left(\frac{\overline{\beta} - \underline{\beta}}{(2\underline{\beta}+1)(2\overline{\beta}+1)} \right)^{1/2}.$$

We also define the optimal index as

$$\check{k} = \max \left\{ 0 \le k \le m \, : \, \check{\omega}_k \le \frac{\check{\lambda}}{N_k} \right\}.$$

The adaptive estimator is now defined as

$$\widehat{S}_{a,n} = S^*_{\check{h}_k} \quad \text{and} \quad \check{h}_k = \check{h}_{\check{k}}\,. \qquad [7.32]$$

REMARK 7.8.– It should be noted that the usual adaptive pointwise estimation (see, for example, Lepskii 1990; Galtchouk and Pergamenshchikov 2001) differs from the threshold $\check{\lambda}$ by the fact that [7.31] does not depend on the parameters $L > 0$ and $K > 0$ of the Hölder class [7.5].

Now we are able to give the main results for the adaptive estimation of the function S at a fixed point $z_0 \in]a, b[$, in the Gaussian model [7.1], i.e. when $p = p_0$ where p_0 is $(0, 1)$–Gaussian density. For any estimate $\widetilde{S}_n$ of $S(z_0)$, we define the adaptive risk as

$$\mathcal{R}_{0,n}(\widetilde{S}_n) = \sup_{\beta\in[\underline{\beta};\overline{\beta}]} \ \sup_{S\in\mathcal{H}^{(\beta)}} \phi_n \, \mathbf{E}_{S,p_0} |\widetilde{S}_n - S(z_0)|\,, \qquad [7.33]$$

where $\mathcal{H}^{(\beta)} = \mathcal{H}^{(\beta)}(z_0, \varepsilon, L, K)$ and ϕ_n is defined in [7.30].

First, we give the lower bound for the minimax risk. More precisely, with the convergence rate ϕ_n, the lower bound for the minimax risk is strictly positive.

THEOREM 7.5.– There exists $K_0 > 0$ such that for all $K > K_0$, the risk [7.33] admits the following lower bound:

$$\liminf_{n\to\infty} \inf_{\widetilde{S}_n} \mathcal{R}_{0,n}(\widetilde{S}_n) > 0\,,$$

where the infimum is taken over all estimators $\widetilde{S}_n$.

To obtain an upper bound for the adaptive risk [7.33] of the procedure [7.32], we need to study the family $(S^*_h)_{\underline{\alpha}\le\alpha\le\overline{\alpha}}$.

THEOREM 7.6.– The sequential procedure [7.25] with the bandwidth h defined in [7.27] for $\kappa_n = \ln n/n$ satisfies the following property:

$$\limsup_{n\to\infty}\ \sup_{\underline{\alpha}\le\alpha\le\overline{\alpha}} \ (\Upsilon_n(h))^{-1}\ \sup_{S\in\mathcal{H}^{(\beta)}}\ \mathcal{R}_n(S^*_h, S) < \infty$$

where $\Upsilon_n(h) = h^\beta + (nh)^{-1/2}$ and the risk $\mathcal{R}_n$ is defined [7.9].

Using this theorem, we can establish the minimax property for the procedure [7.25].

THEOREM 7.7.– The estimation procedure [7.25] satisfies the following asymptotic property:

$$\limsup_{n\to\infty}\ \mathcal{R}_{0,n}(\widehat{S}_{a,n}) < \infty\,.$$

Theorems 7.5–7.7 are shown in Arkoun and Pergamenchtchikov (2016).

REMARK 7.9.– Theorem 7.5 gives the lower bound for the adaptive risk, i.e. the convergence rate ϕ_n is best for the adapted risk. Moreover, by theorem 7.7, the adaptive estimator [7.25] has this convergence rate. In this case, this estimator is called optimal in the sense of the adaptive risk [7.33].

7.4. Estimation with quadratic integral risk

In this section, we still consider the problem of estimating the autoregressive function in the model [7.1] at any point of $[a, b]$ but we will use the quadratic integral risk to measure the mean error of an estimator $\widehat{S}_n$ defined as

$$\overline{\mathcal{R}}_p(\widehat{S}_n, S) = \mathbf{E}_{S,p}\|\widehat{S}_n - S\|^2\,, \quad \|S\|^2 = \int_a^b S^2(x)\mathrm{d}x\,. \qquad [7.34]$$

Recalling that the distribution p of the noises ξ_k in model [7.1] is unknown, we still use the robust non-parametric estimation approach:

$$\overline{\mathcal{R}}^*(\widehat{S}_n, S) = \sup_{p\in\mathcal{P}_\varsigma} \overline{\mathcal{R}}_p(\widehat{S}_n, S)\,, \qquad [7.35]$$

where $\mathcal{P}_\varsigma$ is defined in [7.2].

In this context, we focus here on obtaining an oracle inequality, which is the first step in getting asymptotic efficiency. More precisely, in order to estimate the function

S in model [7.1], we make use of the estimators family $(\widehat{S}_\lambda, \lambda \in \Lambda)$, where $\widehat{S}_\lambda$ is a weighted least square estimator with the Pinsker weights. For this family, similarly to Galtchouk and Pergamenshchikov (2009a) and Konev and Pergamenshchikov (2009a), we construct a special selection rule, i.e. a random variable $\widehat{\lambda}$ with values in Λ, for which we define the selection estimator as $\widehat{S}_* = \widehat{S}_{\widehat{\lambda}}$. Our goal is to establish the non-asymptotic sharp oracle inequality for the robust risk [7.35], i.e. to show that for any $\check{\varrho} > 0$ and $n \geq 1$

$$\overline{\mathcal{R}}^*(\widehat{S}_*, S) \leq (1 + \check{\varrho}) \min_{\lambda \in \Lambda} \overline{\mathcal{R}}^*(\widehat{S}_\lambda, S) + \frac{\mathcal{B}_n}{\check{\varrho} n}, \qquad [7.36]$$

where $\mathcal{B}_n$ is a rest term such that for any $\check{\delta} > 0$,

$$\lim_{n \to \infty} \frac{\mathcal{B}_n}{n^{\check{\delta}}} = 0 .$$

In this case, the estimator $\widehat{S}_*$ is called optimal in the oracle inequality sense.

In order to obtain this inequality for model [7.1], we develop a new model selection method based on the truncated sequential procedure constructed in the previous section for the pointwise efficient estimation. To this purpose, we use the non-asymptotic analysis tool proposed for regression models in Galtchouk and Pergamenshchikov (2009a) based on the non-asymptotic studies from Baron *et al.* (1999) for a family of least-squares estimators and extended in Fourdrinier and Pergamenshchikov (2007) to some other estimators families. Here, we follow the approach proposed by Galtchouk and Pergamenshchikov (2011), i.e. we pass to a discrete time regression model by making use of the truncated sequential procedure introduced previously. To this end, at any point $(z_l)_{1 \leq l \leq d}$ of a partition of the interval $[a, b]$, we define a sequential procedure (τ_l, S_l^*) with a stopping rule τ_l and an estimator S_l^*. For $Y_l = S_l^*$ with $1 \leq l \leq d$, we come to the regression equation on some set $\Gamma \subseteq \Omega$:

$$Y_l = S(z_l) + \zeta_l, \quad 1 \leq l \leq d. \qquad [7.37]$$

Here, in contrast with the classical regression model, the noise sequence $(\zeta_l)_{1 \leq l \leq d}$ has a complex structure, namely,

$$\zeta_l = \xi_l^* + \varpi_l,$$

where $(\xi_l^*)_{1 \leq l \leq d}$ is a "main noise" sequence of uncorrelated random variables and $(\varpi_l)_{1 \leq l \leq n}$ is a sequence of bounded random variables.

It should be emphasized that only sharp oracle inequalities of type [7.36] allow one to propose efficiency property in the adaptive setting.

7.4.1. *Passage to a discrete time regression model*

We will use as a basic procedure the pointwise procedure from Arkoun and Pergamenchtchikov (2016) at the points $(z_l)_{1\le l\le d}$ defined as

$$z_l = a + \frac{l}{d}(b-a) \quad \text{and} \quad d = [\sqrt{n}]\,, \qquad [7.38]$$

where $[a]$ is the integer part of a number a. So we propose to use the first ι_l observations for the auxiliary estimation of $S(z_l)$. We set

$$\widehat{S}_{\iota_l} = \frac{1}{A_{\iota_l}} \sum_{j=1}^{\iota_l} Q_{l,j}\, y_{j-1}\, y_j \quad \text{with} \quad A_{\iota_l} = \sum_{j=1}^{\iota_l} Q_{l,j}\, y_{j-1}^2\,, \qquad [7.39]$$

where $Q_{l,j} = Q(u_{l,j})$ and the kernel $Q(\cdot)$ is the indicator function of the interval $[-1;1]$, i.e. $Q(u) = \mathbf{1}_{[-1,1]}(u)$. The points $(u_{l,j})$ are defined as

$$u_{l,j} = \frac{x_j - z_l}{h}\,.$$

Note that to estimate $S(z_l)$ on the basis of the kernel estimator with the kernel Q, we use only the observations $(y_j)_{k_{1,l}\le j\le k_{2,l}}$ from the h - neighborhood of the point z_l, i.e.

$$k_{1,l} = [n\widetilde{z}_l - n\widetilde{h}] + 1 \quad \text{and} \quad k_{2,l} = [n\widetilde{z}_l + n\widetilde{h}] \wedge n\,,$$

where $\widetilde{z}_l = (z_l - a)/(b-a)$ and $\widetilde{h} = h/(b-a)$. Note that, only for the last point $z_d = b$, we take $k_{2,d} = n$. We choose ι_l in [7.39] as

$$\iota_l = k_{1,l} + \mathbf{q} \quad \text{and} \quad \mathbf{q} = \mathbf{q}_n = [(n\widetilde{h})^{\mu_0}]$$

for some $0 < \mu_0 < 1$. In the sequel for any $0 \le k < m \le n$, we set

$$A_{k,m} = \sum_{j=k+1}^{m} Q_{l,j}\, y_{j-1}^2 \quad \text{and} \quad A_m = A_{0,m}\,.$$

Next, similarly to Arkoun (2011), we use a kernel sequential procedure based on the observations $(y_j)_{\iota_l\le j\le n}$. To transform the kernel estimator in a linear function of observations, we replace the number of observations n by the following stopping time:

$$\tau_l = \inf\{\iota_l + 1 \le k \le k_{2,l} \,:\, A_{\iota_l,k} \ge H_l\}\,, \qquad [7.40]$$

where $\inf\{\emptyset\} = k_{2,l}$ and the positive threshold H_l will be chosen as a positive random variable, which is measurable with respect to the σ-algebra $\{y_1, \ldots, y_{\iota_l}\}$.

Now we define the sequential estimator as

$$S_l^* = \frac{1}{H_l} \left(\sum_{j=\iota_l+1}^{\tau_l-1} Q_{l,j}\, y_{j-1}\, y_j \; + \; \varkappa_l\, Q_{l,\tau_l}\, y_{\tau_l-1}\, y_{\tau_l} \right) \mathbf{1}_{\Gamma_l}\,, \qquad [7.41]$$

where $\Gamma_l = \{A_{\iota_l,k_{2,l}-1} \geq H_l\}$ and the correcting coefficient $0 < \varkappa_l \leq 1$ on this set is defined as

$$A_{\iota_l,\tau_l-1} + \varkappa_l^2\, Q_{l,\tau_l} y_{\tau_l-1}^2 = H_l\,. \qquad [7.42]$$

Note that to obtain the efficient kernel estimator of $S(z_l)$, we need to use all $k_{2,l} - \iota_l - 1$ observations. Similarly to Konev and Pergamenshchikov (1984), one can show that $\tau_l \approx \gamma_l\, H_l$ as $H_l \to \infty$, where

$$\gamma_l = 1 - S^2(z_l)\,.$$

Therefore, one needs to choose H_l as $(k_{2,l} - \iota_l - 1)/\gamma_l$. Taking into account that the coefficients γ_l are unknown, we define the threshold H_l as

$$H_l = \frac{1-\widetilde{\epsilon}}{\widetilde{\gamma}_l}(k_{2,l} - \iota_l - 1) \quad \text{and} \quad \widetilde{\epsilon} = \frac{1}{2+\ln n}\,,$$

where $\widetilde{\gamma}_l = 1 - \widetilde{S}_{\iota_l}^2$ and $\widetilde{S}_{\iota_l}$ is the projection of the estimator $\widehat{S}_{\iota_l}$ on the interval $]-1+\widetilde{\varepsilon}, 1-\widetilde{\varepsilon}[$, i.e.

$$\widetilde{S}_{\iota_l} = \min(\max(\widehat{S}_{\iota_l}, -1+\widetilde{\varepsilon}), 1-\widetilde{\varepsilon})\,. \qquad [7.43]$$

To obtain the uncorrelated stochastic terms in the kernel estimator for $S(z_l)$, we choose the bandwidth h as

$$h = \frac{b-a}{2d}\,.$$

As for the estimator $\widehat{S}_{\iota_l}$, we can show the following property.

PROPOSITION 7.2.– The convergence rate in probability of the estimator [7.43] is more rapid than any power function, i.e. for any $\mathbf{b} > 0$

$$\lim_{n\to\infty} n^{\mathbf{b}} \max_{1\leq l\leq d} \sup_{S\in\Theta_{\varepsilon,L}} \sup_{p\in\mathcal{P}_\varsigma} \mathbf{P}_{S,p}\left(|\widetilde{S}_{\iota_l} - S(z_l)| > \epsilon_0\right) = 0\,, \qquad [7.44]$$

where $\epsilon_0 = \epsilon_0(n) \to 0$ as $n \to \infty$ such that $\lim_{n\to\infty} n^{\check{\delta}} \epsilon_0 = \infty$ for any $\check{\delta} > 0$.

Now we set

$$Y_l = S_l^* \mathbf{1}_\Gamma \quad \text{and} \quad \Gamma = \cap_{l=1}^d \Gamma_l\,. \qquad [7.45]$$

Using the convergence [7.44], we can study the probability properties of the set Γ in the following theorem.

PROPOSITION 7.3.– For any $\mathbf{b} > 0$, the probability of the set Γ^c satisfies the following asymptotic equality:

$$\lim_{n\to\infty} n^{\mathbf{b}} \sup_{S\in\Theta_{\varepsilon,L}} \sup_{p\in\mathcal{P}_\varsigma} \mathbf{P}_{S,p}(\Gamma^c) = 0\,.$$

In view of this theorem, we can shrink the set Γ^c. So, using the estimators [7.45] on the set Γ, we obtain the discrete time regression model [7.37] in which

$$\xi_l^* = \frac{\sum_{j=\iota_l+1}^{\tau_l-1} Q_{l,j}\, y_{j-1}\, \xi_j + \varkappa_l\, Q(u_{l,\tau_l})\, y_{\tau_l-1}\, \xi_{\tau_l}}{H_l} \qquad [7.46]$$

and $\varpi_l = \varpi_{1,l} + \varpi_{2,l}$, where

$$\varpi_{1,l} = \frac{\sum_{j=\iota_l+1}^{\tau_l-1} Q_{l,j}\, y_{j-1}^2 \check{s}_{l,j} + \varkappa_l^2\, Q(u_{l,\tau_l})\, y_{\tau_l-1}^2 \check{s}_{l,\tau_l}}{H_l}\,, \qquad \check{s}_{l,j} = S(x_j) - S(z_l)$$

and

$$\varpi_{2,l} = \frac{(\varkappa_l - \varkappa_l^2)\, Q(u_{l,\tau_l})\, y_{\tau_l-1}^2\, S(x_{\tau_l})}{H_l}\,.$$

Note that in the model [7.37], the random variables $(\xi_j^*)_{1\le j\le d}$ are defined only on the set Γ. For technical reasons, we need to define these variables on the set Γ^c as well. To this end, for any $j \ge 1$ we set

$$\check{Q}_{l,j} = Q_{l,j}\, y_{j-1}\, \mathbf{1}_{\{j<k_{2,l}\}} + \sqrt{H_l}\, Q_{l,j}\, \mathbf{1}_{\{j=k_{2,l}\}}$$

and $\check{A}_{\iota_l,m} = \sum_{j=\iota_l+1}^{m} \check{Q}_{l,j}^2$. Note that for any $j \ge 1$ and $l \ne m$

$$\check{Q}_{l,j}\, \check{Q}_{m,j} = 0\,.$$

and $\check{A}_{\iota_l,k_{2,l}} \ge H_l$. So now we can modify the stopping time [7.40] as

$$\check{\tau}_l = \inf\{k \ge \iota_l + 1 : \check{A}_{\iota_l,k} \ge H_l\}\,.$$

Obviously, $\check{\tau}_l \le k_{2,l}$ and $\check{\tau}_l = \tau_l$ on the set Γ for any $1 \le l \le d$. Now, similarly to [7.42], we define the correction coefficient as

$$\check{A}_{\iota_l,\check{\tau}_l-1} + \check{\varkappa}_l^2\, \check{Q}_{l,\check{\tau}_l}^2 = H_l\,.$$

It is clear that $0 < \check{\varkappa}_l \le 1$ and $\check{\varkappa}_l = \varkappa_l$ on the set Γ for $1 \le l \le d$. Using this coefficient, we set

$$\eta_l = \frac{\sum_{j=\iota_l+1}^{\check{\tau}_l-1} \check{Q}_{l,j}\, \xi_j + \check{\varkappa}_l\, \check{Q}_{l,\check{\tau}_l}\, \xi_{\check{\tau}_l}}{H_l}\,.$$

Note that on the set Γ, for any $1 \le l \le d$, the random variables $\eta_l = \xi_l^*$. Moreover, we can show that for any $1 \le l \le d$ and $p \in \mathcal{P}_\varsigma$

$$\mathbf{E}_{S,p}\left(\eta_l \,|\mathcal{G}_l\right) = 0\,, \quad \mathbf{E}_{S,p}\left(\eta_l^2 \,|\mathcal{G}_l\right) = \sigma_l^2\,, \quad \mathbf{E}_{S,p}\left(\eta_l^4 \,|\mathcal{G}_l\right) \le \check{\mathbf{m}}\sigma_l^4\,, \qquad [7.47]$$

where $\sigma_l = H_l^{-1/2}$, $\mathcal{G}_l = \sigma\{\eta_1, \ldots, \eta_{l-1}, \sigma_l\}$ and $\check{\mathbf{m}} = 4(144/\sqrt{3})^4\,\mathbf{m}_4^*$. It is clear that

$$\sigma_{0,*} \le \min_{1\le l\le d} \sigma_l^2 \le \max_{1\le l\le d} \sigma_l^2 \le \sigma_{1,*}\,, \qquad [7.48]$$

where

$$\sigma_{0,*} = \frac{1-\epsilon^2}{2(1-\widetilde{\epsilon})nh} \quad \text{and} \quad \sigma_{1,*} = \frac{1}{(1-\widetilde{\epsilon})(2nh-\mathbf{q}-3)}\,.$$

Now, taking into account that $|\varpi_{1,l}| \le Lh$, for any $S \in \Theta_{\varepsilon,L}$ we obtain that

$$\sup_{S\in\Theta_{\varepsilon,L}} \mathbf{E}_{S,p}\mathbf{1}_\Gamma \varpi_l^2 \le \left(L^2h^2 + \frac{\check{\upsilon}_n}{(nh)^2}\right)\,, \qquad [7.49]$$

where $\check{\upsilon}_n = \sup_{p\in\mathcal{P}} \sup_{S\in\Theta_{\varepsilon,L}} \mathbf{E}_{S,p} \max_{1\le j\le n} y_j^4$. The behavior of this coefficient is studied in the following proposition.

PROPOSITION 7.4.– For any $\mathbf{b} > 0$, the sequence $(\check{\upsilon}_n)_{n\ge1}$ satisfies the following limiting equality:

$$\lim_{n\to\infty} n^{-\mathbf{b}}\,\check{\upsilon}_n = 0\,. \qquad [7.50]$$

Propositions 7.2–7.4 are shown in Arkoun and Pergamenchtchikov (2016).

REMARK 7.10.– It should be noted that the property [7.50] means that the asymptotic behavior of the upper bound [7.49] is approximately as h^{-2} when $n \to \infty$. We will use this in the oracle inequalities below.

REMARK 7.11.– It should be emphasized that to estimate the function S in [7.1] we use the approach developed in Galtchouk and Pergamenshchikov (2011) for the sequential drift estimation problem in the stochastic differential equation. On the basis of the efficient sequential kernel procedure developed in Galtchouk and Pergamenshchikov (2005, 2006b, 2015) with the kernel-indicator, the stochastic differential equation is replaced by a regression model. It should be noted that to obtain the efficient estimator, one needs to take the kernel-indicator estimator. For this reason, in this section, we use the kernel-indicator in the sequential estimator [7.41]. It also should be noted that the sequential estimator [7.41] has the same form as in Arkoun and Pergamenchtchikov (2016), except the last term, in which the correction coefficient is replaced by the square root of the coefficient used in Konev (2016). We modify this procedure to calculate the variance of the stochastic term [7.46].

7.4.2. *Model selection*

In this section, we consider the non-parametric estimation problem in the non-asymptotic setting for the regression model [7.37] for some set $\Gamma \subseteq \Omega$. The design points $(z_l)_{1\le l\le d}$ are defined in [7.38]. The function $S(\cdot)$ is unknown and has to be estimated from the observations $Y_1,\ldots,Y_d$. Moreover, we assume that the unobserved random variables $(\eta_l)_{1\le l\le d}$ satisfy the properties [7.47] with some non-random constant $\check{\mathbf{m}} > 1$, and the known random positive coefficients $(\sigma_l)_{1\le l\le d}$ satisfy the inequality [7.48] for some non-random positive constants $\sigma_{0,*}$ and $\sigma_{1,*}$. Concerning the random sequence $\varpi = (\varpi_l)_{1\le l\le n}$, we suppose that

$$\mathbf{u}_d^* = \mathbf{E}_{S,p}\mathbf{1}_\Gamma \|\varpi\|_d^2 < \infty\,. \qquad [7.51]$$

The performance of any estimator $\widehat{S}$ will be measured by the empirical squared error

$$\|\widehat{S} - S\|_d^2 = (\widehat{S} - S, \widehat{S} - S)_d = \frac{b-a}{d}\sum_{l=1}^{d}(\widehat{S}(z_l) - S(z_l))^2\,.$$

Now we fix a basis $(\phi_j)_{1\le j\le n}$, which is orthonormal for the empirical inner product:

$$(\phi_i\,,\,\phi_j)_d = \frac{b-a}{d}\sum_{l=1}^{d}\phi_i(z_l)\phi_j(z_l) = \mathbf{1}_{\{i=j\}}\,. \qquad [7.52]$$

For example, we can take the trigonometric basis $(\phi_j)_{j\ge 1}$ in $\mathbf{L}_2[a,b]$ defined as

$$\phi_1 = 1\,, \quad \phi_j(x) = \sqrt{\frac{2}{b-a}}\mathrm{Tr}_j\left(2\pi[j/2]\mathbf{l}_0(x)\right)\,, \quad j\ge 2\,,$$

where the function $\mathrm{Tr}_j(x) = \cos(x)$ for even j and $\mathrm{Tr}_j(x) = \sin(x)$ for odd j, $[x]$ denotes the integer part of x, and $\mathbf{l}_0(x) = (x-a)/(b-a)$. Note that using the orthonormality property [7.52], we can represent for any $1\le l\le d$ the function S as

$$S(z_l) = \sum_{j=1}^{d}\theta_{j,d}\,\phi_j(z_l) \quad \text{and} \quad \theta_{j,d} = \left(S,\phi_j\right)_d\,. \qquad [7.53]$$

So, to estimate the function S we have to estimate the Fourier coefficients $(\theta_{j,d})_{1\le j\le d}$. To this end, we replace the function S by these observations, i.e.

$$\widehat{\theta}_{j,d} = \frac{b-a}{d}\sum_{l=1}^{d} Y_l \phi_j(z_l)\,.$$

From [7.37], we obtain immediately the following regression scheme:

$$\widehat{\theta}_{j,d} = \theta_{j,d} + \zeta_{j,d} \quad \text{with} \quad \zeta_{j,d} = \sqrt{\frac{b-a}{d}}\eta_{j,d} + \varpi_{j,d}\,,$$

where

$$\eta_{j,d} = \sqrt{\frac{b-a}{d}}\sum_{l=1}^{d} \eta_l \phi_j(z_l) \quad \text{and} \quad \varpi_{j,d} = \frac{b-a}{d}\sum_{l=1}^{d} \varpi_l\, \phi_j(z_l)\,.$$

Note that the upper bound [7.48] and the Bounyakovskii–Cauchy–Schwarz inequality imply that

$$|\varpi_{j,d}| \le \|\varpi\|_d\, \|\phi_j\|_d = \|\varpi\|_d\,.$$

We estimate the function S on the grid [7.38] by the weighted least-squares estimator

$$\widehat{S}_\lambda(z_l) = \sum_{j=1}^{d} \lambda(j)\, \widehat{\theta}_{j,d}\, \phi_j(z_l)\, \mathbf{1}_\Gamma\,, \quad 1 \le l \le d\,, \qquad [7.54]$$

where the weight vector $\lambda = (\lambda(1), \ldots, \lambda(d))'$ belongs to some finite set $\Lambda \subset [0,1]^d$, the prime denotes the transposition. We set for any $a \le t \le b$

$$\widehat{S}_\lambda(t) = \sum_{l=1}^{d} \widehat{S}_\lambda(z_l) \mathbf{1}_{\{z_{l-1} < t \le z_l\}}\,. \qquad [7.55]$$

Moreover, denoting $\lambda^2 = (\lambda^2(1), \ldots, \lambda^2(n))'$ we define the following sets

$$\Lambda_1 = \{\lambda^2\,,\, \lambda \in \Lambda\} \quad \text{and} \quad \Lambda_2 = \Lambda \cup \Lambda_1\,.$$

Denote by ν the cardinal number of the set Λ and

$$\nu^* = \max_{\lambda\in\Lambda} \sum_{j=1}^{d} \mathbf{1}_{\{\lambda(j)>0\}}\,.$$

In order to obtain a good estimator, we have to write a rule to choose a weight vector $\lambda \in \Lambda$ in [7.54]. We define the empirical squared risk as

$$\mathrm{Err}_d(\lambda) = \|\widehat{S}_\lambda - S\|_d^2\,.$$

Using [7.53] and [7.54], we can rewrite this risk as

$$\mathrm{Err}_d(\lambda) = \sum_{j=1}^{d} \lambda^2(j)\widehat{\theta}^2_{j,d} - 2\sum_{j=1}^{d} \lambda(j)\widehat{\theta}_{j,d}\,\theta_{j,d} + \sum_{j=1}^{d} \theta^2_{j,d}\,.$$

Since the coefficient $\theta_{j,d}$ is unknown, we need to replace the term $\widehat{\theta}_{j,d}\,\theta_{j,d}$ by some of its estimators which we choose as

$$\widetilde{\theta}_{j,d} = \widehat{\theta}^2_{j,d} - \frac{b-a}{d}\,\mathrm{s}_{j,d} \quad \text{with} \quad \mathrm{s}_{j,d} = \frac{b-a}{d}\sum_{l=1}^{d} \sigma^2_l\,\phi^2_j(z_l)\,.$$

Note that from [7.48]–[7.52], it follows that

$$\mathrm{s}_{j,d} \le \sigma_{1,*}\,.$$

Similarly to Galtchouk and Pergamenshchikov (2011, 2019), one needs to introduce a penalty term in the cost function to compensate such modification of the empirical risk. We choose it as

$$P_d(\lambda) = \frac{b-a}{d}\sum_{j=1}^{d} \lambda^2(j)\mathrm{s}_{j,d}\,.$$

Finally, we define the cost function in the following form:

$$J_d(\lambda) = \sum_{j=1}^{d} \lambda^2(j)\widehat{\theta}^2_{j,d} - 2\sum_{j=1}^{d} \lambda(j)\,\widetilde{\theta}_{j,d} + \delta P_d(\lambda)\,.$$

where $0 < \delta < 1$ is some positive constant, which will be chosen later. We set

$$\widehat{\lambda} = \mathrm{argmin}_{\lambda\in\Lambda}\, J_d(\lambda)$$

and define an estimator of $S(t)$ of the form [7.55]:

$$\widehat{S}_*(t) = \widehat{S}_{\widehat{\lambda}}(t) \quad \text{for} \quad a \le t \le b\,. \qquad [7.56]$$

REMARK 7.12.– We use the procedure [7.56] to estimate the function S in the autoregressive model [7.1] through the regression scheme [7.37] generated by the sequential procedures [7.45].

7.4.3. *Main results*

In this section, we formulate all main results, which are shown in Arkoun *et al.* (2019). First, we obtain the sharp oracle inequality for the selection model procedure [7.56] for the general regression model [7.37].

THEOREM 7.8.– There exists some constant $\mathbf{l}^* > 0$ such that for any weight vectors set Λ, any $p \in \mathcal{P}_\varsigma$, any $n \geq 1$ and $0 < \delta \leq 1/12$, the procedure [7.56], satisfies the following oracle inequality:

$$\mathbf{E}_{S,p}\|\widehat{S}_* - S\|_d^2 \leq \frac{1+4\delta}{1-6\delta} \min_{\lambda \in \Lambda} \mathbf{E}_{S,p}\|\widehat{S}_\lambda - S\|_d^2$$

$$+\mathbf{l}^* \frac{\nu\varsigma^2}{\delta}\left(\frac{\sigma_{1,*}^2}{\sigma_{0,*}d} + \mathbf{u}_d^* + \delta^2\sqrt{\mathbf{P}_{S,p}(\Gamma^c)}\right).$$

Using lemma A.2 from Konev and Pergamenshchikov (2015), we obtain the oracle inequality for the quadratic risk [7.34].

THEOREM 7.9.– There exists some constant $\mathbf{l}^* > 0$ such that for any weight vectors set Λ, any continuously differentiable function S, any $p \in \mathcal{P}_\varsigma$, any $n \geq 1$ and $0 < \delta \leq 1/12$, the procedure [7.56] satisfies the following oracle inequality:

$$\overline{\mathcal{R}}_p(\widehat{S}_*, S) \leq \frac{(1+4\delta)(1+\delta)^2}{1-6\delta} \min_{\lambda \in \Lambda} \overline{\mathcal{R}}_p(\widehat{S}_\lambda, S)$$

$$+\mathbf{l}^* \frac{\varsigma^2\nu}{\delta}\left(\frac{\|\dot{S}\|^2}{d^2} + \frac{\sigma_{1,*}^2}{\sigma_{0,*}d} + \mathbf{u}_d^* + \delta^2\sqrt{\mathbf{P}_{S,p}(\Gamma^c)}\right).$$

Now we assume that the cardinal ν of Λ and the parameter ς in the density family [7.2] are functions of the number observations n, i.e. $\nu = \nu(n)$ and $\varsigma = \varsigma(n)$ such that for any $\check{\delta} > 0$

$$\lim_{n\to\infty} \frac{\nu(n)}{n^{\check{\delta}}} = 0. \qquad [7.57]$$

Using theorems 7.3 and 7.4 and the bounds [7.48] and [7.49], we obtain the oracle inequality for the estimation problem for the model [7.1].

THEOREM 7.10.– Assume that the conditions [7.3] and [7.57] hold. Then for any $p \in \mathcal{P}_\varsigma$, $S \in \Theta_{\varepsilon,L}$, $n \geq 3$ and $0 < \delta \leq 1/12$, the procedure [7.56] satisfies the following oracle inequality:

$$\overline{\mathcal{R}}_p(\widehat{S}_*, S) \leq \frac{(1+4\delta)(1+\delta)^2}{1-6\delta} \min_{\lambda \in \Lambda} \overline{\mathcal{R}}_p(\widehat{S}_\lambda, S) + \frac{\check{\mathbf{B}}_n(p)}{\delta n},$$

where the term $\check{\mathbf{B}}_n(p)$ is such that for any $\check{\delta} > 0$

$$\lim_{n\to\infty} \frac{\check{\mathbf{B}}_n(p)}{n^{\check{\delta}}} = 0.$$

We obtain the same inequality for the robust risks.

THEOREM 7.11.– Assume that the conditions [7.3] and [7.57] hold. Then for any $n \geq 3$, any $S \in \Theta_{\varepsilon,L}$ and any $0 < \delta \leq 1/12$, the procedure [7.56] satisfies the following oracle inequality:

$$\overline{\mathcal{R}}^*(\widehat{S}_*, S) \leq \frac{(1+4\delta)(1+\delta)^2}{1-6\delta} \min_{\lambda \in \Lambda} \overline{\mathcal{R}}^*(\widehat{S}_\lambda, S) + \frac{\mathbf{B}_n^*}{\delta n},$$

where the term $\check{\mathbf{B}}_n$ is such that for any $\check{\delta} > 0$

$$\lim_{n\to\infty} \frac{\mathbf{B}_n^*}{n^{\check{\delta}}} = 0.$$

REMARK 7.13.– It is well known that to obtain the efficiency property, we need to specify the weight coefficients $(\lambda(j))_{1\leq j\leq n}$ (see, for example, Galtchouk and Pergamenshchikov 2009b). Consider for some fixed $0 < \varepsilon < 1$ a numerical grid of the form

$$\mathcal{A} = \{1, \ldots, k^*\} \times \{\varepsilon, \ldots, m\varepsilon\},$$

where $m = [1/\varepsilon^2]$. We assume that both parameters $k^* \geq 1$ and ε are functions of n, i.e. $k^* = k^*(n)$ and $\varepsilon = \varepsilon(n)$, such that

$$\begin{cases} \lim_{n\to\infty} k^*(n) = +\infty, & \lim_{n\to\infty} \dfrac{k^*(n)}{\ln n} = 0, \\ \\ \lim_{n\to\infty} \varepsilon(n) = 0 \quad \text{and} & \lim_{n\to\infty} n^{\check{\delta}}\varepsilon(n) = +\infty \end{cases} \tag{7.58}$$

for any $\check{\delta} > 0$. One can take, for example, for $n \geq 2$

$$\varepsilon(n) = \frac{1}{\ln n} \quad \text{and} \quad k^*(n) = k_0^* + \sqrt{\ln n},$$

where $k_0^* \geq 0$ is some fixed constant. For each $\alpha = (\beta, \mathbf{l}) \in \mathcal{A}$, we introduce the weight sequence

$$\lambda_\alpha = (\lambda_\alpha(j))_{1\leq j\leq n}$$

with the elements

$$\lambda_\alpha(j) = \mathbf{1}_{\{1\leq j<j_*\}} + \left(1 - (j/\omega_\alpha)^\beta\right) \mathbf{1}_{\{j_*\leq j\leq \omega_\alpha\}},$$

where $j_* = 1 + [\ln n]$, $\omega_\alpha = (\mathrm{d}_\beta \mathbf{l}\, n)^{1/(2\beta+1)}$ and

$$\mathrm{d}_\beta = \frac{(\beta+1)(2\beta+1)}{\pi^{2\beta}\beta}.$$

Now we define the set Λ as

$$\Lambda = \{\lambda_\alpha\,, \alpha \in \mathcal{A}\}\,.$$

Note that these weight coefficients are used in Konev and Pergamenshchikov (2012, 2015) for continuous time regression models to show the asymptotic efficiency. It will be noted that in this case the cardinal of the set Λ is $\nu = k^* m$. It is clear that properties [7.58] imply condition [7.57].

7.5. References

Arkoun, O. (2011). Sequential adaptive estimators in nonparametric autoregressive models. *Sequential Analysis*, 44, 228–246.

Arkoun, O., and Pergamenchtchikov, S. (2008). Nonparametric estimation for an autoregressive model. *Tomsk State University Journal of Mathematics and Mechanics*, 2(3), 20–30.

Arkoun, O., and Pergamenchtchikov, S. (2016). Sequential robust estimation for nonparametric autoregressive models. *Sequential Analysis*, 35, 489–515.

Arkoun, O., Brua, J.-Y., and Pergamenchtchikov, S. (2019). Model selection method for nonparametric autoregression. *Sequential Analysis*, 38(4), 437–460.

Baron, A., Birgé, L., and Massart, P. (1999). Risk bounds for model selection via penalization. *Probab. Theory Related Fields*, 113, 301–413.

Belitser, E. (2000). Recursive estimation of a drifted autoregressive parameter. *The Annals of Statistics*, 26, 860–870.

Borisov, V., and Konev, V. (1977). Sequential estimation of parameters of discrete processes. *Automat. and Remote Control*, 10, 58–64.

Dahlhaus, R. (1996a). Maximum likelihood estimation and model selection for locally stationary processes. *Journal of Nonparametric Statistics*, 6, 171–191.

Dahlhaus, R. (1996b). On the Kullback-Leibler information divergence of locally stationary processes. *Stochastic Processes and Their Applications*, 62, 139–168.

Fan, J., and Zhang, W. (2008). Statistical methods with varying coefficient models. *Statistics and Its Interface*, 1, 179–195.

Fourdrinier, D., and Pergamenshchikov, S. (2007). Improved model selection method for a regression function with dependent noise. *Annals of the Institute of Statistical Mathematics*, 59, 435–464.

Galtchouk, L., and Pergamenshchikov, S. (2001). Sequential nonparametric adaptive estimation of the drift coefficient in diffusion processes. *Math. Methods Statist.*, 10(3), 316–330.

Galtchouk, L., and Pergamenshchikov, S. (2005). Nonparametric sequential minimax estimation of the drift coefficient in diffusion processes. *Sequential Analysis*, 24, 303–330.

Galtchouk, L., and Pergamenshchikov, S. (2006a). Asymptotically efficient estimates for nonparametric regression models. *Statistics and Probability Letters*, 76, 852–860.

Galtchouk, L., and Pergamenshchikov, S. (2006b). Asymptotically efficient sequential kernel estimates of the drift coefficient in ergodic diffusion processes. *Statistical Inference for Stochastic Processes*, 9, 1–16.

Galtchouk, L., and Pergamenshchikov, S. (2009a). Sharp non-asymptotic oracle inequalities for nonparametric heteroscedastic regression models. *Journal of Nonparametric Statistics*, 21, 1–16.

Galtchouk, L., and Pergamenshchikov, S. (2009b). Adaptive asymptotically efficient estimation in heteroscedastic nonparametric regression. *Journal of Korean Statistical Society*, 38, 305–322.

Galtchouk, L., and Pergamenshchikov, S. (2011). Adaptive sequential estimation for ergodic diffusion processes in quadratic metric. *Journal of Nonparametric Statistics*, 23, 255–285.

Galtchouk, L., and Pergamenshchikov, S. (2015). Efficient pointwise estimation based on discrete data in ergodic nonparametric diffusions. *Bernoulli*, 21, 2569–2594.

Galtchouk, L., and Pergamenshchikov, S. (2019). Non asymptotic sharp oracle inequalities for high dimensional ergodic diffusion models, Preprint [Online]. Available at: https://hal.archives-ouvertes.fr/hal-02387034.

Galtchouk, L., and Pergamenshchikov, S. (2020). Efficient big data analysis for ergodic diffusion models on the basis of discrete data, Preprint [Online]. Available at: https://hal.archives-ouvertes.fr/hal-02474675.

Ibragimov, I., and Hasminskii, R. (1979). *Statistical Estimation: Asymptotic Theory*. Springer, New York.

Konev, V. (2016). On one property of martingales with conditionally gaussian increments and its application in the theory of nonasymptotic inference. *Doklady Mathematics*, 94, 1–5.

Konev, V., and Pergamenshchikov, S. (1984). Estimate of the number of observations in sequential identification of the parameters of dynamical systems. *Avtomat. i Telemekh*, 12, 56–63.

Konev, V., and Pergamenshchikov, S. (2009a). Non-parametric estimation in a semimartingale regression model. Part 1. Oracle inequalities. *Tomsk State University Journal of Mathematics and Mechanics*, 3(7), 23–41.

Konev, V., and Pergamenshchikov, S. (2009b). Non-parametric estimation in a semimartingale regression model. Part 2. Robust asymptotic efficiency. *Tomsk State University Journal of Mathematics and Mechanics*, 4(8), 31–45.

Konev, V., and Pergamenshchikov, S. (2012). Efficient robust nonparametric estimation in a semimartingale regression model. *Annales de l'Institut Henri Poincaré, Probabilités et Statistiques*, 48, 1244–1277.

Konev, V., and Pergamenshchikov, S. (2015). Robust model selection for a semimartingale continuous time regression from discrete data. *Stochastic Processes and their Applications*, 125, 294–326.

Lepskii, O.V. (1990). A problem of adaptive estimation in gaussian white noise. *Theory of Probability and Its Applications*, 35, 454–466.

Luo, X.H., Yang, Z.H., and Zhou, Y. (2009). Nonparametric estimation of the production function with time-varying elasticity coefficients. *Systems Engineering – Theory and Practice*, 29, 144–149.

Moulines, E., Priouret, P., and Roueff, F. (2005). On recursive estimation for time varying autoregressive processes. *Annals of Statistics*, 33, 2610–2654.

Nussbaum, M. (1985). Spline smoothing in regression models and asymptotic efficiency in L2. *Annals of Statistics*, 13, 984–997.

Pinsker, M.S. (1981). Optimal filtration of square integrable signals in gaussian noise. *Problems Inform. Transmission*, 17, 120–133.

Part 3

Divergence Measures and Entropies

8

Inference in Parametric and Semi-parametric Models: The Divergence-based Approach

This chapter presents a short review of basic features of statistical inference in parametric and semi-parametric models through duality techniques in the realm of divergence-based methods. A simple solution for testing the number of components in a mixture is presented; this is a non-regular model with currently no amenable solution. The interplay between the choice of the divergence and the model is enlightened.

8.1. Introduction

This paper aims to present the main features of Csiszár-type divergences in relation with statistical inference. We first introduce some notations and definitions, and then embed the present setup in the context of decomposable measures of discrepancy between probability measures. A variational representation of Csiszár divergences is the basic ingredient for this embedding. The specificity of the classes of decomposable measures of discrepancy allows us to select such a class for specific inference purposes.

Minimizing a discrepancy between a given distribution P and a class of measures $\mathcal{M}$ occurs in a certain number of problems. On the one hand, when the distribution P is known, then the resulting problem is of analytic nature and bears no statistical

Chapter written by Michel BRONIATOWSKI.

aspect. On the other hand, the measure P might only be known through a sampling of observations $X_1, .., X_n$, whose empirical distribution

$$P_n := \frac{1}{n} \sum_{i=1}^{n} \delta_{X_i}$$

converges to P in some sense. The class $\mathcal{M}$ is then called a model, which may be parametric, therefore indexed by some θ in a subset Θ of a finite dimensional space; the model $\mathcal{M}$ might be semi-parametric, for example described by moment or L-moment conditions; or non-parametric, for example when described by shape conditions, such as unimodality. Inference in those cases aims at answering basic questions:

Does P belong to $\mathcal{M}$?

If yes, and when the model is of parametric or semi-parametric nature, what is the "true" value θ_T of the parameter for which $P = P_{\theta_T}$, a question which should obviously be made more precise according to the context; for example, in semi-parametric models, the class $\mathcal{M}$ is usually described as

$$\mathcal{M} := \bigcup_{\theta} \mathcal{M}_\theta, \qquad [8.1]$$

where $\mathcal{M}_\theta$ represents a whole submodel sharing a common feature θ, and the inference estimation step aims at identifying θ_T such that P belongs to $\mathcal{M}_{\theta_T}$. The case when each of the $\mathcal{M}_\theta$ is defined by specific moment conditions extends the classical empirical likelihood context and will be described in section 8.2. Other fields of application of a minimum divergence approach cover robust inference in parametric models, considering minimum divergence estimation between the distribution of the data and some tubular neighborhoods of the model. Model selection criteria such as AIC or BIC can be considered as minimum divergence estimators. Broadly speaking, divergence-based methods extend the likelihood paradigm in a natural way, paving the way to trade-off between efficiency and robustness requirements. In this chapter, we will show the way in which some non-regular problems can be solved accordingly, in the realm of mixture models.

We now turn to some definitions and notations.

8.1.1. *Csiszár divergences, variational form*

8.1.1.1. *Parametric framework*

Let $\mathcal{P} := \{P_\theta, \theta \in \Theta\}$ be an identifiable parametric model on $\mathbb{R}^d$, where Θ is a subset of $\mathbb{R}^s$. All measures in $\mathcal{P}$ will be assumed to be measure equivalent, therefore sharing the same support. The parameter space Θ does not need to be an open set in the present setting. It may even happen that the model includes measures which would not be probability distributions; cases of interest cover models including mixtures of probability distributions; this will be developed in section 8.3. Let φ be a proper closed convex function from $]-\infty,+\infty[$ to $[0,+\infty]$ with $\varphi(1)=0$ and such that its domain $\text{dom}\varphi := \{x \in \mathbb{R} \text{ such that } \varphi(x)<\infty\}$ is an interval with endpoints $a_\varphi < 1 < b_\varphi$ (which may be finite or infinite). For two measures P_α and P_θ in $\mathcal{P}$, the φ-divergence between P_α and P_θ is defined by

$$\phi(\alpha,\theta) := \int_{\mathcal{X}} \varphi\left(\frac{dP_\alpha}{dP_\theta}(x)\right)\, dP_\theta(x).$$

In a broader context, the φ-divergences were introduced by Csiszár (1964) as "f-divergences", defined between two probability measures Q and P, with Q absolutely continuous (a.c.) with respect to (w.r.t.) P by

$$D_f(Q,P) := \int f\left(\frac{dQ}{dP}\right) dP,$$

where f satisfies the same requirements as φ above. The basic property of φ-divergences states that when φ is strictly convex on a neighborhood of $x=1$, then

$$\phi(\alpha,\theta)=0 \text{ if and only if } \alpha=\theta.$$

We refer to (Liese and Vajda 1987, Chapter 1) for a complete study of those properties. Let us simply quote that in general $\phi(\alpha,\theta)$ and $\phi(\theta,\alpha)$ are not equal. Hence, φ-divergences usually are not distances, but they merely quantify some difference between two measures. A main feature of divergences between distributions of random variables X and Y is the invariance property with respect to a common smooth change of variables.

8.1.1.1.1. Examples of φ-divergences

The Kullback-Leibler (KL), modified Kullback-Leibler (KL_m), χ^2, modified χ^2 (χ^2_m), Hellinger (H) and L_1 divergences are, respectively, associated with the convex functions $\varphi(x)=x\log x - x+1$, $\varphi(x)=-\log x+x-1$, $\varphi(x)=\frac{1}{2}(x-1)^2$, $\varphi(x)=\frac{1}{2}(x-1)^2/x$, $\varphi(x)=2(\sqrt{x}-1)^2$ and $\varphi(x)=|x-1|$. All of these divergences, except the L_1, belong to the class of the so-called "power divergences" introduced in Read and Cressie (1988) (see also Liese and Vajda 1987, Chapter 2), a class that takes its origin from Rényi *et al.* (1961). They are defined through the class of convex functions

$$x \in]0,+\infty[\mapsto \varphi_\gamma(x) := \frac{x^\gamma - \gamma x + \gamma - 1}{\gamma(\gamma-1)} \qquad [8.2]$$

if $\gamma \in \mathbb{R}\setminus\{0,1\}$, $\varphi_0(x) := -\log x + x - 1$ and $\varphi_1(x) := x\log x - x + 1$. So, the KL-divergence is associated with φ_1, the KL_m to φ_0, the χ^2 to φ_2, the χ^2_m to φ_{-1} and the Hellinger distance to $\varphi_{1/2}$.

It may be convenient to extend the definition of the power divergences in such a way that $\phi(\alpha, \theta)$ may be defined (possibly infinite) even when P_α or P_θ is not a probability measure. This is achieved setting

$$x \in]-\infty, +\infty[\mapsto \begin{cases} \varphi_\gamma(x) & \text{if } x \in [0, +\infty[, \\ +\infty & \text{if } x \in]-\infty, 0[. \end{cases} \qquad [8.3]$$

when $\text{dom}\varphi = \mathbb{R}^+/\{0\}$. Note that, for the χ^2-divergence, the corresponding φ function $\phi_2(x) := \frac{1}{2}(x-1)^2$ is defined and convex on whole $\mathbb{R}$.

We will only consider divergences defined through differentiable functions φ, which we assume to satisfy the following condition:

(RC) There exists a positive δ such that, for all c in $[1-\delta, 1+\delta]$, we can find numbers c_1, c_2, c_3 such that $\varphi(cx) \le c_1\varphi(x) + c_2|x| + c_3$, for all real x.

Condition **(RC)** holds for all power divergences including KL and KL_m divergences.

8.1.2. *Dual form of the divergence and dual estimators in parametric models*

Let θ and θ_T be any parameters in Θ. We intend to provide a new expression for $\phi(\theta, \theta_T)$.

As will be considered in statistical applications pertaining to non-regular models, it may be of interest, at times, to consider enhanced statistical models indexed by the parameter θ, and those models may include signed measures, although, obviously, no dataset would be generated by those. In any case, P_{θ_T} clearly will designate a probability measure, under which the data are supposed to be generated. In those cases, the domain of φ will be assumed to contain all possible real values for $(dP_\theta/dP_{\theta_T})(x)$, henceforth a (possibly infinite) interval in $\mathbb{R}$; the χ^2 divergence function

$$\chi^2(x) := \frac{1}{2}(x-1)^2$$

is the prototype for those cases.

By strict convexity, for all a and b the domain of φ it holds

$$\varphi(b) \ge \varphi(b) + \varphi'(a)(b-a) \qquad [8.4]$$

with equality if and only if $a = b$.

Let us denote $\varphi^{\#}(x) := x\varphi'(x) - \varphi(x)$. For any α in Θ, define $a := \frac{dP_\theta}{dP_\alpha}(x)$ and also define $b := \frac{dP_\theta}{dP_{\theta_T}}(x)$. Inserting these values in [8.4] and integrating with respect to P_{θ_T} yields

$$\phi(\theta,\theta_T) \geq \int \left[\varphi'\left(\frac{dP_\theta}{dP_\alpha}\right) dP_\theta - \varphi^{\#}\left(\frac{dP_\theta}{dP_\alpha}\right)\right] dP_{\theta_T}.$$

Assume at present that this entails

$$\phi(\theta,\theta_T) \geq \int \varphi'\left(\frac{dP_\theta}{dP_\alpha}\right) dP_\theta - \int \varphi^{\#}\left(\frac{dP_\theta}{dP_\alpha}\right) dP_{\theta_T} \qquad [8.5]$$

for suitable α's in some set $\mathcal{F}_\theta$ included in Θ.

When $\alpha = \theta_T$, the inequality in [8.5] turns to equality, which yields

$$\phi(\theta,\theta_T) = \sup_{\alpha\in\mathcal{F}_\theta} \int \varphi'\left(\frac{dP_\theta}{dP_\alpha}\right) dP_\theta - \int \varphi^{\#}\left(\frac{dP_\theta}{dP_\alpha}\right) dP_{\theta_T}. \qquad [8.6]$$

Denote

$$h(\theta,\alpha,x) := \int \varphi'\left(\frac{dP_\theta}{dP_\alpha}\right) dP_\theta - \varphi^{\#}\left(\frac{dP_\theta}{dP_\alpha}\right) \qquad [8.7]$$

from which

$$\phi(\theta,\theta_T) = \sup_{\alpha\in\mathcal{F}_\theta} \int h(\theta,\alpha,x) dP_{\theta_T}. \qquad [8.8]$$

Furthermore, by [8.5], for all α in $\mathcal{F}_\theta$

$$\begin{aligned}&\phi(\theta,\theta_T) - \int h(\theta,\alpha,x) dP_{\theta_T}\\ &= \int h(\theta,\theta_T,x) dP_{\theta_T} - \int h(\theta,\alpha,x) dP_{\theta_T} \geq 0\end{aligned}$$

and the function $x \to h(\theta,\theta_T,x) - h(\theta,\alpha,x)$ is non-negative due to [8.4]. It follows that $\phi(\theta,\theta_T) - \int h(\theta,\alpha,x)dP_{\theta_T}$ is zero if and only if (iff) $h(\theta,\alpha,x) = h(\theta,\theta_T,x)-$ P_{θ_T} a.e. Therefore, $\phi(\theta,\theta_T) - \int h(\theta,\alpha,x)dP_{\theta_T} = 0$ iff for any x in the support of P_{θ_T},

$$\begin{aligned}&\left[\int \varphi'\left(\frac{dP_\theta}{dP_{\theta_T}}\right) dP_\theta - \int \varphi'\left(\frac{dP_\theta}{dP_\alpha}\right) dP_\theta\right]\\ &-\varphi^{\#}\left(\frac{dP_\theta}{dP_\alpha}(x)\right) + \varphi^{\#}\left(\frac{dP_\theta}{dP_{\theta_T}}(x)\right) = 0,\end{aligned}$$

which cannot hold when the functions $\varphi^{\#}\left(\frac{dP_\theta}{dP_\alpha}(x)\right)$, $\varphi^{\#}\left(\frac{dP_\theta}{dP_{\theta_T}}(x)\right)$ and 1 are linearly independent, unless $\alpha = \theta_T$. We have proved that θ_T is the unique optimizer in [8.6].

We have skipped some sufficient conditions that ensure that [8.5] holds.

Assume that for all α in $\mathcal{F}_\theta$

$$\int \left|\varphi'\left(\frac{dP_\theta}{dP_\alpha}\right)\right| dP_\theta < \infty. \qquad [8.9]$$

Assume further that $\phi(\theta, \theta_T)$ is finite. Since

$$-\int \varphi^{\#}\left(\frac{dP_\theta}{dP_\alpha}(x)\right) dP_{\theta_T} \leq \phi(\theta,\theta_T) - \int \varphi'\left(\frac{dP_\theta}{dP_{\theta_T}}\right) dP_\theta$$
$$\leq \phi(\theta,\theta_T) + \int \left|\varphi'\left(\frac{dP_\theta}{dP_{\theta_T}}\right)\right| dP_\theta < +\infty$$

we obtain

$$\int \varphi^{\#}\left(\frac{dP_\theta}{dP_\alpha}\right) dP_{\theta_T} > -\infty$$

which entails [8.5]. When $\int \varphi^{\#}\left(\frac{dP_\theta}{dP_\alpha}\right) dP_{\theta_T} = +\infty$, then clearly, under [8.9]

$$\phi(\theta,\theta_T) > \int \varphi'\left(\frac{dP_\theta}{dP_\alpha}\right) dP_\theta - \int \varphi^{\#}\left(\frac{dP_\theta}{dP_\alpha}\right) dP_{\theta_T} = -\infty.$$

We have proved that [8.6] holds when α satisfies [8.9].

Sufficient and simple conditions encompassing [8.9] can be assessed under standard requirements for nearly all divergences. We state the following result (see Liese and Vajda 1987; Broniatowski and Keziou 2006, lemma 3.2).

LEMMA 8.1.– Assume that **RC** holds and $\phi(\theta, \alpha)$ is finite. Then [8.9] holds.

Summing up, we can state the following result.

THEOREM 8.1.– Let θ belong to Θ and let $\phi(\theta, \theta_T)$ be finite. Assume that **RC** holds. Let $\mathcal{F}_\theta$ be the subset of all α's in Θ such that $\phi(\theta, \alpha)$ is finite. Then

$$\phi(\theta,\theta_T) = \sup_{\alpha \in \mathcal{F}_\theta} \int \varphi'\left(\frac{dP_\theta}{dP_\alpha}\right) dP_\theta - \int \varphi^{\#}\left(\frac{dP_\theta}{dP_\alpha}\right) dP_{\theta_T}. \qquad [8.10]$$

Furthermore, the supremum is attained for θ_T and uniqueness holds true.

For the Cressie–Read family of divergences with $\gamma \neq 0, 1$, this representation writes

$$\phi_\gamma(\theta, \theta_T) = \sup_{\alpha \in \mathcal{F}_\theta} \left\{ \frac{1}{\gamma - 1} \int \left(\frac{dP_\theta}{dP_\alpha} \right)^{\gamma - 1} dP_\theta \right.$$
$$\left. - \frac{1}{\gamma} \int \left(\frac{dP_\theta}{dP_\alpha} \right)^{\gamma} dP_{\theta_T} - \frac{1}{\gamma(\gamma - 1)} \right\}.$$

The set $\mathcal{F}_\theta$ may depend on the choice of the parameter θ. Such is the case for the χ^2 divergence, i.e. $\varphi(x) = (x-1)^2/2$, when $p_\theta(x) = \theta \exp(-\theta x) 1_{[0,\infty)}(x)$. In most cases, the difficulty of dealing with a specific set $\mathcal{F}_\theta$ depending on θ can be encompassed when

there exists a neighbourhood $\mathcal{U}$ of θ_T for which [A]

$\phi(\theta, \theta')$ is finite whatever θ and θ' in $\mathcal{U}$

which, for example, holds in the above case for any θ_T. This simplification deserves to be stated in the next result.

THEOREM 8.2.– When $\phi(\theta, \theta_T)$ is finite and **RC** holds, then, under condition [A],

$$\phi(\theta, \theta_T) = \sup_{\alpha \in \mathcal{U}} \int \varphi' \left(\frac{dP_\theta}{dP_\alpha} \right) dP_\theta - \int \varphi^{\#} \left(\frac{dP_\theta}{dP_\alpha} \right) dP_{\theta_T}.$$

Furthermore, the supremum is attained for θ_T and uniqueness holds true.

REMARK 8.1.– Identifying $\mathcal{F}_\theta$ might be cumbersome. This difficulty also appears in the classical MLE case, a special case of the above statement with divergence function φ_0 , for which it is assumed that

$$\int \log p_\theta(x) p_{\theta_T}(x) d\lambda(x)$$

is finite for θ in a neighborhood of θ_T.

Under the above notation and hypotheses, define

$$T_\theta\left(P_{\theta_T}\right) := \arg \sup_{\alpha \in \mathcal{F}_\theta} \int h(\theta, \alpha, x) dP_{\theta_T}. \qquad [8.11]$$

It then holds that $T_\theta\left(P_{\theta_T}\right) = \theta_T$ for all θ_T in Θ. Also let

$$S\left(P_{\theta_T}\right) := \arg \inf_{\theta \in \Theta} \sup_{\alpha \in \mathcal{F}_\theta} \int h(\theta, \alpha, x) dP_{\theta_T}. \qquad [8.12]$$

which also satisfies $S\left(P_{\theta_T}\right) = \theta_T$ for all θ_T in Θ. We thus can state the following result.

THEOREM 8.3.– When $\phi(\theta, \theta_T)$ is finite for all θ in Θ and **RC** holds, both functionals T_θ and S are Fisher consistent for all θ_T in Θ.

8.1.2.1. *Plug in estimators*

From [8.8], simple estimators for θ_T can be defined, plugging any convergent empirical measure in place of P_{θ_T} and taking the infimum w.r.t. θ in the resulting estimator of $\phi(\theta, \theta_T)$.

In the context of simple i.i.d. sampling, introducing the empirical measure

$$P_n := \frac{1}{n} \sum_{i=1}^{n} \delta_{X_i},$$

where the X_i's are i.i.d. r.v's with common unknown distribution P_{θ_T} in $\mathcal{P}$, the natural estimator of $\phi(\theta, \theta_T)$ is

$$\begin{aligned} \phi_n(\theta, \theta_T) &:= \sup_{\alpha \in \mathcal{F}_\theta} \left\{ \int h(\theta, \alpha, x)\, dP_n(x) \right\} \qquad [8.13] \\ &= \sup_{\alpha \in \mathcal{F}_\theta} \int \varphi' \left(\frac{dP_\theta}{dP_\alpha} \right) dP_\theta - \frac{1}{n} \sum_{i=1}^{n} \varphi^{\#} \left(\frac{dP_\theta}{dP_\alpha} (X_i) \right) \\ &= \sup_{\alpha \in \mathcal{U}} \int \varphi' \left(\frac{dP_\theta}{dP_\alpha} \right) dP_\theta - \frac{1}{n} \sum_{i=1}^{n} \varphi^{\#} \left(\frac{dP_\theta}{dP_\alpha} (X_i) \right) \text{ when } A \text{ holds.} \end{aligned}$$

Denote by θ any given value in Θ and call it an escort parameter. Note that for any such θ, the supremum w.r.t. α in [8.10] is achieved at θ_T. A first M-estimator $\widetilde{\theta}(\theta)$ of θ_T with escort parameter θ can then be defined through

$$\widetilde{\theta}(\theta) := \arg\max_{\alpha} \int \varphi' \left(\frac{dP_\theta}{dP_\alpha} \right) dP_\theta - \frac{1}{n} \sum_{i=1}^{n} \varphi^{\#} \left(\frac{dP_\theta}{dP_\alpha} (X_i) \right). \qquad [8.14]$$

Those estimators coincide with the MLE for $\varphi(x) = -\log x + x - 1$ and they can be tuned through the choice of the escort parameter θ. Under condition A, since

$$\inf_{\theta \in \mathcal{U}} \phi(\theta, \theta_T) = \phi(\theta_T, \theta_T) = 0,$$

the resulting estimator of $\phi(\theta_T, \theta_T)$ is

$$\phi_n(\theta_T, \theta_T) := \inf_{\theta \in \Theta} \phi_n(\theta, \theta_T) = \inf_{\theta \in \Theta} \sup_{\alpha \in \mathcal{F}_\theta} \left\{ \int h(\theta, \alpha, x)\, dP_n(x) \right\}. \qquad [8.15]$$

Also the estimator of θ_T is obtained as

$$\widehat{\theta} := \arg \inf_{\theta \in \mathcal{U}} \sup_{\alpha \in \mathcal{U}} \left\{ \int h(\theta, \alpha, x)\, dP_n(x) \right\}. \qquad [8.16]$$

In the case when $\varphi = \varphi_\gamma$ defined by [8.2], it holds true

$$\widehat{\theta} := \arg \inf_{\theta \in \mathcal{U}} \sup_{\alpha \in \mathcal{U}} \left\{ \frac{1}{\gamma - 1} \int \left(\frac{dP_\theta}{dP_\alpha} \right)^{\gamma - 1} dP_\theta \right.$$

$$\left. - \left[\frac{1}{n\gamma} \sum_{i=1}^{n} \left(\frac{dP_\theta}{dP_\alpha} \right)^{\gamma} (X_i) - \frac{1}{n\gamma} \sum_{i=1}^{n} \frac{dP_\theta}{dP_\alpha} (X_i) \right] \right\},$$

for $\gamma \neq 0, 1$. When $\gamma = 0$ the above expression [8.16] boils down to the classical expression

$$\widehat{\theta} = \arg \inf_{\theta \in \mathcal{U}} -\frac{1}{n} \sum_{i=1}^{n} \log \frac{dP_\theta}{d\mu} (X_i)$$

where μ is some dominating measure $\mathcal{P}$. When $\gamma = 1$, then

$$\widehat{\theta} := \arg \inf_{\theta \in \mathcal{U}} \sup_{\alpha \in \mathcal{U}} \left\{ \int \log \frac{dP_\theta}{dP_\alpha} dP_\theta - \left[\frac{1}{n} \sum_{i=1}^{n} \frac{dP_\theta}{d\mu} (X_i) \right] \right\}.$$

The resulting minimum dual divergence estimators [8.15] and [8.16] do not require any smoothing or grouping, in contrast with the classical approach that involves quantization. The paper by Broniatowski and Keziou (2009) provides a complete study of those estimates and subsequent inference tools for the usual i.i.d. sample scheme. For all divergences considered here, these estimators are asymptotically efficient in the sense that they achieve the Cramer–Rao bound asymptotically in regular models. The case when $\varphi = \varphi_0$ leads to θ_{ML} defined as the celebrated maximum likelihood estimator (MLE), in the context of the simple sampling.

8.1.2.2. *Dual form of the divergence and dual estimators in a general framework*

The above derivation clearly leads to a similar description of

$$\phi(Q, P) := \int \varphi \left(\frac{dQ}{dP} \right) dP$$

whenever Q is an a.c. measure w.r.t. P and $\phi(Q, P) = +\infty$ otherwise. Here, Q might be a signed measure and P is a probability measure; therefore, the domain of the convex function φ may at time overlap with the negative axis. The same development as above clearly yields for finite $\phi(Q, P)$

$$\phi(Q, P) = \sup_{f \in \mathcal{F}} \int \varphi'(f) dQ - \int \varphi^{\#}(f) dP, \qquad [8.17]$$

where $\mathcal{F}$ designates any class of function that contains the function $x \to (dQ/dP)(x)$, since indeed the supremum in the RHS of the above display holds when $f = (dQ/dP)(x) - Pa.e.$ Integrability condition for this representation to hold can be written

$$\int |\varphi'(f)| \, dQ < \infty$$

for all f in $\mathcal{F}$, in accordance with [8.9].

Formula [8.17] is of relevance when Q belongs to some class of distributions for which $(dQ/dP)(x)$ has a known functional form; such is the case in many semi-parametric settings, for example for models $\mathcal{M}$ defined through moment conditions, or in odd ratio models (see Broniatowski and Keziou 2012; Keziou and Leoni-Aubin 2008) for which the functional form of the likelihood ratio between the true unknown distribution P and the measures in the model is indeed a specification of the model; the resulting plug in estimator of the ϕ-projection Q^* of P on $\mathcal{M}$ can be obtained through a similar procedure as in the parametric case

$$Q^* := \arg \inf_{Q \in \mathcal{M}} \sup_{f \in \mathcal{F}} \int \varphi'(f) dQ - \frac{1}{n} \sum_{i=1}^{n} \varphi^{\#}(f(X_i)),$$

where $X_1, .., X_n$ are sampled under P.

8.1.3. *Decomposable discrepancies*

We now explore the relation between Csiszár discrepancies and statistical criteria; see Broniatowski and Vajda (2012) for additional insights. Turning back to the parametric setting, the ϕ-divergences $\phi(\theta, \tilde{\theta})$ can be embedded in the broader class of pseudodistances $\mathfrak{D}$, which are defined by the condition that

$$\mathfrak{D}(P_\theta, P_{\tilde{\theta}}) \geq 0 \text{ for all } \theta, \tilde{\theta} \in \Theta \text{ with } \mathfrak{D}(P_\theta, P_{\tilde{\theta}}) = 0 \text{ iff } \theta = \tilde{\theta}.$$

An additional restriction imposed on pseudodistances $\mathfrak{D}(P, Q)$ will be the decomposability property. In the sequel Q would designate a probability measure, which may be an atomic one (typically Q is the empirical measure P_n). Here, $\mathcal{P}$ designates the class of probability measures on some measurable set $\mathcal{X}$.

A pseudodistance $\mathfrak{D}$ is a ***decomposable*** pseudodistance if there exist functionals $\mathfrak{D}^0 : \mathcal{P} \mapsto \mathbb{R}$, $\mathfrak{D}^1 : \mathcal{P} \mapsto \mathbb{R}$ and measurable mappings

$$\rho_\theta : \mathcal{X} \mapsto \mathbb{R}, \quad \theta \in \Theta$$

such that for all $\theta \in \Theta$ and $Q \in \mathcal{P}$ the expectations $\int \rho_\theta dQ$ exist and

$$\mathfrak{D}(P_\theta, Q) = \mathfrak{D}^0(P_\theta) + \mathfrak{D}^1(Q) + \int \rho_\theta dQ.$$

We say that a functional $T_{\mathfrak{D}} : \mathcal{Q} \mapsto \Theta$ defines a ***minimum pseudodistance estimator*** (briefly, $\min \mathfrak{D}$-estimator) if $\mathfrak{D}(P_\theta, Q)$ is a decomposable pseudodistance and the parameters $T_{\mathfrak{D}}(Q) \in \Theta$ minimize $\mathfrak{D}^0(P_\theta) + \int \rho_\theta dQ$ on Θ, in symbols

$$T_{\mathfrak{D}}(Q) = \text{argmin}_\theta \left[\mathfrak{D}^0(P_\theta) + \int \rho_\theta dQ \right].$$

In particular, for $Q = P_n$

$$\theta_{\mathfrak{D},n} := T_{\mathfrak{D}}(P_n) = \text{argmin}_\theta \left[\mathfrak{D}^0(P_\theta) + \frac{1}{n} \sum_{i=1}^{n} \rho_\theta(X_i) \right] .$$

Every $\min \mathfrak{D}$-estimator $\theta_{\mathfrak{D},n}$ is Fisher consistent in the sense that

$$T_{\mathfrak{D}}(P_{\theta_0}) = \text{argmin}_\theta \mathfrak{D}(P_\theta, P_{\theta_0}) = \theta_0 \quad \text{for all } \theta_0 \in \Theta.$$

Clearly, minimal ϕ-divergence estimators with escort θ, as defined in [8.14], are $\min \mathfrak{D}$-estimators for the decomposable pseudodistance

$$\mathfrak{D}(P_\theta, Q) := \mathfrak{D}^0_{\phi,\theta}(P_\theta) + \mathfrak{D}^1_{\phi,\theta}(Q) + \int \rho_{\phi,\theta,\alpha} dQ$$

and the maximum subdivergence estimator

$$\tilde{\theta}(\theta) = \arg\max_\alpha \left[\int \phi' \left(\frac{p_\theta}{p_\alpha} \right) p_\theta d\lambda + \frac{1}{n} \sum_{i=1}^{n} \phi^\sharp \left(\frac{p_\theta(X_i)}{p_\alpha(X_i)} \right) \right]$$

with the divergence ϕ and escort parameter $\theta \in \Theta$ is the $\min \mathfrak{D}$-estimator for the decomposable pseudodistance.

8.1.3.1. *Power pseudodistance estimators*

A special class of pseudodistances $\mathfrak{D}_\psi(P_\theta, Q)$ is defined on $\mathcal{P} \otimes \mathcal{P}$ by the integral formula

$$\mathfrak{D}_\psi(P_\theta, Q) = \int \psi(p_\theta, q)\, \mathrm{d}\lambda \quad \text{for } p_\theta = \frac{\mathrm{d}P_\theta}{\mathrm{d}\lambda}, q = \frac{\mathrm{d}Q}{\mathrm{d}\lambda},$$

where $\psi(s,t)$ are *reflexive* in the sense that they are non-negative functions of arguments $s, t > 0$ with $\psi(s,t) = 0$ iff $s = t$. If a function ψ is reflexive and also *decomposable* in the sense

$$\psi(s,t) = \psi^0(s) + \psi^1(t) + \rho(s)t, \quad s, t \geq 0 \qquad [8.18]$$

for some $\psi^0, \psi^1, \rho : (0,\infty) \to \mathbb{R}$, then the corresponding ψ-pseudodistance $\mathfrak{D}_\psi(P_\theta, Q)$ is a *decomposable pseudodistance* satisfying

$$\mathfrak{D}_\psi(P_\theta, Q) = \mathfrak{D}^0_\psi(P_\theta) + \mathfrak{D}^1_\psi(Q) + \int \rho_\theta dQ$$

for

$$\mathfrak{D}^0_\psi(P_\theta) = \int \psi^0(p_\theta)\,\mathrm{d}\lambda, \quad \mathcal{D}^1_\psi(Q) = \int \psi^1(q)\,\mathrm{d}\lambda \text{ and } \rho_\theta = \rho(p_\theta).$$

8.1.3.1.1. ϕ-divergences $D_\phi(P_\theta, Q)$

The ϕ-divergences $D_\phi(P_\theta, Q)$ are special ψ-pseudodistances for the functions

$$\psi(s,t) = \phi(s/t)\,t - \phi'(1)(s-t), \quad s,t > 0 \tag{8.19}$$

since they are non-negative and reflexive, and [8.19] implies $\mathfrak{D}_\psi(P_\theta, Q) = D_\phi(P_\theta, Q)$ for all $P \in \mathcal{P}, Q \in \mathcal{P}$ when ϕ and ψ are related by [8.19]. However, the functions [8.19] in general do not satisfy the decomposability condition [8.18] so that the ϕ-divergences are not in general decomposable pseudodistances. An exception is the logarithmic function $\phi = \phi_0 = -\log x + x - 1$ for which the $\min \mathfrak{D}_{\phi_0}$-estimator is the MLE.

8.1.3.1.2. L_2-estimator

The quadratic function $\psi(s,t) = (s-t)^2$ is reflexive and also decomposable. Thus, it defines the decomposable pseudodistance

$$\mathfrak{D}_\psi(P_\theta, Q) = \int (p_\theta - q)^2\,\mathrm{d}\lambda = \|p_\theta - q\|^2$$

on $\mathcal{P} \otimes \mathcal{P}$. It is easy to verify that

$$\mathfrak{D}^0_\psi(P_\theta) = \int p_\theta^2\,\mathrm{d}\lambda \quad \mathcal{D}^1_\psi(Q) = \int q^2\,\mathrm{d}\lambda, \text{ and } \rho_\theta = -2p_\theta.$$

The corresponding $\min \mathfrak{D}_\psi$-estimator is in this case the L_2-estimator

$$\theta_n = \mathrm{argmin}_\theta \left[\int p_\theta^2\,\mathrm{d}\lambda - \frac{2}{n}\sum_{i=1}^{n} p_\theta(X_i) \right]$$

which is known to be robust but not efficient (see for example Hampel *et al.* 1988).

To build a smooth bridge between the robustness and efficiency, one needs to replace the reflexive and decomposable functions ψ by families $\{\psi_\alpha : \alpha \geq 0\}$ of reflexive functions decomposable in the sense

$$\psi_\alpha(s,t) = \psi^0_\alpha(s) + \psi^1_\alpha(t) + \rho_\alpha(s)\,t \quad \text{for all } \alpha \geq 0$$

where the two functions ψ_α^0 and ψ_α^1 are defined at $\alpha = 0$ through the limits

$$\psi_0^0(s) = \lim_{\alpha \downarrow 0} \psi_\alpha^0(s) = \varkappa\, s \text{ and } \lim_{\alpha \downarrow 0} \rho_\alpha(s) = \rho_0(s) = -\ln s$$

for some constant $\varkappa$ all $s > 0$. Then for all $\alpha \geq 0$ and $(P_\theta, Q) \in \mathcal{P} \otimes \mathcal{P}$, the family of ψ_α-pseudodistances

$$\mathfrak{D}_\alpha(P_\theta, Q) := \mathfrak{D}_{\psi_\alpha}(P_\theta, Q), \quad \alpha \geq 0$$

satisfies the decomposability condition

$$\mathfrak{D}_\alpha(P_\theta, Q) = \mathfrak{D}_\alpha^0(Q) + \mathfrak{D}_\alpha^1(P_\theta) + Q \cdot \rho_{\alpha,\theta}$$

for

$$\mathfrak{D}_\alpha^0(P_\theta) = \int \psi_\alpha^0(p_\theta)\, \mathrm{d}\lambda, \quad \mathfrak{D}_\alpha^1(Q) = \int \psi_\alpha^1(q)\, \mathrm{d}\lambda \text{ and } \rho_{\alpha,\theta} = \rho_\alpha(p_\theta).$$

8.1.3.2. $\mathfrak{D}_\alpha$ *estimators*

The pseudodistances $\mathfrak{D}_\alpha(P_\theta, Q)$ are decomposable and define the family of $\min \mathfrak{D}_\alpha$-estimators

$$\begin{aligned} \theta_{\alpha,n} &= \arg\min_\theta \left[\mathfrak{D}_{\psi_\alpha}^0(P_\theta) + P_n \cdot \rho_{\alpha,\theta}\right] \\ &= \arg\min_\theta \left[\int \psi_\alpha^0(p_\theta)\, \mathrm{d}\lambda + \frac{1}{n}\sum_{i=1}^n \rho_\alpha(p_\theta(X_i))\right], \quad \alpha \geq 0. \end{aligned}$$

This family contains as a special case for $\alpha = 0$ the efficient but non-robust MLE

$$\theta_{0,n} = \arg\min_\theta \left[\text{const} - \frac{1}{n}\sum_{i=1}^n \ln p_\theta(X_i)\right],$$

while for $\alpha > 0$ the $\theta_{\alpha,n}$'s are expected to be less efficient but more robust than $\theta_{0,n}$.

Decomposable pseudodistances $\mathfrak{D}_\alpha(P_\theta, Q)$ defined by the functions

$$\psi_\alpha(s,t) = t^{1+\alpha}\left[\alpha\phi_{1+\alpha}\left(\frac{s}{t}\right) + (1-\alpha)\phi_\alpha\left(\frac{s}{t}\right)\right], \quad \alpha \geq 0$$

of variables s, $t > 0$ where $\phi_{1+\alpha}$ and ϕ_α are the power functions defined through

$$\psi_\alpha^0(s) = \frac{s^{1+\alpha}}{1+\alpha}, \quad \psi_\alpha^1(t) = \begin{cases} t\left[\frac{t^\alpha - 1}{\alpha} - \frac{t^\alpha}{1+\alpha}\right] \\ t\ln t - t \end{cases} \text{ and } \rho_\alpha(s) = \begin{cases} -\frac{s^\alpha - 1}{\alpha} & \text{if } \alpha > 0 \\ -\ln s & \text{if } \alpha = 0. \end{cases}$$

Moreover, this family is continuous in the parameter $\alpha \downarrow 0$ and leads to MLE as its limit.

8.1.3.3. *Relation with other estimation criteria*

Let us call the pseudodistances $\mathfrak{D}_\alpha(P, Q)$ simply ***power distances*** of orders $\alpha \geq 0$.

8.1.3.3.1. L_2-estimator revisited

The $\min \mathfrak{D}_\alpha$-estimator of order $\alpha = 1$ is defined by

$$\theta_{1,n} = \text{argmin}_\theta \left[\int p_\theta^2 \, \mathrm{d}\lambda - \frac{2}{n} \sum_{i=1}^{n} p_\theta(X_i) \right]$$

so that it is nothing but the L_2-estimator θ_n. The family of estimators $\theta_{n,\alpha}$ smoothly connects this robust estimator with the efficient MLE $\theta_{n,0}$ when the parameter α decreases from 1 to 0.

The special class of the $\min \mathfrak{D}_\alpha$-estimators $\theta_{\alpha,n}$ given above was proposed by Basu *et al.* (1998) who confirmed their efficiency for $\alpha \approx 0$ and their intuitively expected robustness for $\alpha > 0$. These authors called $\theta_{\alpha,n}$ *minimum density power divergence estimators* without actual clarification of the relation of the "density power divergences" $\mathfrak{D}_\alpha(P, Q)$ to the standard power divergences $D_\alpha(P, Q)$ studied in Liese and Vajda (1987) and Read and Cressie (1988).

$$\theta_{\alpha,n} = \text{argmax}_\theta \begin{cases} \frac{1}{n} \sum_{i=1}^{n} \frac{1}{\alpha} \left(p_\theta^\alpha(X_i) - 1\right) - \frac{1}{1+\alpha} \int p_\theta^{1+\alpha} \, \mathrm{d}\lambda & \text{if } \alpha > 0 \\ \frac{1}{n} \sum_{i=1}^{n} \ln p_\theta(X_i) - 1 & \text{if } \alpha = 0. \end{cases} \quad [8.20]$$

This subclass of general $\min \mathfrak{D}_\alpha$-estimators [8.20] was included in a wider family of generalized MLE's introduced and studied previously in Vajda (1984 and 1986).

8.1.3.4. *Invariance*

If the observations $x \in \mathcal{X}$ are replaced by $y = T(x)$, where $T : (\mathcal{X}, \mathcal{A}) \mapsto (\mathcal{Y}, \mathcal{B})$ is a measurable statistic with the inverse T^{-1} then the densities

$$\tilde{p}_\theta = \frac{\mathrm{d}\tilde{P}_\theta}{\mathrm{d}\tilde{\lambda}}$$

in the transformed model $\langle (\mathcal{Y}, \mathcal{B}); \tilde{\mathcal{P}} = (\tilde{P}_\theta = P_\theta T^{-1} : \theta \in \Theta) \rangle$ w.r.t. the σ-finite dominating measure $\tilde{\lambda} = \lambda T^{-1}$

$$\tilde{p}_\theta(y) = p_\theta(T^{-1}y) \, \mathcal{J}_T(y),$$

where $\mathcal{J}_T(y)$ is the determinant$\mathcal{J}_T(y) = \left| \frac{\mathrm{d}}{\mathrm{d}y} T^{-1}(y) \right|$.

The $\min \mathcal{D}_\alpha$-estimators are in general not equivariant w.r.t. invertible transformations of observations T, unless $\alpha = 0$.

However, the $\min \mathfrak{D}_\alpha$-estimates $\tilde{\theta}_{\alpha,n}$ in the above-considered transformed model coincide with the original $\min \mathfrak{D}_\alpha$-estimates $\theta_{\alpha,n}$ if the Jacobian $\mathcal{J}_T$ of transformation is a non-zero constant on the transformed observation space $\mathcal{Y}$. Thus, if $\mathcal{X}, \mathcal{Y}$ are Euclidean spaces, then the $\min \mathfrak{D}_\alpha$-estimators are equivariant under linear statistics $Tx = ax + b$.

8.2. Models and selection of statistical criteria

We now turn back to the Csiszár class of divergences and illustrate the role of this class in statistical inference.

Semi-parametric models defined by moment conditions provide a simple case where the question of the selection of statistical criterion appears in a natural way. Such models $\mathcal{M}$ satisfy [8.1] where the submodel $\mathcal{M}_\theta$ is defined as the collection of all distributions Q such that, for some function $(x, \theta) \to g(x, \theta) \in \mathbb{R}^d$, the moment condition

$$\int g(x, \theta)\, dQ = 0 \qquad [8.21]$$

holds. The parameter of interest is θ and an extensive literature is devoted to inference about it in many contexts. The Csiszár divergence approach to it consists of finding the ϕ-projection of P_n on $\mathcal{M}_\theta$

$$\widehat{Q}_\theta := \arg \min_{Q \in \mathcal{M}_\theta} \phi(Q, P_n) \qquad [8.22]$$

and then to estimate θ_T through

$$\widehat{\theta} := \arg \min_{\theta} \phi\left(\widehat{Q}_\theta, P_n\right). \qquad [8.23]$$

The optimization in [8.22] is easy; indeed, $\mathcal{M}_\theta$ can be divided into the class of all distributions with support included in the support of P_n (hence supported by the sample points), denoted $\mathcal{M}_\theta(P_n)$, and its complementary set. On $\mathcal{M}_\theta(P_n)$, the divergence $\phi(Q, P_n)$ is finite, and this divergence is infinite on the complementary set. Therefore, the ϕ-projection of P_n on $\mathcal{M}_\theta$ is a point in the simplex of $\mathbb{R}^n$. The optimization step in [8.22] is a finite dimensional problem. This is the core argument underlying the empirical likelihood approach and its generalizations, and this gives a sound motivation for the Csiszár divergence approach in this setting. However, in some cases we may be interested in introducing some additional assumptions in the model, specifying some smoothness property for the distributions in $\mathcal{M}_\theta$, typically pertaining to their density with respect to the Lebesgue measure. The approach through Csiszár divergences does not allow for such a requirement since definition [8.22] cannot include any additional information on $\widehat{Q}_\theta$ but the moment condition [8.21]. On the contrary, the density power divergences introduced in Basu *et al.* (1998), which are not of Csiszár type, allow us to handle those cases, at the cost of some enhanced optimization approaches in appropriate function spaces (see Broniatowski and Moutsouka 2020). Similar approaches can be handled for models defined by constraints pertaining to expectations of L-statistics (linear functions of order statistics); see Broniatowski and Decurninge (2016) for inference in models defined by conditions on L-moments.

8.3. Non-regular cases: the interplay between the model and the criterion

This section illustrates the flexibility of the Csiszár divergence approach to handle inference in non-regular models in a simple and commonly met situation. We consider the test of the number of components in a mixture. For the sake of clearness, the model at hand consists of a mixture of two components with given densities; therefore, the test consists of assessing whether the mixing coefficient θ is 0. There exists a large literature devoted to such questions either when the components of the mixture are known or when their own parameters are unknown; a classical result quoted in Titterington *et al.* (1985) provides the limit distribution of the likelihood ratio statistics under the null hypothesis, which includes a Dirac measure; such results have been extended to mixtures with more than two compnents; however, the asymptotic distribution is quite complex and the problem is therefore still quite open. In the present context, we will embed the mixture model in the space of the signed measures, allowing the mixing coefficient to take negative values; the coefficient θ is then an interior point of the parameter set, and we make use of divergences whose domain extends on $\mathbb{R}^-$. The extended model is regular and the divergence between the empirical measure and the model can be analyzed under the null hypothesis, leading to a classical χ^2 distribution, as expected.

Let us thus consider a set of signed measures defined by

$$p_\theta = (1-\theta)p_0 + \theta p_1, \quad \theta \in \mathbb{R} \qquad [8.24]$$

where p_0 and p_1 are two known densities (belonging, or not, to the same parametric family).

We observe a random sample $X_1, \ldots, X_n$ of distribution p_T. We are willing to test:

$$H_0 : p_T = p_0 \text{ vs } H_1 : p_T = p_\theta \neq p_0 \qquad [8.25]$$

which can be reduced to

$$H_0 : \theta_T = 0 \text{ vs } H_1 : \theta_T \neq 0 \qquad [8.26]$$

whenever $p_0 \neq p_1$ is met. The latter condition ensures the identifiability of the model and enables us to consider different parametric families for p_0 and p_1. Conversely, Chen *et al.* (2001), for instance, assume that $0 < \theta < 1$, and tests the equality of the parameters of p_0 and p_1 inside a unique family $\mathcal{F}$.

In the following, we thus assume that $p_0 \neq p_1$. The result proposed in Titterington *et al.* (1985) adapts the Wilks statistics (likelihood ratio test) for this situation; however, the limit distribution is not standard, even in this simple case, since it is a mixture of a Dirac distribution at 0 and a χ^2 distribution with 1df; this procedure seems untractable in cases with more than two components, adding the difficulty due to the estimation of the parameters of the components; note that in the present derivation we assumed that the parameters of the components are known, which is clearly not the case in practice. So any global inference on the number of components adds an extra derivation of the test statistics taking into account the distribution of the estimators of the parameters of the components; this does not play in favor of methods that would involve non-classical limit laws for the Wilks statistics. This explains the amount of efforts toward a classical solution for this non-regular problem, still quite commonly met.

8.3.1. *Test statistics*

For adequate divergence ϕ, the estimate of $\phi(0, \theta_T)$ provides a test statistic for the test defined by [8.25]; indeed, it quantifies the discrepancy between the actual distribution p_{θ_T} and p_0. Let ϕ be any Csiszár divergence with 0 an interior point in the domain of φ. Then the statistic $2n\phi_n(0, \theta_T)$ can be used as a test statistic for [8.25] and

$$2n\phi_n(0, \theta_T) \longrightarrow \chi^2_{(1)} \text{ when } H_0 \text{ holds.} \qquad [8.27]$$

This result also holds when testing whether the true distribution is a k_0-component mixture or a k-component mixture, which can be stated as follows. Consider a k-component parametric mixture model P_θ ($k \geq 2$) defined by

$$P_\theta := \sum_{i=1}^{k} w_i P_{a_i}^{(i)} \qquad [8.28]$$

where $\left\{P_{a_1}^{(1)}; a_1 \in A_1\right\}, \ldots, \left\{P_{a_k}^{(k)}; a_k \in A_k\right\}$ are k-parametric models and $A_1, \ldots, A_k$ are k sets in $\mathbb{R}^{d_1}, \ldots, \mathbb{R}^{d_k}$ with $d_1, \ldots, d_k \in \mathbb{N}^*$ and $0 \leq w_i \leq 1$,

$\sum w_i = 1$. Note that we consider a non-standard framework in which the weights are allowed to be equal to 0. Note Θ the parameter space

$$\theta \in \Theta :=$$
$$\left\{ (w_1, \ldots, w_k, a_1, \ldots, a_k)^T \in [0,1]^k \times A_1 \times \cdots \times A_k \text{ such that } \sum_{i=1}^{k} w_i = 1 \right\},$$

and assume that the model is identifiable. Let $k_0 \in \{1, \ldots, k-1\}$.

We are willing to test if $(k - k_0)$ components in [8.28] have null coefficients. We assume that their labels are $k_0 + 1, ..., k$. Denote Θ_0 the subset of Θ defined by

$$\Theta_0 := \{\theta \in \Theta \text{ such that } w_{k_0+1} = \cdots = w_k = 0\}.$$

In this case, the test statistic $2n\phi_n(\Theta_0, \theta_T) := 2n \inf_{\theta \in \Theta_0} \phi_n(\theta, \theta_T)$ converges to a $\chi^2_{(k-k_0)}$ distribution when H_0 holds.

While many divergences meet the former properties, we restrict ourselves to two generators in the following.

8.3.1.1. *Chi-square divergence*

The first divergence that we consider is the χ^2-divergence. The corresponding φ function $\varphi_2(x) := \frac{1}{2}(x-1)^2$ is defined and convex on whole $\mathbb{R}$; an example when $\mathcal{P}$ may contain signed finite measures and not be restricted to probability measures is considered in Broniatowski and Keziou (2006) in relation with a two-component mixture model defined in [8.24] and where θ is allowed to assume values in an open neighborhood Θ of 0, in order to provide a test for [8.26], with θ an interior point of Θ.

8.3.1.2. *Modified Kullback–Leibler divergence*

The second divergence that we retain is generated by a function described below, namely

$$\varphi_c(x) := (x + e^c - 1) \cdot \log(x + e^c - 1) + 1 - (x + e^c - 1) \qquad [8.29]$$
$$+(1-c) \cdot (e^c - 1) - c \cdot x \geq 0 \text{ for } x \in]1 - e^c, \infty[, \text{ and } \quad c \in \mathbb{R}, \qquad [8.30]$$

which has been derived within the recent general framework of Broniatowski and Stummer (2019). It is straightforward to see that φ_c is strictly convex and satisfies $\varphi_c(1) = 0 = \varphi_c'(1)$. For the special choice $c = 0$, [8.29] reduces to the omnipresent Kullback–Leibler divergence generator

$$\varphi_0(x) := x \log x - x + 1 \geq 0, \qquad x \in]0, \infty[.$$

According to [8.29], in the case of $c > 0$ the domain $]1 - e^c, \infty[$ of φ_c also covers negative numbers (see Broniatowski and Stummer (2019) for insights on divergence-generators with general real-valued domain); thus, the same facts hold for the new generator as for the χ^2 and this opens the gate to considerable comfort in testing mixture-type hypotheses against corresponding marginal-type alternatives, as we derive in the following. We denote KL_c the corresponding divergence functional for which $KL_c(Q, P)$ is well defined, whenever P is a probability measure and Q is a signed measure.

It can be noted that, depending on the type of model considered, the validity of the test can be subject to constraints over the parameters of the densities. Indeed, the convergence of $I = \int \mid \phi'\left(\frac{p_0}{p_\theta}\right) \mid dP_0$ is not always guaranteed. This kind of consideration may guide the choice of the test statistic. For instance, in some cases, including scaling models, conditions that are required for the χ^2-divergence, do not apply to the KL_c-divergence. For instance, consider a Gaussian mixture model with different variances

$$p_0 \sim \mathcal{N}(\mu, \sigma_0^2), \qquad p_1 \sim \mathcal{N}(\mu, \sigma_1^2).$$

The convergence of I with the χ^2 requires either $\sigma_1^2 > \sigma_0^2$ or $\sigma_0^2 > \sigma_1^2 > \frac{1}{2}\sigma_0^2$. On the other hand, the convergence is always ensured with the KL_c-divergence.

The same observations can be made for the lognormal, exponential and Weibull densities. This fact enlightens the necessary interplay between the model and the statistical criterion, which should impose as little restriction as possible on the unknown part of the inference process. We refer to Broniatowski *et al.* (2019) for extensions to models with estimated parameters and numerical results.

8.4. References

Basu, A., Harris, I.R., Hjort, N.L., and Jones, M. (1998). Robust and efficient estimation by minimising a density power divergence. *Biometrika*, 85(3), 549–559.

Broniatowski, M., and Moutsouka, J. (2020). Smooth min-divergence inference in semi parametric models. [Online]. Available at: https://hal.archives-ouvertes.fr/hal-02586204.

Broniatowski, M., and Decurninge, A. (2016). Estimation for models defined by conditions on their L-moments. *IEEE Transactions on Information Theory*, 62(9), 5181–5198.

Broniatowski, M., and Keziou, A. (2006). Minimization of φ-divergences on sets of signed measures. *Studia Scientiarum Mathematicarum Hungarica*, 43(4), 403–442.

Broniatowski, M., and Keziou, A. (2009). Parametric estimation and tests through divergences and the duality technique. *Journal of Multivariate Analysis*, 100(1), 16–36.

Broniatowski, M., and Keziou, A. (2012). Divergences and duality for estimation and test under moment condition models. *Journal of Statistical Planning and Inference*, 142(9), 2554–2573.

Broniatowski, M., Miranda, E., and Stummer, W. (2019). Testing the number and the nature of the components in a mixture distribution. Lecture Notes in Computer Science 11712. *Geometric Science of Information - 4th International Conference*. Springer, Cham.

Broniatowski, M., and Stummer, W. (2019). Some universal insights on divergences for statistics, machine learning and articial intelligence. In Nielsen, F. (ed.). *Geometric Structures of Information*. Springer, Cham.

Broniatowski, M., and Vajda, I. (2012). Several applications of divergence criteria in continuous families. *Kybernetika*, 48(4), 600–636.

Chen, H., Chen, J., and Kalbfleisch, J.D. (2001). A modified likelihood ratio test for homogeneity in finite mixture models. *Journal of the Royal Statistical Society: Series B (Statistical Methodology)*, 63(1), 19–29.

Csiszár, I. (1964). Eine informationstheoretische Ungleichung und ihre Anwendung auf Beweis der Ergodizitaet von markoffschen Ketten. *Magyer Tud. Akad. Mat. Kutato Int. Koezl.*, 8, 85–108.

Hampel, F.R., Ronchetti, E.M., Rousseeuw, P.J., Stahel, W.A. (1986). *Robust statistics. The approach based on influence functions*. John Wiley & Sons, Inc., New York.

Keziou, A., and Leoni-Aubin, S. (2008). On empirical likelihood for semiparametric two-sample density ratio models. *Journal of Statistical Planning and Inference*, 138(4), 915–928.

Liese, F., and Vajda, I. (1987). *Convex Statistical Distances*. BSB B.G. Teubner Verlagsgesellschaft, Leipzig.

Read, T.R.C., Cressie, N.A.C. (1988). *Goodness-of-Fit Statistics for Discrete Multivariate Data*. Springer-Verlag, New York.

Rényi, A. (1961). On measures of entropy and information. *Proc. 4th Berkeley Sympos. Math. Statist. and Prob.* University of California Press, Berkeley.

Titterington, D.M., Smith, A.F.M., and Makov, U.E. (1985). *Statistical Analysis of Finite Mixture Distributions*. John Wiley & Sons Ltd., Chichester.

Vajda, I. (1984). Minimum divergence principle in statistical estimation. *Statistics and Decisions*, 1, 239–261.

Vajda, I. (1986). Efficiency and robustness control via distorted maximum likelihood estimation. *Kybernetika*, 22(1), 47–67.

9

Dynamics of the Group Entropy Maximization Processes and of the Relative Entropy Group Minimization Processes Based on the Speed-gradient Principle

The maximum or minimum entropy principles answer the question, "Where does a system tend to?", while the speed-gradient (SG) principle answers the question, "How does a system reach its steady-state mode?". By combining these principles, the equilibrium state distribution of a non-stationary process, which follows a maximum entropy principle, and the dynamic equations of the process are obtained. SG principle is a local principle that investigates the transient modes of a physical/mechanical system, either stationary or non-stationary. In this chapter, we consider non-stationary systems, which we assume have steady-state distributions, given by the maximum entropy principle (various entropies are investigated).

Some other applications of the SG principle include the construction of statistical dynamics equations for finite systems of particles that obey the principle of maximum entropy, molecular dynamics based on the equations of classical mechanics, hock-wave transient processes in condensed matter, the Rayleigh problem on the time evolution of the shear medium flow along the plane rigid surface and microscopic description of the time evolution of a system's internal structure.

In this chapter, we investigate the non-stationary process that obeys the group entropy maximization principle or the relative entropy group minimization principle and obtains its dynamic equations by applying the SG principle. Group entropies satisfy one important axiom, apart from the classical Shannon–Khinchin axioms an entropy

Chapter written by Vasile PREDA and Irina BĂNCESCU.

should satisfy. The fourth axiom, called the composability axiom, tells us that the entropy of a macroscopic system obtained through the composition of two statistically independent systems should depend only on the entropy of the two systems. The composability axiom is an alternative to the additivity axiom, extensivity of systems being crucial for an entropy to have applicability in thermodynamics.

We also introduce new entropies called (G, a) power relative entropy group measures. Some possible future directions of the present work are the analysis of the continuous case for the relative entropy groups and (G, a) power relative entropy groups, and also considering new entropies and divergence measures.

9.1. Introduction

The evolution of physical, chemical, biological and socioeconomic processes has attracted the attention of the scientific community since the very beginning. Many discussions and questions involving the concept of entropy were debated, entropy being a basic tool of thermodynamics and modern statistical physics.

Nowadays, entropies play an essential role in many fields such as physics, information theory, analysis of complex systems, electronics and engineering, but what is entropy? Concepts of the term started with the discovery of atoms, with many researchers trying to grasp the intuitive nature of entropy. The question that leads to the statistical interpretation of entropy in thermodynamics is, "Are all molecules moving at the same speed in a gas?". The answer was given by Boltzmann (Cercignani 1998). Entropy tells us the story of atoms that can, just as probably, all stay in the same space or some atoms can stay in a space while the remaining stay in another space, or other possibilities.

Important entropic principles are the maximum and minimum entropy principles that originated in statistical physics. The first to formulate the maximum entropy principle (MaxEnt) was E.T. Jaynes in 1957 (Jaynes 1957, 1980, 1979). Jaynes implied that "the entropy of any physical system tends to increase until it achieves its maximum value under some constraints imposed by other physical laws". Nowadays, applications of maximum and minimum principles can also be found in other fields such as economics, biodiversity, pattern recognition and chemistry, just to mention a few (Kapur and Kesavan 1992; Preda and Bălcău 2011). In economics, the MaxEnt principle was applied to obtain new Lorenz curves, which maximize Tsallis entropy under equality and inequality constraints of the mean and Gini index (Preda *et al.* 2015).

The MaxEnt principle has the ability to expand the research of non-equilibrium systems, which can be found in non-equilibrium phase transitions (Haken 2006), various physical, biological and chemical systems or in relaxation processes Jaynes (1980, 1979); Dougherty (1994); Grandy Jr. (1980). The question that the MaxEnt or

MinEnt principles answer to is, "Where does the system tend to?"; it does not tell us how it reaches that state or what is the speed to reach that state. Hence, some other principle was needed and the answer was given by Alexander Fradkov.

9.1.1. *The SG principle*

Alexander Fradkov proposed a new local evolution principle, namely the SG principle, which answers the question, "How does a system reach its steady-state mode?" (Fradkov 2005, 2007). This principle originates from nonlinear control theory (Fradkov 1979; Fradkov and Pogromsky 1998).

The SG principle is a local principle that investigates the transient modes of a physical/mechanical system which is either stationary or non-stationary (Fradkov and Krivtsov 2011). In this chapter, we consider non-stationary systems that we assume have steady-state distributions given by the maximum entropy principle (various entropies are investigated). Therefore, we assume that their limit laws will maximize their entropy considered (Jaynes 1957). According to the SG principle, the physical meaning of the transient dynamics of the system obtained with this method suggest that the system moves along the "direction of the maximum entropy production rate" (Fradkov and Krivtsov 2011). Hence, knowing the steady-state distribution of a system given by the maximum entropy principle, we also know how the system reaches its state of maximum entropy by applying the SG principle.

Consider a physical system described by systems of differential equations of the form

$$\dot{x} = f(x, u, t), \tag{9.1}$$

where $x = (x_1, x_2, ..., x_n)^T$ represents the systems' states, $u = (u_1, u_2, ..., u_m)^T$ are the free (input) variables and $\dot{x} = \frac{dx}{dt}$, $t \geq 0$.

According to the SG principle, "among all possible motions of the system only those are realized for which the input variables change proportionally to the SG $\nabla_u \dot{W}(x, u, t)$ of an appropriate goal functional $W(x, t)$". Therefore, to describe the dynamic equations of a system, one needs to choose an appropriate goal function $W(x, t)$. We search for the law of change of $u(t)$ subject to some criterion of behavior of the system given by the tendency to achieve a goal functional $W(x, t)$ set *a priori* (or, more precisely, the behavior is given by the decreasing of the goal functional).

The solution of the problem is obtained by deriving the speed of the goal functional $W(x, t)$, evaluating the gradient of the speed $\nabla_u \dot{W}$ with respect to input vector u (SG vector), and formulate the law of dynamics as

$$u = -\gamma \nabla_u \dot{W}(x, u, t), \text{ where } \gamma > 0 \text{ is a scalar.} \tag{9.2}$$

By taking the goal function $W(x,t)$ to be any given entropy, we obtain the transient dynamics equations of systems having their limited law given by a distribution that maximizes the given entropy.

If we consider a system also having an additional total energy conservation constraint (Shalymov and Fradkov 2016a) given by

$$\int_{\Omega} p(t,x)h(x)dx = E, \quad \forall t \geq 0, \qquad [9.3]$$

where $p(\cdot, x)$ is the probability distribution of the system, while x is the state of the system, E is the common energy of a system and h is the density of energy, $h \in L_2(\Omega)$, h is not a constant function, then in this case we can also obtain an equilibrium distribution law of a system that obeys maximum entropy principle. The dynamic equation of this system is defined as (Shalymov and Fradkov 2016a)

$$u = -\gamma \nabla_u \dot{S_G} + \lambda_1 h + \lambda_2, \qquad [9.4]$$

where λ_1 and λ_2 are Lagrange multipliers. Solutions of problems that employ the SG and maximum entropy principles are obtained under conditions of regularity as given by Cover and Thomas (2006).

Examples of SG principle laws of dynamics include the motion of a particle in the potential field and wave, diffusion and heat transfer equations (Fradkov 2008). For the first example, the goal function $W(x)$ is chosen as the potential energy of a particle while, for the second example, the goal function is given by

$$W_t(x) = \frac{1}{2}\int_{\Omega} |\nabla_r x(r,t)|^2 dr.$$

Many studies of dynamics of systems have considered the application of variational principles, for example, the Gauss principle of least constraint, the Pontryagin maximum principle and the minimum energy dissipation principle. The SG principle and the dynamics of discrete systems for the Tsallis entropy are discussed in Fradkov and Shalymov (2015b), while for Rényi entropy this principle is investigated in Shalymov and Fradkov (2016a). Some continuous probability distributions and the SG principle are considered in Fradkov and Shalymov (2015a) and Shalymov (2016). In Shalymov and Fradkov (2016b), the f-divergence minimization and the SG principle are discussed along with the particular cases of Kullback–Leibler divergence, Tsallis relative entropy, Burg entropy, quadratic term of the BGS entropy and Cressie–Read entropy.

As an observation, the SG principle can be applied not only to finite dimensional systems but also to infinite dimensional ones (Fradkov 2008).

Some other applications of the SG principle include the construction of statistical dynamics equations for finite systems of particles that obey the principle of maximum entropy (Fradkov 2008, 2007), molecular dynamics based on the equations of classical mechanics (Fradkov and Krivtsov 2011), hock-wave transient processes in condensed matter (Khantuleva 2005; Khantuleva and Abdullaev 2014; Khantuleva 2003), the Rayleigh problem on the time evolution of the shear medium flow along the plane rigid surface (Rayleigh 1911) and microscopic description of the time evolution of a system's internal structure (Khantuleva and Shalymov 2017).

In this chapter, we consider some equations that describe the evolution of non-stationary processes that maximize the group entropy or minimize the relative entropy group. The SG principle is employed in studying these type of non-stationary processes. Finally, new dynamic equations for many entropic measures are obtained as special cases.

In the following sections of this chapter, three types of entropies are considered and explored, namely group entropies in section 9.2, relative entropy groups in section 9.3 and then a new class of power relative entropy groups in section 9.4.

9.1.2. *Entropy groups*

Entropies play an essential role in many fields such as information theory, physics, electronics, engineering and analysis of complex systems. Some generalized classes of entropy are the entropy groups introduced by Tempesta in 2011 and again in 2016 (Tempesta 2016b, 2011, 2016a, 2015). These types of entropies were introduced due to the need for a unified theoretical framework suitable for constructing generalized entropies for a wide class of systems (Jeldtoft Jensen and Tempesta 2018). Among various entropies introduced in the specialized literature, we mention Kaniadakis entropy (Kaniadakis 2009; Ourabah *et al.* 2015; Abul-Magd 2007), relative entropies (Kleeman 2002; Liu *et al.* 2006; Furuichi *et al.* 2004), Tsallis entropy along with Tsallis statistics (De Albuquerque *et al.* 2004; Bhandari *et al.* 2015; Plastino and Plastino 1993; Băncescu 2018), weighted entropies (Belis and Guiaşu 1968; Guiaşu 1977, 1971; Dial and Taneja 1981) and cumulative entropies (Kumar and Taneja 2011). Some well-known entropies are Shannon and Rényi entropies, the latter being a generalization of the former with both entropies having many applications, for example, in classifying emergent behaviors in a simulation of laser dynamics and ultrasonic molecular imaging (Guisado *et al.* 2005; Hughes *et al.* 2009).

Various forms of group entropies were proposed, but the most generalized one is the universal group entropy.

The universal group entropy (discrete case), denoted by s_G, is defined by

$$s_G(p) = \sum_{i=1}^{n} p_i G\Big(\log \frac{1}{p_i}\Big) = \sum_{i=1}^{n} p_i G\Big(-\log p_i\Big), \qquad [9.5]$$

where $p = (p_1, p_2, ..., p_n)^T$ is a discrete probability distribution, and $G(z)$ is an arbitrary real analytic invertible function such that $G(0) = 0$, $\dot{G}(0) = 1$.

In the continuous case, the group entropy takes the form

$$S_G(p) = \int_\Omega p(t,x) G\Big(\log \frac{1}{p(t,x)}\Big) dx = \int_\Omega p(t,x) G\Big(-\log p(t,x)\Big) dx, \quad [9.6]$$

where $p(t, x)$ is a probability density function in x, where $t \geq 0$.

These type of entropies can classify entropies "according to how the phase space grows with number of particles" (Jeldtoft Jensen and Tempesta 2018). Hence, we have entropies classified by non-exponential phase spaces and exponential phase spaces. Also, constructing Riemannian structures in statistical manifolds can be done by group entropies (Amari 2016). Some quantum versions of these entropies have also be obtained so far (Tempesta 2016b).

Group entropies satisfy one important axiom apart from the classical Shannon–Khinchin axioms that an entropy $H(p)$ should satisfy (continuity, concavity and $H(0) = 0$, $H(1) = 0$) (Shannon 1948; Shannon and Weaver 1949; Khinchin 1957). The fourth axiom, called the composability axiom, tells us that the entropy of a macroscopic system obtained through the composition of two statistically independent systems should depend only on the entropy of the two systems (Tsallis 2009; Callen 1985; Gell-Mann 1995).

The composability axiom is an alternative to the additivity axiom, extensivity of systems being crucial for an entropy to have applicability in thermodynamics. It has been observed that additivity does not imply extensivity and that extensivity does not imply additivity (Tempesta 2016a). As an example, Tsallis entropy, which is non-additive, is extensive for some special values of the q parameter. In 2017, the domain of applicability of the group entropy in the perspective of the third law of thermodynamics has been investigated, which also ensures the extensivity of the group entropy (Canturk *et al.* 2017).

Some special cases of entropy can be obtained by choosing different G functions.

For $G(z) = z$, we get the Shannon entropy

$$S(p) = \sum_{i=1}^{n} p_i \log \frac{1}{p_i}. \quad [9.7]$$

For $G(z) = \frac{\exp\left[(1-q)z\right]-1}{1-q}$, $0 \leq q < 1$, the universal group entropy reduces to Tsallis entropy

$$S_q(p) = \frac{\sum_{i=1}^{n} p_i^q - 1}{1-q}, \qquad [9.8]$$

while for $q > 1$, we choose $G(z) = \frac{\exp\left[(q-1)z\right]-1}{q-1}$.

Kaniadakis entropy is obtained by choosing $G(z) = \frac{\exp(kz)-\exp(-kz)}{2k}$, given by

$$S_k(p) = \sum_{i=1}^{n} p_i \frac{p_i^{-k} - p_i^k}{2k}, \quad k \neq 0. \qquad [9.9]$$

For $G(z) = \frac{\exp(az)-\exp(bz)}{a-b}$, $a \neq b$, we get Abe–Borges–Roditi entropy.

For

$$G(z) = \frac{1}{\sigma}\sum_{i=l}^{m} k_i \exp(i\sigma z), \quad l, m \in \mathbb{Z}, \quad m - l = r > 0,$$

subject to constraint

$$\sum_{i=l}^{m} k_i = 0, \quad \sum_{i=l}^{m} ik_i = 1, \quad k_m \neq 0, k_l \neq 0,$$

we get the original group entropy from Tempesta (2011):

$$S_G(p) = \sum_{i=1}^{n} p_i Log_G\left(\frac{1}{p_i}\right), \qquad [9.10]$$

where

$$Log_G(x) = \frac{1}{\sigma}\sum_{i=l}^{m} k_i x^{\sigma i}, \quad x > 0,$$

where Log_G is the generalized logarithm. The question Tempesta answered when he introduced the original group entropy is how the joint entropy of a system $S_1 \cup S_2$, a system obtained by composition of two independent systems S_1 and S_2, can be expressed as a polynomial function of only the individual entropies of S_1 and S_2. To answer that question, Tempesta investigated the Lazard universal formal groups, which originate from algebraic topology Bukhshtaber *et al.* (1971).

If we denote by H an entropy measure, then the original group entropy mentioned above satisfies the following property:

$$H(S_1 \cup S_2) = \Phi(H(S_1), H(S_2)), \qquad [9.11]$$

where $\Phi(x, y)$ is a polynomial in the variables (x, y) defined as

$$\Phi(x, y) = G(F(x) + F(y)), \quad G(z) = \sum_{i=0}^{\infty} c_i \frac{z^{i+1}}{i+1}, \quad c_0 = 1, c_i \in \mathbb{Q},$$

$G(z)$ being the formal group exponential,

$$F(s) = \sum_{i=0}^{\infty} \gamma_i \frac{s^{i+1}}{i+1},$$

with $\gamma_0 = 1$, $\gamma_1 = -c_1$, $\gamma_2 = 3/2c_1^2 - c_2$,... and $G(F(z)) = z$.

Main properties of universal group entropy include composability (weakly composable), the first three Shannon–Khinchin axioms and extensivity in suitable regimes (Tempesta 2016a). Hence, the universal group entropy can be applied in many scientific fields having many important properties.

9.2. Group entropies and the SG principle

Based on the SG principle, we derive the laws of motion for a non-stationary process, which obeys the maximum group entropy principle. Similar to Shalymov and Fradkov (2016a), we consider a system having a continuous distribution of possible states over a compact carrier Ω.

Let $(X_t)_{t\geq 0}$ be a non-stationary stochastic process defined over a probability space, where X_t, $t \geq 0$ is a random variable. Suppose that Ω is the state space of the process and that $p(\cdot, t)$ is an absolute continuous density for every $t \geq 0$ in the continuous case, while in the discrete case it is a mass probability function. We also assume that $p(t, x)$ satisfies some continuity and/or differentiability hypothesis required with respect to t for any $x \in \Omega$ and $t \geq 0$.

The density function of the system satisfies the following condition:

$$\int_{\Omega} p(t, x) dx = 1, \quad \forall t \geq 0, \qquad [9.12]$$

and it is continuous over Ω, except for a set having zero measure. From equation [9.12], we get the following restriction for the law of motion u

$$\int_{\Omega} u(t, x) dx = 0, \quad \forall t \geq 0, \qquad [9.13]$$

where $u(t,x) = \dot{p}(t,x) = \frac{dp}{dt}$. Applying the SG principle for non-stationary processes as described above, we calculate the speed of the goal functional $\dot{\widetilde{S_G}} = \frac{d\widetilde{S_G}}{dt}$ as

$$\dot{\widetilde{S_G}}(X_t) = \int_\Omega \dot{p}(t,x)\Big[\dot{G}(-\log p(t,x)) - G(-\log p(t,x))\Big]dx, \qquad [9.14]$$

where $\widetilde{S_G} = -S_G$. The gradient of $\dot{\widetilde{S_G}}$ by u is equal to

$$\nabla_u \dot{\widetilde{S_G}}(X_t) = \dot{G}(-\log p(t,x)) - G(-\log p(t,x)). \qquad [9.15]$$

According to the SG principle, we obtain the following evolution law:

$$u(t,x) = \dot{p}(t,x) = -\gamma\Big[\dot{G}(-\log p(t,x)) - G(-\log p(t,x))\Big] + \lambda_t', \qquad [9.16]$$

where λ_t' is the Lagrange multiplier, which obeys the following condition:

$$\int_\Omega \Big[-\gamma\Big[\dot{G}(-\log p(t,x)) - G(-\log p(t,x))\Big] + \lambda_t'\Big]dx = 0.$$

From this, we obtain the Lagrange multiplier as

$$\lambda_t' = \frac{\gamma}{mes(\Omega)}\int_\Omega \Big[\dot{G}(-\log p(t,x)) - G(-\log p(t,x))\Big]dx, \qquad [9.17]$$

where $mes(\Omega) = \int_\Omega dx$. The dynamic equation of the system has the following form:

$$\begin{aligned}\dot{p}(t,x) &= -\gamma\Big[\dot{G}(-\log p(t,x)) - G(-\log p(t,x))\Big]\\ &\quad + \frac{\gamma}{mes(\Omega)}\int_\Omega \Big[\dot{G}(-\log p(t,x)) - G(-\log p(t,x))\Big]dx. \qquad [9.18]\end{aligned}$$

The general form of the evolution of the system is

$$\dot{p}(t,x) = -\gamma(I - \Psi)\Big(\dot{G}\big(-\log p(t,x)\big) - G\big(-\log p(t,x)\big)\Big), \qquad [9.19]$$

where $\Psi = \int_\Omega(\cdot)dx/mes(\Omega)$ is a linear operator, which is invariant to $\dot{G}\big(-\log p(t,x)\big) - G\big(-\log p(t,x)\big)$ and I is an identity operator. The stability of the obtained equilibrium equation [9.18] is derived considering the following Lyapunov function $M(p) = \widetilde{S_G}(p) - \widetilde{S_{Gmin}} \geq 0$, where $\widetilde{S_{Gmin}}$ is the minimum of $\widetilde{S_G}(X_t)$ corresponding to equation [9.14]. We have the following:

$$\dot{M}(p) = \dot{\widetilde{S_G}}(X_t) = \int_\Omega u(t,x)\Big[\dot{G}(-\log p(t,x)) - G(-\log p(t,x))\Big]dx.$$

We substitute the expression for u from [9.18] and obtain

$$\dot{M}(p) = -\gamma \int_{\Omega} \Big[\dot{G}(-\log p(t,x)) - G(-\log p(t,x))\Big]^2 dx$$
$$+ \frac{\gamma}{mes(\Omega)} \left(\int_{\Omega} \Big[\dot{G}(-\log p(t,x)) - G(-\log p(t,x))\Big] dx \right)^2. \qquad [9.20]$$

Applying the Cauchy–Bunyakovsky–Schwarz (CBS) inequality

$$|\int_{\Omega} f(x)g(x)dx|^2 \leq \Big(\int_{\Omega} |f(x)|^2 dx\Big)\Big(\int_{\Omega} |g(x)|^2 dx\Big), \qquad [9.21]$$

for $f = \dot{G}(-\log p(t,x)) - G(-\log p(t,x))$ and $g = 1$, since $\gamma > 0$, we obtain $\dot{M}(p) \leq 0$. We have that $\dot{M}(p) = 0$ is true when $\dot{G}(-\log p(t,x)) - G(-\log p(t,x)) = \sigma_t$=constant for all $t \geq 0$, which is possible only if we take $p(t,x) = \exp\Big(-\big(\dot{G} - G\big)^{-1}(\sigma_t)\Big)$, where $\big(\dot{G} - G\big)^{-1}$ is the inverse function of $\dot{G} - G$, which we assume exists. Applying equation [9.12], we get $\sigma_t = (\dot{G} - G)(-\log(mes(\Omega))^{-1})$ and prove that σ_t does not depend on t. From the same equation [9.12], we derive p to be $p(t,x) = (mes(\Omega))^{-1}$ constant for all t. Therefore, there is only one unique density function $p^* = (mes(\Omega))^{-1}$ for the equilibrium state of the system, which evolves by evolution law given by [9.18].

Asymptotic convergence

Asymptotic convergence of all solutions to p^* can be proved by applying Barbalat's lemma (Lyapunov 1966) given below.

LEMMA 9.1 (Barbalat's lemma).– If differentiable function $f(t)$ has a finite limit for $t \to \infty$ and its derivative $\dot{f}(t)$ is uniformly continuous, then $\dot{f}(t) \to 0$ for all $t \to 0$.

Asymptotic convergence of all solutions to p^* is given in the following result:

THEOREM 9.1.– Assuming that function $\ddot{G} = \frac{\partial^2 G}{\partial t^2}$ exists, if $p(t,x)$ is a density function defined by equation [9.18] and functions $|G - \dot{G}|$ and $|\dot{G} - \ddot{G}|$ are bounded, then for any x we have $\lim_{t\to\infty} p(t,x) = p^*$.

PROOF.– Similar to Shalymov and Fradkov (2016a), the theorem can be proven by calculating $\ddot{M}(p)$ and applying Barbalat's lemma. We denote by $m(t)$ Lyapunov function $M(p(t,x))$, i.e. $m(t) = M(p(t,x))$. We proved that $m(t) \geq 0$ and $\dot{m}(t) \leq 0$ for all $t \geq 0$, hence we have that limit $\lim_{t\to\infty} m(t)$ is finite.

For the next step of the proof, we calculate $\dddot{m}(t)$ given by

$$\begin{aligned}\dddot{m}(t) &= -2\gamma^2 \int_\Omega \Big\{\frac{[G(-\log p(t,x)) - \dot{G}(-\log p(t,x))]^2}{p(t,x)} \\ &\times \Big[\dot{G}(-\log p(t,x)) - \ddot{G}(-\log p(t,x))\Big]\Big\}dx \\ &+ 3\frac{\gamma^2}{mes(\Omega)} \int_\Omega \Big[\dot{G}(-\log p(t,x)) - G(-\log p(t,x))\Big]dx \\ &\times \int_\Omega \Big\{\frac{[G(-\log p(t,x)) - \dot{G}(-\log p(t,x))]}{p(t,x)} \\ &\times \Big[\dot{G}(-\log p(t,x)) - \ddot{G}(-\log p(t,x))\Big]\Big\}dx \\ &- \lambda^2 \int_\Omega \frac{\Big[\dot{G}(-\log p(t,x)) - \ddot{G}(-\log p(t,x))\Big]}{p(t,x)} dx.\end{aligned}$$

Because functions $|G - \dot{G}|$ and $|\dot{G} - \ddot{G}|$ are bounded, and due to constraint [9.18], it can be shown that function $|\dddot{m}(t)|$ is bounded over compact carrier Ω, hence $\ddot{m}(t)$ is uniformly continuous. Applying Barbalat's lemma, we have $\ddot{m}(t) \to 0$ for $t \to 0$. We rewrite $\ddot{m}(t)$ as

$$\begin{aligned}\ddot{m}(t) &= -\gamma||\dot{G}(-\log p(t,x)) - G(-\log p(t,x))||^2 \\ &\times \left[1 - \frac{(1, \dot{G}(-\log p(t,x)) - G(-\log p(t,x)))^2}{mes(\Omega)||\dot{G}(-\log p(t,x)) - G(-\log p(t,x))||^2}\right] \\ &= -\gamma||\dot{G}(-\log p(t,x)) - G(-\log p(t,x))||^2(1 - cos^2(r(t))),\end{aligned}$$

where $(1, f) = \int_\Omega f(x)dx$, $||f||^2 = \int_\Omega f^2(x)dx$ and $r(t) = \cos\big(\angle(1, \dot{G}(-\log p(t,x)) - G(-\log p(t,x)))\big)$. If $||(\dot{G} - G)(-\log p(t,x))||^2 \to 0$, then $mes\{x|p(t,x) \neq 0\} \to 0$. However, this case contradicts condition [9.12]. Hence, we obtain $r(t) \to 1$ for $t \to 0$, which means that $\dot{G}(-\log p(t,x)) - G(-\log p(t,x)) \to 1$, hence $p(t,x) \to p^*$ for $t \to \infty$ for all $p(t,x)$ that follow evolution law [9.12]. □

9.2.1. *Total energy constraint*

Consider a system having a total energy conservation constraint given by equations [9.3] and [9.4]. The constraints [9.12] and [9.3] are equivalent to

$$\int_\Omega u(t,x)dx = 0 \quad \text{and} \quad \int_\Omega u(t,x)h(x)dx = 0, \quad \forall t \geq 0, \quad \text{respectively.}$$

Based on the above constraints, the Lagrange multipliers λ_1 and λ_2 are given as

$$\lambda_1 = \gamma\Big[mes(\Omega)\int_\Omega h^2(x)dx - \Big(\int_\Omega h(x)dx\Big)^2\Big]^{-1}$$

$$\times\Big\{mes(\Omega)\int_\Omega h(x)\big[\dot{G}(-\log p(t,x)) - G(-\log p(t,x))\big]dx$$

$$-\int_\Omega h(x)\big[\dot{G}(-\log p(t,x)) - G(-\log p(t,x))\big]dx\Big\}, \qquad [9.22]$$

$$\lambda_2 = \gamma\Big[mes(\Omega)\int_\Omega h^2(x)dx - \Big(\int_\Omega h(x)dx\Big)^2\Big]^{-1}$$

$$\times\Big\{\int_\Omega h^2(x)dx\int_\Omega\big[\dot{G}(-\log p(t,x)) - G(-\log p(t,x))\big]dx$$

$$-\int_\Omega h(x)dx\int_\Omega h(x)\big[\dot{G}(-\log p(t,x)) - G(-\log p(t,x))\big]dx\Big\}. \qquad [9.23]$$

Since h is not a constant function, the quantity $mes(\Omega)\int_\Omega h^2(x)dx - \Big(\int_\Omega h(x)dx\Big)^2$ is different from zero.

Equilibrium stability

Using the same Lyapunov function M used in the previous section, for $\widetilde{S_G}$ and Lagrange multipliers $\lambda_1' = -\lambda_1$ and $\lambda_2' = -\lambda_2$, we obtain the following expression:

$$\dot{M}(p) = \gamma\Big[mes(\Omega)\int_\Omega h^2(x)dx - \Big(\int_\Omega h(x)dx\Big)^2\Big]^{-1}$$

$$\times\Big\{-mes(\Omega)\int_\Omega\Big[\dot{G}(-\log p(t,x)) - G(-\log p(t,x))\Big]^2dx\int_\Omega h^2(x)dx$$

$$+\int_\Omega\Big[G(-\log p(t,x)) - \dot{G}(-\log p(t,x))\Big]^2\Big(\int_\Omega h(x)dx\Big)^2dx\Big\}$$

$$+mes(\Omega)\Big(\int_\Omega h(x)\Big[G(-\log p(t,x)) - \dot{G}(-\log p(t,x))\Big]^2dx\Big)^2$$

$$-2\int_\Omega h(x)\int_\Omega\Big[G(-\log p(t,x)) - \dot{G}(-\log p(t,x))\Big]dx$$

$$\times \int_\Omega h(x)\Big[G(-\log p(t,x)) - \dot{G}(-\log p(t,x))\Big]dx$$

$$+ \int_\Omega h^2(x)dx\Big(\int_\Omega \Big[G(-\log p(t,x)) - \dot{G}(-\log p(t,x))\Big]dx\Big)^2. \qquad [9.24]$$

Applying the inequality $< f,g >^2 \;\leq\; < f,f > \times < g,g >$ from Shalymov and Fradkov (2016a) where the functional $<,>$: $L_2(\Omega) \times L_2(\Omega) \to \mathbb{R}$, $f,g \in L_2(\Omega)$ is defined as

$$< f,g >= mes(\Omega)\int_\Omega fgdx - \int_\Omega fdx \int_\Omega gdx,$$

for $f = G(-\log p(t,x)) - \dot{G}(-\log p(t,x))$ and $g = h$, we get

$$\Big(mes(\Omega)\int_\Omega h(x)\Big[G(-\log p(t,x)) - \dot{G}(-\log p(t,x))\Big]dx$$

$$- \int_\Omega h(x)dx \int_\Omega \Big[G(-\log p(t,x)) - \dot{G}(-\log p(t,x))\Big]\Big)^2$$

$$\leq \Big(mes(\Omega)\int_\Omega \Big[G(-\log p(t,x)) - \dot{G}(-\log p(t,x))\Big]^2 dx$$

$$- \Big(\int_\Omega \Big[G(-\log p(t,x)) - \dot{G}(-\log p(t,x))\Big]dx\Big)^2\Big)$$

$$\times \Big(mes(\Omega)\int_\Omega h^2(x)dx - \Big(\int_\Omega h(x)dx\Big)^2\Big).$$

Hence, we obtain $\dot{M}(p) \;\leq\; 0$. $\dot{M}(p) \;=\; 0$ holds if and only if $\exists\sigma,\rho \in \mathbb{R}$ such that

$$G(-\log p(t,x)) - \dot{G}(-\log p(t,x)) = \sigma h(x) + \rho.$$

According to the SG principle, the density equilibrium is given by

$$G(-\log p(t,x)) - \dot{G}(-\log p(t,x)) = \frac{\lambda_1' h(x)}{\gamma} + \frac{\lambda_2'}{\gamma}.$$

Applying equations [9.12] and [9.3], in this case, we have an unique distribution for the equilibrium state $p^*(x)$.

Asymptotic convergence

Asymptotic convergence is proven in this case in a similar manner as in the previous section by applying Barbalat's lemma.

THEOREM 9.2.– If $p(t, x)$ is a density function defined by $u = -\gamma \nabla_u \dot{\widetilde{S_G}} + \lambda_1' h + \lambda_2'$, then for any x we have $\lim_{t \to \infty} p(t, x) = p^*(x)$.

9.3. Relative entropy group and the SG principle

The relative entropy group was studied in 2018 by Gomez *et al.* (2018), having been introduced by Tempesta (2011) as a generalization of the well-known Kullback–Leibler divergence, defined by:

$$D_{KL}(p||q) = \sum_{i=1}^{n} p_i \log\left(\frac{p_i}{g_i}\right), \qquad [9.25]$$

where $p = (p_i)_{i \in \overline{1,n}}$ and $g = (g_i)_{i \in \overline{1,n}}$ are probability mass functions. This relative entropy measures the information gain achieved if p is used instead of g.

Other relative entropies are the Jensen–Shannon divergence, f-divergences, Bregman divergences and Tsallis relative entropy.

Tsallis relative entropy is a non-negative, symmetric, which can be extensive, pseudoadditive, joint convex, strong additive entropy. The expression for Tsallis relative entropy is given as

$$D_q(p||g) = -\sum_{i=1}^{n} p_i \log_q \frac{g_i}{p_i}, \quad q > 0, q \neq 1. \qquad [9.26]$$

Some applications of relative entropies include anomalous diffusion problems, quantum correlation, measuring anonymity, the problem of information entanglement, magnetic resonance imaging and atomic dissimilarity (Prehl *et al.* 2012; Misra *et al.* 2015; Deng *et al.* 2006; Czinner and Mena 2016; Cabella *et al.* 2009; Antolín *et al.* 2010).

The relative entropy group is defined as

$$D_G(p||p^*) = \sum_{i=1}^{n} p_i G\Big(\log \frac{p_i}{p_i^*}\Big), \qquad [9.27]$$

where $p = (p_i)_{i=\overline{1,n}}$ is a discrete probability distribution, $p^* = (p_i^*)_{i=\overline{1,n}}$, $p_i^* > 0$ is a positive "equilibrium distribution".

Evolution of systems toward a "equilibrium distribution" can be studied applying the SG principle described in section 9.1. In the case of relative entropy, we choose the Lyapunov function so that the relative entropy is minimized instead of maximized. By minimization of the relative entropy, the system tends to the "equilibrium state". This principle is discussed next for the relative entropy group. In Shalymov and Fradkov (2016b), the SG principle was investigated in the case of f-divergence.

Some applications of minimum relative entropy include deep clustering, calibrating volatility surfaces, calibration of asset-pricing models, spectrum analysis and groundwater hydrology problems (Ghasedi Dizaji *et al.* 2017; Avellaneda *et al.* 1997; Avellaneda 1998; Johnson 1987; Woodbury and Ulrych 1998).

Let us consider a discrete system with a set of n possible states $s_1, \cdots, s_n$. The system evolves somehow with probability distribution over states $P(t) = \Big(p_1(t), \cdots, p_n(t)\Big)$, where $p_i(t)$ is a probability of the system location in state i. $P^* = \Big(p_1^*, \cdots, p_n^*\Big)$ is a desired distribution. Both $P(t)$ and P^* satisfy the normalization conditions:

$$\sum_{i=1}^{n} p_i(t) = 1, \quad \forall t \geq 0, \quad \sum_{i=1}^{n} p_i^* = 1. \qquad [9.28]$$

This constraint is equivalent to

$$\sum_{i=1}^{n} \dot{p}_i(t) = \sum_{i=1}^{n} u_i(t) = 0, \quad \forall t \geq 0. \qquad [9.29]$$

We choose the goal functional W to be the relative entropy group $D_G(P||P^*)$ given by equation [9.27].

Similar to Shalymov and Fradkov (2016b), we define the law evolution of a system as

$$\dot{x} = u(t), \quad x = P(t).$$

According to the SG principle, first we have to calculate the rate of change for W given by

$$\begin{aligned}\dot{W} &= \sum_{i=1}^{n} \dot{p}_i(t)\Big[G(\log\frac{p_i(t)}{p_i^*}) + \dot{G}(\log\frac{p_i(t)}{p_i^*})\Big] \\ &= \sum_{i=1}^{n} u_i\Big[G(\log\frac{p_i(t)}{p_i^*}) + \dot{G}(\log\frac{p_i(t)}{p_i^*})\Big].\end{aligned}$$

The SG of function W is given as

$$\frac{\partial \dot{W}}{\partial u_i} = G\Big(\log\frac{p_i(t)}{p_i^*}\Big) + \dot{G}\Big(\log\frac{p_i(t)}{p_i^*}\Big).$$

The SG principle tells us that the form of u_i is given by

$$u_i(t) = -\gamma \frac{\partial \dot{W}}{\partial u_i} + \lambda', \qquad [9.30]$$

where $\lambda' = \lambda'(t)$ is a Lagrange multiplier that can be obtained by applying constraint [9.29] and $\gamma > 0$ is a scalar value. The Lagrange multiplier λ' is written as

$$\lambda'(t) = \frac{\gamma}{n} \sum_{i=1}^{n} \left[G\left(\log \frac{p_i(t)}{p_i^*} \right) + \dot{G}\left(\log \frac{p_i(t)}{p_i^*} \right) \right].$$

Hence, the final system dynamics is given by

$$\dot{p}_i(t) = \gamma \left\{ - \left[G\left(\log \frac{p_i(t)}{p_i^*} \right) + \dot{G}\left(\log \frac{p_i(t)}{p_i^*} \right) \right] \right.$$
$$\left. + \frac{1}{n} \sum_{i=1}^{n} \left[G\left(\log \frac{p_i(t)}{p_i^*} \right) + \dot{G}\left(\log \frac{p_i(t)}{p_i^*} \right) \right] \right\}. \qquad [9.31]$$

The general form of the system's evolution is given as

$$\dot{P}(t) = -\gamma(I - \Psi)(A(t)), \qquad [9.32]$$

where

$$A(t) = \left(G\left(\log \frac{p_1(t)}{p_1^*} \right) + \dot{G}\left(\log \frac{p_1(t)}{p_1^*} \right), \cdots, G\left(\log \frac{p_n(t)}{p_n^*} \right) + \dot{G}\left(\log \frac{p_n(t)}{p_n^*} \right) \right)^T$$

and $p_i(t)$ is the probability that the system is in state i, p_i^* is the desired probability for state i, I is an identity matrix and Ψ is a $n \times n$ symmetric matrix where each element is equal to $\frac{1}{n}$, i.e. $\Psi_{ij} = \frac{1}{n}$.

As an observation, in equilibrium state the expression $u_i = 0$, for any i, is true, which is equivalent to

$$\sum_{i=1}^{n} \left[G\left(\log \frac{p_i(t)}{p_i^*} \right) + \dot{G}\left(\log \frac{p_i(t)}{p_i^*} \right] = n \left[G\left(\log \frac{p_j(t)}{p_j^*} \right) + \dot{G}\left(\log \frac{p_j(t)}{p_j^*} \right) \right],$$

for all $j \in \overline{1,n}$. Hence, we obtain

$$(G + \dot{G})\left(\log \frac{p_i(t)}{p_i^*} \right) = (G + \dot{G})\left(\log \frac{p_j(t)}{p_j^*} \right), \quad \forall j.$$

We have

$$(G + \dot{G})\left(\log \frac{p_j(t)}{p_j^*} \right) = (G + \dot{G})\left(\log \frac{p_k(t)}{p_k^*} \right) = \alpha_t, \quad \forall k, j.$$

Hence, $p_j(t) = p_j^* \exp\left[(G+\dot{G})^{-1}(\alpha_t)\right]$, meaning

$$\sum_{j=1}^{n} p_j^*(t) = \sum_{j=1}^{n} p_j^* \exp\left[(G+\dot{G})^{-1}(\alpha_t)\right],$$

which leads to $\exp\left[(G+\dot{G})^{-1}(\alpha_t)\right] = 1$ equivalent to $\alpha_t = (G+\dot{G})(0), \quad t \geq 0$, therefore α_t does not depend on t.

9.3.1. *Equilibrium stability*

Consider the following Lyapunov function:

$$V(P) = W(P, P^*) - W_{min}(P, P^*) \geq 0, \qquad [9.33]$$

where $W_{min}(P, P^*)$ is the minimum of $W(P, P^*)$. We have

$$\dot{V}(P) = \dot{W}(P, P^*) = \sum_{i=1}^{n} u_i \left[G\Big(\log \frac{p_i(t)}{p_i^*}\Big) + \dot{G}\Big(\log \frac{p_i(t)}{p_i^*}\Big)\right].$$

Substituting the expression for u_i from [9.31], we get the following:

$$\dot{V}(P) = -\gamma \sum_{i=1}^{n} \left[G\Big(\log \frac{p_i(t)}{p_i^*}\Big) + \dot{G}\Big(\log \frac{p_i(t)}{p_i^*}\Big)\right]^2 + \frac{\gamma}{n}\Big(\sum_{i=1}^{n} \left[G\Big(\log \frac{p_i(t)}{p_i^*}\Big) + \dot{G}\Big(\log \frac{p_i(t)}{p_i^*}\Big)\right]\Big)^2.$$

We apply the CBS inequality [9.21] for

$$f = \left(G\Big(\log \frac{p_1(t)}{p_1^*}\Big) + \dot{G}\Big(\log \frac{p_1(t)}{p_1^*}\Big), \cdots, G\Big(\log \frac{p_n(t)}{p_n^*}\Big) + \dot{G}\Big(\log \frac{p_n(t)}{p_n^*}\Big)\right)^T$$

and $g = (1, 1, .., 1, 1)^T$. Hence, we get $\dot{V}(P) \leq 0$. We have $\dot{V}(P) = 0$ when $G\Big(\log \frac{p_i(t)}{p_i^*}\Big) + \dot{G}\Big(\log \frac{p_i(t)}{p_i^*}\Big) = \alpha_t$, for any i which corresponds to the state of equilibrium. Applying 9.29, we conclude that α_t does not depend on t, meaning $\alpha_t = \alpha$, for all $t \geq 0$. Thus, law [9.31] gives us a global asymptotic stability for the relative entropy group minimization.

9.3.2. *Total energy constraint*

Consider a new constraint for the system in the form of a energy constraint given by

$$E = \sum_{i=1}^{n} p_i(t) E_i, \quad E = \sum_{i=1}^{n} p_i^* E_i, \quad \forall t \geq 0, \qquad [9.34]$$

where E_i is the energy of a system in the state i and E is the common energy of a system. We have the following constraints:

$$\sum_{i=1}^{n} u_i(t)E_i = 0, \quad \sum_{i=1}^{n} u_i(t) = 0, \quad \forall t \geq 0. \tag{9.35}$$

We are looking for u_i that satisfies constraints [9.28] and [9.34]. Similar to Shalymov and Fradkov (2016b), the evolution law should have the form

$$u_i(t) = -\gamma\Big[G\Big(\log\frac{p_i(t)}{p_i^*}\Big) + \dot{G}\Big(\log\frac{p_i(t)}{p_i^*}\Big)\Big] + \lambda_1 E_i + \lambda_2, \quad i \in \overline{1,n}, \tag{9.36}$$

where λ_1 and λ_2 are Lagrange multipliers, which can be obtained by constraint [9.35]. Hence, we have

$$\lambda_1 = \gamma\Big[n\sum_{i=1}^{n} E_i^2 - \Big(\sum_{i=1}^{n} E_i\Big)^2\Big]^{-1}\Big\{n\sum_{i=1}^{n} E_i\Big[G\Big(\log\frac{p_i(t)}{p_i^*}\Big) + \dot{G}\Big(\log\frac{p_i(t)}{p_i^*}\Big)\Big]$$
$$-\sum_{i=1}^{n}\Big[G\Big(\log\frac{p_i(t)}{p_i^*}\Big) + \dot{G}\Big(\log\frac{p_i(t)}{p_i^*}\Big)\Big]\sum_{i=1}^{n} E_i\Big\},$$

$$\lambda_2 = \gamma\Big[n\sum_{i=1}^{n} E_i^2 - \Big(\sum_{i=1}^{n} E_i\Big)^2\Big]^{-1}\Big[\sum_{i=1}^{n}\Big[G\Big(\log\frac{p_i(t)}{p_i^*}\Big) + \dot{G}\Big(\log\frac{p_i(t)}{p_i^*}\Big)\Big]$$
$$\times\sum_{i=1}^{n} E_i^2 - \sum_{i=1}^{n} E_i\sum_{i=1}^{n} E_i\Big[G\Big(\log\frac{p_i(t)}{p_i^*}\Big) + \dot{G}\Big(\log\frac{p_i(t)}{p_i^*}\Big)\Big]\Big].$$

We do not consider the case where E_i are equal, so $n\sum_{i=1}^{n} E_i^2 \neq \Big(\sum_{i=1}^{n} E_i\Big)^2$. The existence of a solution in this case is possible to show in similar ways as it was shown in the previous case with one constraint.

9.4. A new (G, a) power relative entropy group and the SG principle

In 2012, Vonta *et al.* (2012) introduced the (Φ, a)-power divergence family as a general framework for different relative entropies such as Csiszár's family and BHHJ power divergence, the latter being given by (Basu *et al.* 1998; Mattheou and Karagrigoriou 2010; Mattheou *et al.* 2009),

$$d^a(P^*, P) = \sum_{i=1}^{n} p_i^{1+a} - \Big(1+\frac{1}{a}\Big)\sum_{i=1}^{n} p_i^* p_i^a + \frac{1}{a}\sum_{i=1}^{n}(p_i^*)^{1+a} = \sum_{i=1}^{n}(p_i^*)^{1+a}\Phi\Big(\frac{p_i}{p_i^*}\Big),$$

where $\Phi(z) = z^{1+a} - (1 + 1/a)z^a + 1/a$ and $P = (p_i)_{i \in \overline{1,n}}$ and $P^* = (p_i^*)_{i \in \overline{1,n}}$ are two discrete probability distributions. The (Φ, a) discrete power relative entropy is given as

$$d_\Phi^a(P, P^*) = \sum_{i=1}^{n} (p_i^*)^{1+a} \Phi\Big(\frac{p_i}{p_i^*}\Big), \quad a > 0, \qquad [9.37]$$

where Φ is a strictly convex function, $\Phi(x)$, $x \geq 0$, such that $0\Phi(0/0) = 0$, $\Phi(1) = \dot{\Phi}(1) = 0$ and $0\Phi(p/0) = \lim_{n \to \infty} \Phi(u)/u$. For $a = 0$ and

$$\Phi(x) = \frac{x^{\mu+1} - x - \mu(u-1)}{\mu(\mu+1)}, \quad \mu \neq 0, -1,$$

the (Φ, a)- power divergence reduces to Cressie and Read (CR) power divergence.

Another relative entropy that is a submodel of the (Φ, a) divergence is the CR family of power relative entropy. Some theoretical properties for Markov chains considering different relative entropies including CR case are considered in Barbu *et al.* (2017, 2018a,b).

Some properties of (Φ, a) relative entropy include the order preserving property, the limiting property and the quadratic convergence.

Similar to the (Φ, a)- power divergence, we introduce the following relative entropy, denoted by (G, a), which we call (G, a)- power relative entropy group:

$$D_G^a(P||P^*) = \sum_{i=1}^{n} p_i^a(t) G\Big(\log \frac{p_i(t)}{p_i^*}\Big), \quad a > 0, \qquad [9.38]$$

where probability distribution $P(t) = \Big(p_1(t), \cdots, p_n(t)\Big)$ represents the evolution law of a system, where $p_i(t)$ is a probability of the system location in state i, while $P^* = \Big(p_1^*, \cdots, p_n^*\Big)$ is a desired distribution. Function $G(z)$ is an arbitrary real analytic invertible function such that $G(0) = 0, \dot{G}(0) = 1$. For $a = 0$, we obtain the relative entropy group.

Next, we discuss the case of this new relative entropy in relation to the SG principle. Hence, we choose the goal functional W to be the relative entropy given by equation [9.38]. Probability laws P and P^* satisfy constraints [9.28] and [9.29].

According to the SG principle, we determine the rate of change of W given by

$$\dot{W} = \sum_{i=1}^{n} p_i^a(t)\dot{p}_i(t)\Big[(a+1)G\big(\log\frac{p_i(t)}{p_i^*}\big) + \dot{G}\big(\log\frac{p_i(t)}{p_i^*}\big)\Big]$$

$$= \sum_{i=1}^{n} p_i^a(t)u_i(t)\Big[(a+1)G\big(\log\frac{p_i(t)}{p_i^*}\big) + \dot{G}\big(\log\frac{p_i(t)}{p_i^*}\big)\Big].$$

The SG of function W is

$$\frac{\partial \dot{W}}{\partial u_i} = p_i^a(t)\Big[(a+1)G\big(\log\frac{p_i(t)}{p_i^*}\big) + \dot{G}\big(\log\frac{p_i(t)}{p_i^*}\big)\Big].$$

The evolution law of the system is given by equation [9.30] where $\lambda' = \lambda'(t)$ is a Lagrange multiplier given by

$$\lambda' = \lambda'(t) = \frac{\gamma}{n}\sum_{i=1}^{n} p_i^a(t)\Big[(a+1)G\big(\log\frac{p_i(t)}{p_i^*}\big) + \dot{G}\big(\log\frac{p_i(t)}{p_i^*}\big)\Big],$$

and $\gamma > 0$ is a scalar value. The final system dynamics is given by

$$\dot{p}_i(t) = \gamma\Big\{ - p_i^a(t)\Big[(a'+1)G\big(\log\frac{p_i(t)}{p_i^*}\big) + \dot{G}\big(\log\frac{p_i(t)}{p_i^*}\big)\Big]$$

$$+ \frac{1}{n}\sum_{i=1}^{n} p_i^a(t)\Big[(a+1)G\big(\log\frac{p_i(t)}{p_i^*}\big) + \dot{G}\big(\log\frac{p_i(t)}{p_i^*}\big)\Big]\Big\}. \quad [9.39]$$

The general form of the system's equation is

$$\dot{P}(t) = -\gamma(I - \Psi)(A(t)), \quad [9.40]$$

where

$$A(t) = \Big(p_1^a(t)\Big[(a+1)G\big(\log\frac{p_1(t)}{p_1^*}\big) + \dot{G}\big(\log\frac{p_1(t)}{p_1^*}\big)\Big], \cdots,$$

$$p_n^a(t)\Big[(a+1)G\big(\log\frac{p_n(t)}{p_n^*}\big) + \dot{G}\big(\log\frac{p_n(t)}{p_n^*}\big)\Big]\Big)^T. \quad [9.41]$$

In the equilibrium state, the expression $u_i = 0$, for any i, is true and it is equivalent to

$$\sum_{i=1}^{n} p_i^a(t)\Big[(a+1)G\big(\log\frac{p_i(t)}{p_i^*}\big) + \dot{G}\big(\log\frac{p_i(t)}{p_i^*}\big)\Big] = np_j^a(t)$$

$$\times \Big[(a+1)G\big(\log\frac{p_j(t)}{p_j^*}\big) + \dot{G}\big(\log\frac{p_j(t)}{p_j^*}\big)\Big], \quad \forall j \in \overline{1,n}.$$

Hence,

$$p_j^a(t)\Big[(a+1)(G+\dot{G})\big(\log\frac{p_j(t)}{p_j^*}\big)\Big] = p_k^a(t)\Big[(a+1)(G+\dot{G})\big(\log\frac{p_k(t)}{p_k^*}\big)\Big], \quad \forall k, j,$$

equation is satisfied by the relationship between $p_j(t)$ and p_j^*, and by the state equilibrium distribution p_j^*.

Equilibrium stability

We consider the Lyapunov function given by [9.33]. We have

$$\dot{V}(P) = \sum_{i=1}^{n} p_i^a(t)u_i(t)\Big[(a+1)G\big(\log\frac{p_i(t)}{p_i^*}\big) + \dot{G}\big(\log\frac{p_i(t)}{p_i^*}\big)\Big].$$

Substituting the expression for u_i from [9.39], we get the following:

$$\dot{V}(P) = -\gamma\sum_{i=1}^{n} p_i^{2a}(t)\Big[(a+1)G\big(\log\frac{p_i(t)}{p_i^*}\big) + \dot{G}\big(\log\frac{p_i(t)}{p_i^*}\big)\Big]^2$$
$$+\frac{\gamma}{n}\left(\sum_{i=1}^{n} p_i^a(t)\Big[(a+1)G\big(\log\frac{p_i(t)}{p_i^*}\big) + \dot{G}\big(\log\frac{p_i(t)}{p_i^*}\big)\Big]\right)^2.$$

By applying the CBS inequality given in [9.21] with $f = A(t)$ given in [9.41] and $g = (1, 1, \cdots, 1, 1)^T$, we get $\dot{V}(P) \le 0$. We have $\dot{V}(P) = 0$ when $p_i^a(t)\Big[(a+1)G\big(\log\frac{p_i(t)}{p_i^*}\big) + \dot{G}\big(\log\frac{p_i(t)}{p_i^*}\big)\Big] = \alpha_t$, $\forall i$ which corresponds to the state of equilibrium. Thus, law [9.39] gives us a global asymptotic stability.

Total energy constraint

Consider a new constraint for the system in the form of a energy constraint given by equations [9.34] and [9.35]. We are looking for u_i, which satisfies constraints [9.28] and [9.34]. The evolution law should have the form

$$u_i(t) = -\gamma p_i^a(t)\Big[(a+1)G\big(\log\frac{p_i(t)}{p_i^*}\big) + \dot{G}\big(\log\frac{p_i(t)}{p_i^*}\big)\Big] + \lambda_1 E_i + \lambda_2, \quad i \in \overline{1,n},$$

where λ_1 and λ_2 are Lagrange multipliers that can be obtained by constraint [9.35]. We have

$$\lambda_1 = \gamma\Big[n\sum_{i=1}^{n} E_i^2 - \big(\sum_{i=1}^{n} E_i\big)^2\Big]^{-1}$$

$$\times\left\{n\sum_{i=1}^{n}E_i p_i^a(t)\Big[(a+1)G\big(\log\frac{p_i(t)}{p_i^*}\big)+\dot{G}\big(\log\frac{p_i(t)}{p_i^*}\big)\Big]\right.$$

$$\left.-\sum_{i=1}^{n}p_i^a(t)\Big[(a+1)G\big(\log\frac{p_i(t)}{p_i^*}\big)+\dot{G}\big(\log\frac{p_i(t)}{p_i^*}\big)\Big]\sum_{i=1}^{n}E_i\right\},$$

$$\lambda_2=\gamma\Big[n\sum_{i=1}^{n}E_i^2-\Big(\sum_{i=1}^{n}E_i\Big)^2\Big]^{-1}$$

$$\times\left\{\sum_{i=1}^{n}p_i^a(t)\Big[(a+1)G\big(\log\frac{p_i(t)}{p_i^*}\big)+\dot{G}\big(\log\frac{p_i(t)}{p_i^*}\big)\Big]\sum_{i=1}^{n}E_i^2\right.$$

$$\left.-\sum_{i=1}^{n}E_i\sum_{i=1}^{n}E_i p_i^a(t)\Big[(a+1)G\big(\log\frac{p_i(t)}{p_i^*}\big)+\dot{G}\big(\log\frac{p_i(t)}{p_i^*}\big)\Big]\right\}.$$

9.5. Conclusion

This chapter addressed the question, "How does a physical system reach its equilibrium stability?", considering group entropies, relative entropy groups, new entropies called (G, a) power relative entropy group measures and the SG principle. Dynamic evolution equations that describe the evolution of systems in each case were obtained. Some possible future directions of the present work are the analysis of the continuous case for the relative entropy groups and (G, a) power relative entropy groups, and also considering new entropies and divergence measures.

9.6. References

Abul-Magd, A. (2007). Nonextensive random-matrix theory based on Kaniadakis entropy. *Physics Letters A*, 361(6), 450–454.

Amari, S. (2016). *Information Geometry and Its Applications*. Springer, Berlin

Antolín, J., López-Rosa, S., Angulo, J., and Esquivel, R. (2010). Jensen-Tsallis divergence and atomic dissimilarity for position and momentum space electron densities. *The Journal of Chemical Physics*, 132(4), 044105.

Avellaneda, M. (1998). Minimum-relative-entropy calibration of asset-pricing models. *International Journal of Theoretical and Applied Finance*, 1(4), 447–472.

Avellaneda, M., Friedman, C., Holmes, R., and Samperi, D. (1997). Calibrating volatility surfaces via relative-entropy minimization. *Applied Mathematical Finance*, 4(1), 37–64.

Barbu, V.S., Karagrigoriou, A., and Preda, V. (2017). Entropy and divergence rates for Markov chains: I. The alpha-gamma and beta-gamma case. *Proceedings of the Romanian Academy-Series A*, 18(4), 293–301.

Barbu, V.S., Karagrigoriou, A., and Preda, V. (2018a). Entropy and divergence rates for Markov chains: II. The weighted case. *Proceedings of the Romanian Academy-Series A*, 19(1), 3–10.

Barbu, V.S., Karagrigoriou, A., and Preda, V. (2018b). Entropy and divergence rates for Markov chains: III. The Cressie and Read case and applications. *Proceedings of the Romanian Academy-Series A*, 19(3), 413–421.

Basu, A., Harris, I.R., Hjort, N.L., and Jones, M. (1998). Robust and efficient estimation by minimising a density power divergence. *Biometrika*, 85(3), 549–559.

Belis, M., and Guiaşu, S. (1968). A quantitative-qualitative measure of information in cybernetic systems. *IEEE Transactions on Information Theory*, 14(4), 593–594.

Bhandari, A.K., Kumar, A., and Singh, G.K. (2015). Tsallis entropy based multilevel thresholding for colored satellite image segmentation using evolutionary algorithms. *Expert Systems with Applications*, 42(22), 8707–8730.

Băncescu, I. (2018). q-Log-distributions: Log-concavity and log-convexity. *European Physical Journal Plus*, 133, 1–18.

Bukhshtaber, V.M., Mishchenko, A.S., and Novikov, S.P. (1971). Formal groups and their role in the apparatus of algebraic topology. *Russian Mathematical Surveys*, 26(2), 63–90.

Cabella, B.C.T., Sturzbecher, M.J., de Araujo, D.B., and Neves, U.P.C. (2009). Generalized relative entropy in functional magnetic resonance imaging. *Physica A: Statistical Mechanics and its Applications*, 388(1), 41–50.

Callen, H.B. (1985). *Thermodynamics and an Introduction to Thermostatistics*, 2nd edition. John Wiley & Sons, New York.

Canturk, B., Oikonomou, T., and Bagci, G.B. (2017). Group theory, entropy and the third law of thermodynamics. *Annals of Physics*, 377, 62–70.

Cercignani, C. (1998). *Ludwig Boltzmann: The Man Who Trusted Atoms*. Oxford University Press, New York.

Cover, T.M., and Thomas, J.A. (2006). *Elements of Information Theory*. Wiley, Hoboken.

Czinner, V.G., and Mena, F.C. (2016). Relative information entropy in cosmology: The problem of information entanglement. *Physics Letters B*, 758, 9–13.

De Albuquerque, M.P., Esquef, I.A., and Mello, A.G. (2004). Image thresholding using Tsallis entropy. *Pattern Recognition Letters*, 25(9), 1059–1065.

Deng, Y., Pang, J., and Wu, P. (2006). *Measuring Anonymity with Relative Entropy*. Springer, Berlin.

Dial, G., and Taneja, I.J. (1981). On weighted entropy of type (α, β) and its generalizations. *Aplikace Matematiky*, 26(6), 418–425.

Dougherty, J.P. (1994). Foundations of non-equilibrium statistical mechanics. *Philosophical Transactions of the Royal Society of London. Series A: Physical and Engineering Sciences*, 346(1680), 259–305.

Fradkov, A. (2007). *Cybernetical physics: From Control of Chaos to Quantum Control*. Springer, Berlin, Heidelberg, New York.

Fradkov, A. (2008). Speed-gradient entropy principle for nonstationary processes. *Entropy*, 10(4), 757–764.

Fradkov, A., and Krivtsov, A. (2011). Speed-gradient principle for description of transient dynamics in systems obeying maximum entropy principle. *AIP Conference Proceedings*, 1305(1), 399–406.

Fradkov, A.L. (1979). Speed-gradient scheme and its applications in adaptive control. *Automation and Remote Control*, 40, 1333–1342.

Fradkov, A.L. (2005). Application of cybernetic methods in physics. *Physics-Uspekhi*, 48, 103–148.

Fradkov, A.L., and Pogromsky, A.Y. (1998). *Introduction to Control of Oscillations and Chaos*. World Scientific Publishing, Singapore.

Fradkov, A.L., and Shalymov, D.S. (2015a). Dynamics of non-stationary nonlinear processes that follow the maximum of differential entropy principle. *Communications in Nonlinear Science and Numerical Simulation*, 29(1–3), 488–498.

Fradkov, A., and Shalymov, D. (2015b). Speed gradient and MaxEnt principles for Shannon and Tsallis entropies. *Entropy*, 17(3), 1090–1102.

Furuichi, S., Yanagi, K., and Kuriyama, K. (2004). Fundamental properties of Tsallis relative entropy. *Journal of Mathematical Physics*, 45(12), 4868–4877.

Gell-Mann, M. (1995). *The Quark and the Jaguar: Adventures in the Simple and the Complex*. Macmillan, New York.

Ghasedi Dizaji, K., Herandi, A., Deng, C., Cai, W., and Huang, H. (2017). Deep clustering via joint convolutional autoencoder embedding and relative entropy minimization. *Proceedings of the IEEE Computer Society Conference on Computer Vision and Pattern Recognition*, 22–29 October, Venice, Italy, 5736–5745.

Gomez, I.S., Borges, E.P., and Portesi, M. (2018). Fisher metric from relative entropy group. *Cornell University*, arXiv:1805.1115.

Grandy Jr., W. (1980). Principle of maximum entropy and irreversible processes. *Physics Reports*, 62(3), 175–266.

Guiaşu, S. (1971). Weighted Entropy. *Reports on Mathematical Physics*, 2(3), 165–179.

Guiaşu, S. (1977). *Information Theory with New Applications*. McGraw-Hill, New York.

Guisado, J.L., Jiménez-Morales, F., and Guerra, J.M. (2005). Application of Shannon's entropy to classify emergent behaviors in a simulation of laser dynamics. *Mathematical and Computer Modelling*, 42(7–8), 847–854.

Haken, H. (2006). *Information and Self-organization: A Macroscopic Approach to Complex Systems*. Springer, Berlin, Heidelberg, New York.

Hughes, M., Marsh, J., Arbeit, J., Neumann, R., Fuhrhop, R., Wallace, K., Thomas, L., Smith, J., Agyem, K., Lanza, G., and Wickline, S. (2009). Application of Rényi entropy for ultrasonic molecular imaging. *The Journal of the Acoustical Society of America*, 125(5), 3141–3245.

Jaynes, E. (1979). Where do we stand on maximum entropy? In *The Maximum Entropy Formalism*, Levine, R.D. and Tribus, M. (eds). MIT Press, Cambridge.

Jaynes, E.T. (1957). Information theory and statistical mechanics. *Physical Review*, 106(4), 620–630.

Jaynes, E.T. (1980). The minimum entropy production principle. *Annual Review of Physical Chemistry*, 31(1), 579–601.

Jeldtoft Jensen, H., and Tempesta, P. (2018). Group entropies: From phase space geometry to entropy functionals via group theory. *Entropy*, 20(10), 804.

Johnson, R.W. (1987). Relative-entropy minimization with uncertain constraints: Theory and application to spectrum analysis. In *Maximum-Entropy and Bayesian Spectral Analysis and Estimation Problems*, Smith, C.R. and Erickson G.J. (eds), Springer, Dordrecht, 57–73.

Kaniadakis, G. (2009). Relativistic entropy and related Boltzmann kinetics. *The European Physical Journal A*, 40(3), 275.

Kapur, J.N., and Kesavan, H.K. (1992). Entropy optimization principles and their applications. In *Entropy and Energy Dissipation in Water Resources*, Singh, V.P. and Fiorentino, M. (eds), Springer, Dordrecht, 3–20.

Khantuleva, T. A. (2003). The shock wave as a nonequilibrium transport process. In *High-Pressure Shock Compression of Solids VI: Old Paradigms and New Challenges*, Horie, Y., Davison, L., and Thadhani, N.N. (eds), Springer, New York, 215–254.

Khantuleva, T. A. (2005). Internal control in nonequilibrium transport. *Proceedings of the 2005 International Conference Physics and Control*, 24–26 August, Saint Petersburg, Russia, 41–46.

Khantuleva, T.A., and Shalymov, D.S. (2017). Modelling non-equilibrium thermodynamic systems from the speed-gradient principle. *Philosophical Transactions of the Royal Society A: Mathematical, Physical and Engineering Sciences*, 375(2088), 1–22.

Khantuleva, T., and Abdullaev, K. (2014). Internal control in nonequilibrium physical processes. *Cybernetics and Physics*, 3(1), 21–27.

Khinchin, A.Y. (1957). *Mathematical Foundations of Information Theory*. Dover Publications, New York.

Kleeman, R. (2002). Measuring dynamical prediction utility using relative entropy. *Journal of the Atmospheric Sciences*, 59(13), 2057–2072.

Kumar, V., and Taneja, H. (2011). Some characterization results on generalized cumulative residual entropy measure. *Statistics and Probability Letters*, 81(8), 1072–1077.

Liu, H., Chen, W., and Sudjianto, A. (2006). Relative entropy based method for probabilistic sensitivity analysis in engineering design. *Journal of Mechanical Design*, 128(2), 326–336.

Lyapunov, A.M. (1966). *Stability of Motion*. Academic Press, New York.

Mattheou, K., and Karagrigoriou, A. (2010). A new family of divergence measures for tests of fit. *Australian & New Zealand Journal of Statistics*, 52(2), 187–200.

Mattheou, K., Lee, S., and Karagrigoriou, A. (2009). A model selection criterion based on the BHHJ measure of divergence. *Journal of Statistical Planning and Inference*, 139(2), 228–235.

Misra, A., Biswas, A., Pati, A.K., Sen, A., and Sen, U. (2015). Quantum correlation with sandwiched relative entropies: Advantageous as order parameter in quantum phase transitions. *Physical Review E*, 91(5), 052125.

Ourabah, K., Hamici-Bendimerad, A.H., and Tribeche, M. (2015). Quantum entanglement and Kaniadakis entropy. *Physica Scripta*, 90(4), 045101.

Plastino, A.R., and Plastino, A. (1993). Tsallis' entropy, Ehrenfest theorem and information theory. *Physics Letters A*, 177(3), 177–179.

Preda, V., and Bălcău, C. (2011). *Entropy Optimization with Applications*. The Publishing House of the Romanian Academy, Bucharest.

Preda, V., Dedu, S., and Gheorghe, C. (2015). New classes of Lorenz curves by maximizing Tsallis entropy under mean and Gini equality and inequality constraints. *Physica A: Statistical Mechanics and its Applications*, 436, 925–932.

Prehl, J., Essex, C., and Hoffmann, K.H. (2012). Tsallis relative entropy and anomalous diffusion. *Entropy*, 14(4), 701–716.

Rayleigh, L. (1911). On the motion of solid bodies through viscous liquid. *The London, Edinburgh, and Dublin Philosophical Magazine and Journal of Science*, 21(126), 697–711.

Shalymov, D. (2016). Dynamics of non-stationary processes that follow the maximum of continuous Tsallis entropy. *Cybernetics and Physics*, 5, 59–66.

Shalymov, D.S., and Fradkov, A.L. (2016a). Dynamics of non-stationary processes that follow the maximum of the Rényi entropy principle. *Proceedings of the Royal Society A: Mathematical, Physical and Engineering Sciences*, 472(2185), 1–18.

Shalymov, D.S., and Fradkov, A.L. (2016b). Dynamics of the f-divergence minimization processes based on the speed-gradient principle. *2016 IEEE Conference on Norbert Wiener in the 21st Century (21CW)*, 13–16 July, Melbourne, Australia, 1–5.

Shannon, C.E. (1948). A mathematical theory of communication. *Bell System Technical Journal*, 27(3), 379–423.

Shannon, C.E., and Weaver, W. (1949). *The Mathematical Theory of Communication*. University of Illinois Press, Urbana.

Tempesta, P. (2011). Group entropies, correlation laws, and zeta functions. *Physical Review E*, 84(2), 1–17.

Tempesta, P. (2015). A theorem on the existence of trace-form generalized entropies. *Proceedings of the Royal Society A: Mathematical, Physical and Engineering Sciences*, 471(2183), 1–13.

Tempesta, P. (2016a). Beyond the Shannon-Khinchin formulation: The composability axiom and the universal-group entropy. *Annals of Physics*, 365, 180–197.

Tempesta, P. (2016b). Formal groups and Z-entropies. *Proceedings of the Royal Society A: Mathematical, Physical and Engineering Sciences*, 472(2195), 1–17.

Tsallis, C. (2009). *Introduction to Nonextensive Statistical Mechanics: Approaching a Complex World*. Springer, Berlin.

Vonta, F., Mattheou, K., and Karagrigoriou, A. (2012). On properties of the (ϕ, a)-power divergence family with applications in goodness of fit tests. *Methodology and Computing in Applied Probability*, 14(2), 335–356.

Woodbury, A.D., and Ulrych, T.J. (1998). Minimum relative entropy and probabilistic inversion in groundwater hydrology. *Stochastic Hydrology and Hydraulics*, 12(5), 317–358.

10

Inferential Statistics Based on Measures of Information and Divergence

In this chapter, we attempt an overview of measures of information and divergence. Many of the most known measures and some of their most important properties are briefly presented. We discuss the use of measures of divergence in model selection criteria and in goodness-of-fit tests for multinomial distributions. Emphasis is given to the newly proposed double index (Φ, α) test of fit, which through the two indices involved provides greater flexibility to the researcher for inferential purposes and combines under a single umbrella most of the well-known test statistics previously investigated.

10.1. Introduction

Divergence measures have numerous applications in statistical inference and can be applied in various fields such as signal processing and pattern recognition (Bekara *et al.* 2006), speech recognition (Qiao and Minematsu 2010), clustering (Fischer 2010), finance and economics (Toma and Leoni-Aubin 2014), analysis of contingency tables (Alin and Kurt 2008; Kateri 2018), model selection (Akaike 1973; Cavanaugh 2004; Shang and Cavanaugh 2008; Mattheou *et al.* 2009; Mantalos *et al.* 2010; Toma 2014), parameter estimation (Neath *et al.* 2015; Jiménez-Gamero and Batsidis 2017), goodness-of-fit tests (Toma and Leoni-Aubin 2010; Stehlík *et al.* 2014, 2018), etc.

Although, in special cases, physical data generating mechanisms suggest a good probability model, in general, we do not have sufficient knowledge of the machinery generating the data to convert it into a probability distribution. A way to resolve the issue of model identification is to estimate the underlying mechanism by approaches

Chapter written by Alex KARAGRIGORIOU and Christos MESELIDIS.

based on information or divergence measures, which are used for measuring the closeness or the distance between each candidate model and the true but unknown model.

This work focuses on measures of divergence and distance providing an overview of their properties and their uses in statistical inference. The most known measures are briefly presented in section 10.2, while some of their properties are provided in section 10.3. Model selection criteria and how they are derived via divergence measures are discussed in section 10.4, while in section 10.5 we provide the basic theoretical results about goodness-of-fit tests for multinomial distributions based on divergence measures. Among the goodness-of-fit tests discussed is the newly proposed double index (Φ, α) test of fit (Meselidis and Karagrigoriou 2020), which through the two indices involved provides greater flexibility to the researcher for inferential purposes. Finally, in section 10.6, we investigate the behavior of the proposed test in terms of size and power through a simulation study.

10.2. Divergence measures

10.2.1. φ-Divergences

Let f and g be two probability density functions that may depend on an unknown parameter of fixed finite dimension. The φ-divergence family of measures (Csiszár 1963; Ali and Silvey 1966) is defined by

$$I_X^{\varphi}(f, g) = \int g(z)\varphi\left(\frac{f(z)}{g(z)}\right) dz \qquad [10.1]$$

where φ is a convex function in $[0, \infty)$ such that $0\varphi(u/0) = \lim_{u\to\infty}[\varphi(u)/u]$ and $0\varphi(0/0) = 0$. Many of the well-known measures of divergence can be considered as particular cases of the φ-divergence measure. For instance, Pearson's chi-squared divergence, also known as Kagan's divergence (Kagan 1963), which is given by

$$I_X^{Ka}(f, g) = \frac{1}{2}\int g(z)\left(1 - \frac{f(z)}{g(z)}\right)^2 dz$$

can be derived from [10.1] for $\varphi(u) \doteq \varphi_1(u) = 1/2(1-u)^2$. Hellinger distance is another important divergence measure that can be derived from [10.1] for $\varphi(u) = 2(\sqrt{x} - 1)$ and is given by

$$I_X^{He}(f, g) = \left(\int \left(\sqrt{f(z)} - \sqrt{g(z)}\right)^2 dz\right)^{1/2}.$$

Note that Hellinger distance is the square root of Matusita's divergence (Matusita 1967). For $\varphi(u) \doteq \varphi_2(u) = u \log u + 1 - u$, the φ-divergence measure yields the Kullback–Leibler divergence (Kullback and Leibler 1951) given by

$$I_X^{KL}(f, g) = \int f(z) \log \left(\frac{f(z)}{g(z)} \right) dz$$

and for $\varphi(u) \doteq \varphi_3(u) = (u^{\lambda+1} - u - \lambda(u-1))/(\lambda(\lambda+1))$, $\lambda \neq 0, -1$ yields the family of power divergences introduced by Cressie and Read (1984), which is given by

$$I_X^{CR}(f, g) = \frac{1}{\lambda(\lambda+1)} \int f(z) \left[\left(\frac{f(z)}{g(z)} \right)^{\lambda} - 1 \right] dz, \quad \lambda \in \mathbb{R} \qquad [10.2]$$

where for $\lambda = 0, -1$ is defined by continuity. Note that for $\lambda \to 0$, the Kullback–Leibler divergence is obtained, while for $\lambda = 1$, [10.2] reduces to the Pearson's chi-squared divergence.

10.2.2. α-*Divergences*

Another class of divergence measures is the class of α-divergences. The additive and non-additive directed divergences of order α were introduced in the 1960s and the 1970s (Rényi 1961; Csiszár 1963; Rathie and Kannappan 1972). The so-called order α information measure of Rényi (1961) is given by

$$I_X^{R}(f, g) = \frac{1}{\alpha - 1} \log \left(\int f^{\alpha}(z) g^{1-\alpha}(z) \; dz \right), \quad \alpha > 0, \; \alpha \neq 1.$$

It should be noted that for $\alpha \to 1$ the above measure becomes the Kullback–Leibler divergence.

One of the most recently proposed measures is the family of pseudodistances of order $\alpha > 0$ (see Jones *et al.* 2001; Broniatowski *et al.* 2012), which is defined by

$$I_X^{P}(f, g) = \frac{1}{\alpha+1} \log \left(\int g^{1+\alpha}(z) dz \right) + \frac{1}{\alpha(\alpha+1)} \log \left(\int f^{1+\alpha}(z) dz \right) - \frac{1}{\alpha} \log \left(\int f(z) g^{\alpha}(z) dz \right)$$

where for α tending to 0 reduces to the Kullback–Leibler divergence $I_X^{KL}(f, g)$.

The BHHJ (Basu, Harris, Hjort and Jones) power divergence (Basu *et al.* 1998) between f and g is another measure indexed by a positive parameter α and defined as

$$I_X^{BHHJ}(f,g)$$
$$= \int \left\{ g^{1+\alpha}(z) - \left(1+\frac{1}{\alpha}\right) f(z) g^{\alpha}(z) + \frac{1}{\alpha} f^{1+\alpha}(z) \right\} dz, \quad \alpha > 0.$$

Note that for $\alpha \to 0$ the BHHJ measure reduces to the Kullback–Leibler divergence and for $\alpha = 1$ to the square of the L_2 distance between f and g. Mattheou *et al.* (2009) proposed a general class of divergence measures; the (Φ, α)-power divergence family is given by

$$I_\Phi^\alpha(f,g) = \int g^{1+\alpha}(z) \Phi\left(\frac{f(z)}{g(z)}\right) dz \qquad [10.3]$$

where Φ is a convex function such that $0\Phi(u/0) = \lim_{u\to\infty}[\Phi(u)/u]$ and $0\Phi(0/0) = 0$. For $\Phi(u) = 1 - (1+\alpha^{-1})u + \alpha^{-1}u^{1+\alpha}$, [10.3] reduces to the BHHJ power divergence. The (Φ, α)-measure is a broad class of measures that also covers the Cressie and Read power divergence family and the φ-divergence family. Indeed, for $\alpha = 0$ if we take $\Phi = \varphi$, then [10.3] coincides with [10.1] while for $\Phi = \varphi_3$ the (Φ, α)-measure reduces to [10.2].

10.2.3. *Bregman divergences*

Finally, the Bregman divergence (Bregman 1967) between the probability density functions f and g is given by

$$I_X^{Breg}(f,g) = \int \Big(\phi(f(z)) - \phi(g(z)) - \phi'(g(z))(f(z) - g(z)) \Big) dz$$

for any differentiable convex function ϕ with $\phi(0) = \lim_{u\to 0}\phi(u) \in (-\infty, \infty)$. Observe that for $\phi(u) = u \log u$, the Bregman divergence reduces to the Kullback–Leibler divergence and for $\phi(u) = \alpha^{-1}(u^{\alpha+1} - u)$, $\alpha > 0$ reduces to the BHHJ power divergence.

The above divergence measures can also be defined for discrete settings. Let $\mathbf{p} = (p_1, \ldots, p_m)^\top$ and $\mathbf{q} = (q_1, \ldots, q_m)^\top$ be two discrete finite probability distributions. Then the discrete version of the φ-divergence measure is given by

$$d_\varphi(\mathbf{p}, \mathbf{q}) = \sum_{i=1}^{m} q_i \varphi\left(\frac{p_i}{q_i}\right)$$

while the discrete version of the (Φ, α)-power divergence measure is given by

$$d_\Phi^\alpha(\mathbf{p}, \mathbf{q}) = \sum_{i=1}^{m} q_i^{1+\alpha} \Phi\left(\frac{p_i}{q_i}\right), \quad \alpha > 0. \qquad [10.4]$$

The other measures of divergence can be defined in a similar fashion.

10.3. Properties of divergence measures

The measures of divergence are not formal distance functions. Any bivariate function $I_X(\cdot,\cdot)$ that satisfies the non-negativity property, namely $I_X(\cdot,\cdot) \geq 0$ with equality if and only if its two arguments are equal can be possibly used as a measure of divergence. Several properties of divergence measures have been investigated over the years and some of them are presented in the sequel.

φ-divergences are invariant with respect to common changes of variables. Also, generally, the divergences are not symmetric. Indeed, $I^{\varphi}(f,g)$ and $I^{\varphi}(g,f)$ are, in general, different. They are equal when the convex function φ verifies the condition $\varphi(x) - x\varphi(1/x) = c(x-1)$, where c is a constant (Liese and Vajda 1987). Also, $I_X^{\varphi}(f,g) = 0$ if and only if the two arguments are equal, provided that the function $x \to \varphi(x)$ is strictly convex in a neighborhood of $x = 1$. Otherwise, only "$f = g$ implies $I_X^{\varphi}(f,g) = 0$" occurs.

The order-preserving property has been introduced by Shiva *et al.* (1973) and shows that the relation between the amount of information contained in an r.v. (random variable) X_1 and that contained in another r.v. X_2 remains intact irrespective of the measure of divergence used. In particular, if the superscripts 1 and 2 represent two different measures of divergence, then

$$I^1_{X_1}(f_1,g_1) \leq I^1_{X_2}(f_2,g_2) \Leftrightarrow I^2_{X_1}(f_1,g_1) \leq I^2_{X_2}(f_2,g_2).$$

The limiting property that refers to the $\mu-$almost everywhere convergence states that a sequence of probability density functions f_n converges to a probability density function f if and only if the corresponding measure of divergence tends to zero, i.e.

$$f_n \to f \text{ iff } I_X(f_n,f) \to 0.$$

Note that the $\mu-$almost everywhere convergence is a weakened version of pointwise convergence according to which, for a measure space X, a sequence $g_n(x)$ tends to $g(x)$ for all $x \in Y$, a measurable subset of X such that, $\mu(X|Y) = 0$ (see Vonta *et al.* 2012).

Another property is the Ali–Silvey property, where if $f(x,\theta)$ (or simply f_θ) has the monotone likelihood ratio property in x, then

$$\theta_1 < \theta_2 < \theta_3 \Rightarrow I_X(f_{\theta_1},f_{\theta_2}) < I_X(f_{\theta_1},f_{\theta_3}).$$

Note that a family of density functions f_θ has the monotone likelihood ratio property in x if for any $\theta_1 < \theta_2$, $f_{\theta_2}/f_{\theta_1}$ is a non-decreasing function of x.

Recently, Vonta *et al.* (2012) proved the limiting property and the order preserving property for the (Φ, α)-power divergence family of measures [10.3]. As they state, the order preserving property holds for [10.3] provided that the parameter α belongs to a specific interval, which is different for each measure. They also showed the quadratic convergence of the (Φ, α)-measure. In other words, they showed that the discretized version of the (Φ, α)-measure converges quadratically to the theoretical (Φ, α)-measure.

10.4. Model selection criteria

A model selection criterion can be considered as an approximately unbiased estimator of the expected overall discrepancy, a non-negative quantity that measures the distance between the true unknown model and a fitted approximating model. If the value of the criterion is small, then the approximated model is good.

Let $\mathbf{Y} = (Y_1, \ldots, Y_n)^\top$ be a random vector from the distribution g (the true model) and f_θ a candidate model from a parametric family of models $\{f_\theta\}$, indexed by an unknown parameter $\theta \in \Theta \subset \mathbb{R}^p$. Akaike (1973) proposed the Akaike information criterion (AIC), which is based on the Kullback–Leibler measure of divergence between the true and the candidate models. For a realization $\mathbf{y} = (y_1, \ldots, y_n)^\top$ of $\mathbf{Y} = (Y_1, \ldots, Y_n)^\top$, the AIC is given by

$$AIC = -2l(\hat{\theta}_{MLE}; \mathbf{y}) + 2p$$

where $l(\theta; \mathbf{y})$ is the log-likelihood and $\hat{\theta}_{MLE}$ is the maximum likelihood estimator (MLE) of the unknown parameter, i.e.

$$l(\hat{\theta}_{MLE}; \mathbf{y}) = \sum_{i=1}^{n} \log(f_{\hat{\theta}_{MLE}}(y_i)) = \max_{\theta \in \Theta} l(\theta; \mathbf{y}).$$

Applying the same methodology used for AIC, Mattheou *et al.* (2009) developed the divergence information criterion (DIC), which is based on the quantity

$$W_\theta = \int \left\{ f_\theta^{1+\alpha}(z) - (1 + \alpha^{-1}) g(z) f_\theta^{\alpha}(z) \right\} dz, \quad \alpha > 0. \qquad [10.5]$$

Note that [10.5] is the same as the BHHJ power divergence $I_X^{BHHJ}(g, f_\theta)$ without the last term that remains constant irrespectively of the model f_θ used. The expected overall discrepancy that is the target theoretical quantity for which Mattheou *et al.* (2009) provided an asymptotically unbiased estimator is given by

$$E(W_{\hat{\theta}}) = E\Big(W_\theta \big| \theta = \hat{\theta}\Big)$$

where $\hat{\theta}$ is any consistent and asymptotically normal estimator of θ. This quantity can be viewed as the average distance between g and f_θ up to a constant. For a set of observations $y_1, \dots, y_n$, the resulting criterion is given by

$$DIC = nQ_{\hat{\theta}} + (2\pi)^{-\alpha/2}(1+\alpha)^{2+p/2}p$$

where

$$Q_{\hat{\theta}} = \int f_{\hat{\theta}}^{1+\alpha}(z)dz - (1+\alpha^{-1})\frac{1}{n}\sum_{i=1}^{n} f_{\hat{\theta}}^{\alpha}(y_i).$$

Because of the fact that the integral $\int f_{\hat{\theta}}^{1+\alpha}(z)dz$ in the first part of $Q_{\hat{\theta}}$ is approximately constant for the various candidate models, Mantalos *et al.* (2010) proposed a modified version of DIC, the modified divergence information criterion (MDIC), which is given by

$$MDIC = nMQ_{\hat{\theta}} + (2\pi)^{-\alpha/2}(1+\alpha)^{2+p/2}p$$

where

$$MQ_{\hat{\theta}} = -(1+\alpha^{-1})\frac{1}{n}\sum_{i=1}^{n} f_{\hat{\theta}}^{\alpha}(y_i)$$

and the expected overall discrepancy up to a constant is given by

$$E(V_{\hat{\theta}}) = E\Big(V_\theta \big| \theta = \hat{\theta}\Big) \text{ with } V_\theta = -\int (1+\alpha^{-1})g(z)f_\theta^{\alpha}(z)dz.$$

Recently, Toma *et al.* (2020) proposed a new model selection criterion, the so-called pseudodistance information criterion (PIC). For this new criterion, the family of pseudodistances of order $\alpha > 0$ was considered. The quantity that have been used for the construction of the criterion is given by

$$M_\theta = \frac{1}{\alpha+1}\log\left(\int f_\theta^{1+\alpha}(z)dz\right) - \frac{1}{\alpha}\log\left(\int g(z)f_\theta^{\alpha}(z)dz\right).$$

Observe that M_θ is the same as the pseudodistance $I_X^P(g, f_\theta)$ without the middle term, which is constant irrespectively of the model f_θ used. The expected overall discrepancy is given by $E(M_{\hat{\theta}_n}) = E(M_\theta | \theta = \hat{\theta}_n)$, where $\hat{\theta}_n$ is the minimum pseudodistance estimator (Broniatowski *et al.* 2012) defined as

$$\hat{\theta}_n = \arg\inf_{\theta\in\Theta}\left\{\frac{1}{\alpha+1}\log\left(\int f_\theta^{1+\alpha}(z)dz\right) - \frac{1}{\alpha}\log\left(\frac{1}{n}\sum_{i=1}^{n} f_\theta^{\alpha}(y_i)\right)\right\}.$$

The resulting criterion for the univariate normal case (Toma *et al.* 2020) is given by

$$PIC = nQ^*_{\hat{\theta}_n} + \frac{(\alpha+1)^2}{(2\alpha+1)^{3/2}}p + \frac{1}{2\alpha}\left(1 - \frac{\alpha+1}{\sqrt{2\alpha+1}}\right)$$

where

$$Q^*_{\hat{\theta}_n} = \frac{1}{\alpha+1}\log\left(\int f^{1+\alpha}_{\hat{\theta}_n}(z)dz\right) - \frac{1}{\alpha}\log\left(\frac{1}{n}\sum_{i=1}^{n} f^{\alpha}_{\hat{\theta}_n}(y_i)\right).$$

10.5. Goodness of fit tests

10.5.1. *Simple null hypothesis*

Let $\mathbf{X} = (X_1, \ldots, X_m)^\top$ be a random vector from the multinomial distribution $M(n, \mathbf{p})$ with parameters n and $\mathbf{p} = (p_1, \ldots, p_m)^\top$. For a realization $\mathbf{x} = (x_1, \ldots, x_m)^\top$ of $\mathbf{X} = (X_1, \ldots, X_m)^\top$, we have that $\hat{\mathbf{p}} = (\hat{p}_1, \ldots, \hat{p}_m)^\top$ with $\hat{p}_i = x_i/n$, $i = 1, \ldots, m$ is the non-parametric maximum likelihood estimator of the true probabilities p_i, $i = 1, \ldots, m$ and $n = \sum_{i=1}^{m} x_i$. If we want to test the simple null hypothesis that the data $(x_1, \ldots, x_m)^\top$ come from a known multinomial distribution $M(n, \mathbf{p}_0)$, where $\mathbf{p}_0 = (p_{10}, \ldots, p_{m0})^\top$ is the vector of hypothesized probabilities, namely

$$H_0 : \mathbf{p} = \mathbf{p}_0 \quad [10.6]$$

the most commonly used test statistics are Pearson's or chi-squared test statistic given by

$$X^2 = \sum_{i=1}^{m} \frac{(x_i - np_{i0})^2}{np_{i0}} \quad [10.7]$$

and the likelihood ratio test statistic is given by

$$G^2 = 2\sum_{i=1}^{m} x_i \log\left(\frac{x_i}{np_{i0}}\right). \quad [10.8]$$

In the domain of hypothesis testing, divergence measures have a pivotal role. Traditionally, for testing hypothesis we use the chi-squared test and the likelihood ratio tests mentioned above, but Kupperman (1957, 1958) proposed a new ground-breaking way. He established a test statistic based on the Kullback–Leibler divergence, which is asymptotically distributed as a chi-squared random variable. The connection between divergence measure theory and hypothesis testing, continued with several alternative approaches. Indeed, Simon (1973) connected the likelihood ratio test statistic for hypothesis testing under the exponential distribution family with the Kullback–Leibler divergence. Nayak (1983, 1985) used entropy measures while Basu *et al.* (2010) used the Hellinger distance for the construction of statistical tests for discrete populations.

Most of the above test statistics are special cases of the family of power divergence test statistics introduced by Cressie and Read (1984), which is given by

$$T_n^{\lambda}(\hat{\mathbf{p}}, \mathbf{p}_0) = \frac{2n}{\lambda(\lambda+1)} \sum_{i=1}^{m} \hat{p}_i \left[\left(\frac{\hat{p}_i}{p_{i0}} \right)^{\lambda} - 1 \right], \quad \lambda \in \mathbb{R} \setminus \{0, -1\} \qquad [10.9]$$

where specific values of λ correspond to well-known test statistics. Particularly for $\lambda = 1$ and for $\lambda \to 0$, the chi-squared and the likelihood ratio test statistics are obtained, respectively.

Recently, Mattheou and Karagrigoriou (2010) proposed the more general (Φ, α)-power divergence family of test statistics, for testing the simple null hypothesis [10.6], which is defined by

$$T_{\Phi}^{\alpha}(\hat{\mathbf{p}}, \mathbf{p}_0) = \frac{2n}{\Phi''(1)} \sum_{i=1}^{m} p_{i0}^{1+\alpha} \Phi\left(\frac{\hat{p}_i}{p_{i0}} \right) \qquad [10.10]$$

for any convex function Φ such that $\Phi(1) = \Phi'(1) = 0$ and $\Phi''(1) \neq 0$. Observe that for $\alpha = 0$ and Φ equal to φ_1, φ_2, and φ_3 the (Φ, α)-power divergence test statistic reduces to chi-squared, likelihood ratio and power divergence test statistics, respectively. Note also that for $\alpha = 0$ and Φ as in [10.1], the test statistic [10.10] reduces to the traditional tests of fit based on the φ-divergence family of measures.

Under the simple null hypothesis [10.6], the asymptotic distribution of the tests [10.7], [10.8], [10.9] and the φ-divergence test statistic is chi-squared with $m - 1$ degrees of freedom ($\mathcal{X}_{m-1}^2$). These results were established by Pearson (1900), Neyman and Pearson (1928), Cressie and Read (1984) and Zografos *et al.* (1990), respectively. Mattheou and Karagrigoriou (2010) proved that under [10.6] the asymptotic distribution of the (Φ, α)-power divergence test statistic is estimated to be approximately $c\mathcal{X}_{m-1}^2$, where $c = 0.5(\min_i p_{i0}^{\alpha} + \max_i p_{i0}^{\alpha})$.

10.5.2. *Composite null hypothesis*

Now consider a random vector $\mathbf{X} = (X_1, \ldots, X_m)^\top$ from the multinomial distribution $M(n, \mathbf{p}(\boldsymbol{\theta}_0))$, where $\mathbf{p}(\boldsymbol{\theta}_0) = (p_1(\boldsymbol{\theta}_0), \ldots, p_m(\boldsymbol{\theta}_0))^\top$ is the vector of true probabilities and n is the number of the experiments. The vector $\mathbf{p}(\boldsymbol{\theta}_0)$ is assumed to be unknown but belonging to a known family

$$\mathcal{P} = \left\{ \mathbf{p}(\boldsymbol{\theta}) = (p_1(\boldsymbol{\theta}), \ldots, p_m(\boldsymbol{\theta}))^\top : \boldsymbol{\theta} = (\theta_1, \ldots, \theta_s)^\top \in \Theta \right\}$$

with $\Theta \subset \mathbb{R}^s$ $(s < m - 1)$. In other words, the true value $\boldsymbol{\theta}_0$ of the parameter $\boldsymbol{\theta} = (\theta_1, \ldots, \theta_s)^\top$ is assumed to be unknown. For a realization $(x_1, \ldots, x_m)^\top$ of $\mathbf{X} = (X_1, \ldots, X_m)^\top$, we have $n = \sum_{i=1}^m x_i$, $x_i \geq 0$. Then $\hat{\boldsymbol{p}} = (\hat{p}_1, \ldots, \hat{p}_m)^\top$ with $\hat{p}_i = x_i/n$ is the non-parametric maximum likelihood estimator of the true probabilities $p_i(\boldsymbol{\theta}_0), i = 1, \ldots, m$.

In order to estimate the unknown parameter $\boldsymbol{\theta}$, one could use the maximum likelihood method, where in the multinomial setting it suffices to find a value $\hat{\boldsymbol{\theta}}$ over Θ that maximizes $\sum_{i=1}^m \hat{p}_i \log p_i(\boldsymbol{\theta})$. Then $\hat{\boldsymbol{\theta}}$ is the point MLE of the parameter $\boldsymbol{\theta}$. Equivalently the MLE can be defined as the argument that minimizes the Kullback–Leibler divergence:

$$\hat{\boldsymbol{\theta}} = \arg \inf_{\boldsymbol{\theta} \in \Theta} d_{\mathrm{KL}}(\hat{\boldsymbol{p}}, \mathbf{p}(\boldsymbol{\theta})).$$

Cressie and Read (1984) proposed the minimum power divergence estimator

$$\hat{\boldsymbol{\theta}}_{\mathrm{CR}}^{\lambda} = \arg \inf_{\boldsymbol{\theta} \in \Theta} d_{\mathrm{CR}}^{\lambda}(\hat{\boldsymbol{p}}, \mathbf{p}(\boldsymbol{\theta}))$$

and studied its properties. As they argue (Read and Cressie 1988, p. 30), this estimator is a natural generalization of the MLE. Furthermore, Morales *et al.* (1995) considered the more general minimum φ-divergence estimator, defined as

$$\hat{\boldsymbol{\theta}}_{\varphi} = \arg \inf_{\boldsymbol{\theta} \in \Theta} d_{\varphi}(\hat{\boldsymbol{p}}, \mathbf{p}(\boldsymbol{\theta})). \qquad [10.11]$$

In the sequel, we assume that the model is correct, so that there exists a value $\boldsymbol{\theta}_0$ that leads to the true probabilities that generate the observations. Under the regularity conditions introduced by Birch (1964) and the assumption that the function $\mathbf{p}(\boldsymbol{\theta})$ has continuous second partial derivatives in a neighbourhood of $\boldsymbol{\theta}_0$, it holds

$$\begin{aligned}\hat{\boldsymbol{\theta}}_{\varphi} =& \boldsymbol{\theta}_0 + \left(\mathbf{A}(\boldsymbol{\theta}_0)^\top \mathbf{A}(\boldsymbol{\theta}_0)\right)^{-1} \mathbf{A}(\boldsymbol{\theta}_0)^\top diag\left(\mathbf{p}(\boldsymbol{\theta}_0)^{-1/2}\right)\left(\hat{\boldsymbol{p}} - \mathbf{p}(\boldsymbol{\theta}_0)\right) \\ &+ o\left(\left\|\hat{\boldsymbol{p}} - \mathbf{p}(\boldsymbol{\theta}_0)\right\|\right)\end{aligned} \qquad [10.12]$$

and

$$\sqrt{n}(\hat{\boldsymbol{\theta}}_{\varphi} - \boldsymbol{\theta}_0) \xrightarrow[n \to \infty]{L} N_s\left(\mathbf{0}, \left(\mathbf{A}(\boldsymbol{\theta}_0)^\top \mathbf{A}(\boldsymbol{\theta}_0)\right)^{-1}\right) \qquad [10.13]$$

where

$$\mathbf{A}(\boldsymbol{\theta}_0) = diag\left(\mathbf{p}(\boldsymbol{\theta}_0)^{-1/2}\right) \left(\frac{\partial p_i(\boldsymbol{\theta}_0)}{\partial \theta_j}\right)_{\substack{i=1,\ldots,m \\ j=1,\ldots,s}}$$

and $\left(\mathbf{A}(\boldsymbol{\theta}_0)^\top \mathbf{A}(\boldsymbol{\theta}_0)\right)$ is the Fisher information matrix associated with the multinomial model.

For testing the composite null hypothesis

$$H_0\colon \mathbf{p} = \mathbf{p}(\boldsymbol{\theta}_0),\ \boldsymbol{\theta}_0 \in \Theta \subset \mathbb{R}^s \qquad [10.14]$$

the family of φ-divergence test statistics can be used where the unknown parameter $\boldsymbol{\theta}_0$ is estimated by the minimum φ-divergence estimator given in [10.11], for which we have that (Menéndez *et al.* 2001)

$$T_\varphi\left(\hat{\boldsymbol{\theta}}_{\varphi*}\right) = \frac{2n}{\varphi''(1)} d_\varphi\left(\hat{\boldsymbol{p}}, \mathbf{p}(\hat{\boldsymbol{\theta}}_{\varphi*})\right) \xrightarrow[n\to\infty]{L} \mathcal{X}^2_{m-1-s}$$

where φ, φ^* as in [10.1]. Observe that functions φ, φ^* are not necessarily the same.

Here in order to test the composite null hypothesis [10.14], we propose the new double index (Φ, α)-test statistic, which is associated with the (Φ, α)-power divergence measure [10.4]. This general family of test statistics involves two indices and two functions that are not necessarily equal and this feature provides greater flexibility in terms of statistical inference. The new double index (Φ, α)-test statistic is given by

$$T^{\alpha_1}_{\Phi_1}\left(\hat{\boldsymbol{\theta}}^{\alpha_2}_{\Phi_2}\right) = \frac{2n}{\Phi''_1(1)} d^{\alpha_1}_{\Phi_1}\left(\hat{\boldsymbol{p}}, \mathbf{p}(\hat{\boldsymbol{\theta}}^{\alpha_2}_{\Phi_2})\right) \qquad [10.15]$$

where Φ_1 and Φ_2 as in [10.10] and $\hat{\boldsymbol{\theta}}^{\alpha}_{\Phi}$ is the minimum (Φ, α)-power divergence estimator, defined as

$$\hat{\boldsymbol{\theta}}^{\alpha}_{\Phi} = \arg \inf_{\boldsymbol{\theta}\in\Theta} d^{\alpha}_{\Phi}(\hat{\boldsymbol{p}}, \mathbf{p}(\boldsymbol{\theta})).$$

Under the same assumptions used for [10.12] and [10.13], it can be proved (Meselidis and Karagrigoriou 2020) that

$$\hat{\boldsymbol{\theta}}^{\alpha}_{\Phi} = \boldsymbol{\theta}_0 + \left(\mathbf{B}(\boldsymbol{\theta}_0)^\top \mathbf{B}(\boldsymbol{\theta}_0)\right)^{-1} \mathbf{B}(\boldsymbol{\theta}_0)^\top diag(\mathbf{p}(\boldsymbol{\theta}_0)^{\alpha/2}) \times \\ \times\, diag(\mathbf{p}(\boldsymbol{\theta}_0)^{-1/2})(\hat{\boldsymbol{p}} - \mathbf{p}(\boldsymbol{\theta}_0)) + o(\|\hat{\boldsymbol{p}} - \mathbf{p}(\boldsymbol{\theta}_0)\|) \qquad [10.16]$$

and

$$\sqrt{n}(\hat{\boldsymbol{\theta}}^{\alpha}_{\Phi} - \boldsymbol{\theta}_0) \xrightarrow[n\to\infty]{L} N_s(\mathbf{0}, \mathbf{W}(\boldsymbol{\theta}_0)\boldsymbol{\Sigma}_{\mathbf{p}(\boldsymbol{\theta}_0)}\mathbf{W}(\boldsymbol{\theta}_0)^\top) \qquad [10.17]$$

where

$$\mathbf{W}(\boldsymbol{\theta}_0) = \left(\mathbf{B}(\boldsymbol{\theta}_0)^\top \mathbf{B}(\boldsymbol{\theta}_0)\right)^{-1} \mathbf{B}(\boldsymbol{\theta}_0)^\top diag(\mathbf{p}(\boldsymbol{\theta}_0)^{\alpha/2}) diag(\mathbf{p}(\boldsymbol{\theta}_0)^{-1/2})$$

$$\mathbf{B}(\boldsymbol{\theta}_0) = diag(\mathbf{p}(\boldsymbol{\theta}_0)^{\alpha/2})\mathbf{A}(\boldsymbol{\theta}_0)$$

and

$$\boldsymbol{\Sigma}_{\mathbf{p}(\boldsymbol{\theta}_0)} = diag(\mathbf{p}(\boldsymbol{\theta}_0)) - \mathbf{p}(\boldsymbol{\theta}_0)\mathbf{p}(\boldsymbol{\theta}_0)^\top.$$

We should remark that when $\alpha = 0$ or if we assume that the true model $\mathbf{p}(\boldsymbol{\theta}_0)$ is the equiprobable, then the decomposition of $\hat{\boldsymbol{\theta}}_\Phi^\alpha$ given in [10.16] coincides with the decomposition of $\hat{\boldsymbol{\theta}}_\varphi$ given in [10.12] and the variance–covariance matrix in [10.17] reduces to the inverse of the Fisher information matrix associated with the multinomial model. Note further that for values of α close to zero (as is the case in most applications), the above two results hold approximately true.

Under the same assumptions used for [10.12] and [10.13], the composite null hypothesis [10.14] and assuming further that the true model $\mathbf{p}(\boldsymbol{\theta}_0)$ is the equiprobable we have (Meselidis and Karagrigoriou 2020) that

$$T_{\Phi_1}^{\alpha_1}(\hat{\boldsymbol{\theta}}_{\Phi_2}^{\alpha_2}) \xrightarrow[n\to\infty]{L} c\mathcal{X}_{m-1-s}^2$$

where $c = m^{-\alpha_1}$. We should remark here that when the true model is not the equiprobable but $\alpha_2 = 0$ or the value of α_2 is close to zero, we can reformulate theorem 2 of Mattheou and Karagrigoriou (2010) and set

$$Z = \sum_{i=1}^{m} \frac{n}{p_i(\boldsymbol{\theta}_0)}(\hat{p}_i - p_i(\hat{\boldsymbol{\theta}}_{\Phi_2}^{\alpha_2}))^2$$

in order to have

$$(\min_i p_i(\boldsymbol{\theta}_0)^{\alpha_1})Z \prec_{st} \sum_{i=1}^{m} \frac{np_i(\boldsymbol{\theta}_0)^{\alpha_1}}{p_i(\boldsymbol{\theta}_0)}(\hat{p}_i - p_i(\hat{\boldsymbol{\theta}}_{\Phi_2}^{\alpha_2}))^2 \prec_{st} (\max_i p_i(\boldsymbol{\theta}_0)^{\alpha_1})Z.$$

Then, the asymptotic distribution of $T_{\Phi_1}^{\alpha_1}(\hat{\boldsymbol{\theta}}_{\Phi_2}^{\alpha_2})$ is estimated to be approximately $c\mathcal{X}_{m-1-s}^2$ where

$$c = 0.5(\min_i p_i(\boldsymbol{\theta}_0)^{\alpha_1} + \max_i p_i(\boldsymbol{\theta}_0)^{\alpha_1}) \qquad [10.18]$$

and $\prec_{st}$ is the symbol for stochastic ordering. Observe that under the equiprobable model, the value of c in [10.18] equals $m^{-\alpha_1}$. We note further that for values of α_1 close to 0 the asymptotic distribution of the test statistic [10.15], under the equiprobable model, coincides with the $\mathcal{X}_{m-1-s}^2$, while for bigger values of α_1 coincides with the $c\mathcal{X}_{m-1-s}^2$. However, when the true model is not the equiprobable, as α_2 deviates from 0 the asymptotic $c\mathcal{X}_{m-1-s}^2$ fails to describe properly the underlying distribution of the new double index (Φ, α)-test statistic. Also for small departures from the equiprobable model, the asymptotic distribution of [10.15] coincides with the $c\mathcal{X}_{m-1-s}^2$, even for values of α_2 not close to zero.

The new double index (Φ, α)-test statistic combines under a single umbrella most of the well-known test statistics previously investigated. For instance, when $\alpha_1 = 0$ the chi-squared test statistic can be derived from [10.15] for $\Phi_1 = \varphi_1$, the likelihood ratio statistic for $\Phi_1 = \varphi_2$ and the Cressie and Read power divergence test statistic for $\Phi_1 = \varphi_3$. Observe also that for $\alpha_1 = 0$ and $\Phi_1 = \varphi$, we have the family of test statistics that is associated with the φ-divergence measure. The BHHJ-test statistic can be derived from [10.15] for $\Phi_1(u) = 1 - (1 + \alpha_1^{-1})u + \alpha_1^{-1}u^{1+\alpha_1}$ and $\alpha_1 > 0$.

10.6. Simulation study

Let $\mathbf{Y} = (Y_1, \ldots, Y_n)^\top$ be a random vector from the distribution F. We can test the composite null hypothesis, $H_0\colon F = F_{\boldsymbol{\theta}_0}$, that the distribution function F is a member of a parametric family $\{F_{\boldsymbol{\theta}}\}_{\boldsymbol{\theta}\in\Theta}$, $\Theta \subset \mathbb{R}^s$ by partitioning the data range into m disjoint intervals and testing the hypothesis $H_0\colon \mathbf{p} = \mathbf{p}(\boldsymbol{\theta}_0)$ about the parameter-vector of the resulting multinomial distribution, for some unknown $\boldsymbol{\theta}_0 \in \Theta \subset \mathbb{R}^s$ and $s < m - 1$.

Let $P = \{E_i\}_{i=1,\ldots,m}$ be a partition of the real line $\mathbb{R}$ into m intervals. The probabilities of the intervals E_i, $i = 1, \ldots, m$ depend on the unknown parameter $\boldsymbol{\theta}_0$, in such a way that $p_i(\boldsymbol{\theta}_0) = \Pr_{F_{\boldsymbol{\theta}_0}}(E_i) = \int_{E_i} dF_{\boldsymbol{\theta}_0}$. Let $x_i = \sum_{j=1}^{n} I_{E_i}(Y_j)$ where

$$I_{E_i}(Y_j) = \begin{cases} 1, & \text{if } Y_j \in E_i \\ 0, & \text{otherwise} \end{cases},$$

with $\sum x_i = n$ and $\hat{p}_i$ the non-parametric MLE of the true probability of the E_i interval, defined at the beginning of the previous section.

In order to understand the behavior of the double index (Φ, α)-test statistic given in [10.15], we compare it with three tests based on the φ-divergence family of measures, namely the tests of fit based on the Kullback–Leibler measure $(G^2(\hat{\boldsymbol{\theta}}))$, the Pearson's chi-squared measure $(X^2(\hat{\boldsymbol{\theta}}_{X^2}))$ and the Cressie and Read measure with $\lambda = 2/3$ $(CR(\hat{\boldsymbol{\theta}}^{\lambda}_{\mathrm{CR}}))$. Observe that for every test of fit, the estimator of the unknown parameter is based on the same measure, although this is not mandatory in general. The proposed test $(T^{\alpha_1}_{\Phi_1}(\hat{\boldsymbol{\theta}}^{\alpha_2}_{\Phi_2}))$ is applied for $\Phi_1(u) = 1 - (1 + \alpha_1^{-1})u + \alpha_1^{-1}u^{1+\alpha_1}$ and $\Phi_2(u) = 1 - (1 + \alpha_2^{-1})u + \alpha_2^{-1}u^{1+\alpha_2}$ and three different values of α_1 and α_2, namely $\alpha_1, \alpha_2 = 0.01, 0.10$ and 0.50.

For the null distribution, we focus on the Gamma distribution where the shape parameter is taken to be equal to 0.44, 1 and 2. These values of the shape parameter correspond to skewness equal to 3, 2 and 1.414, respectively. Distributions with the same skewness values have been used as possible alternatives in each case, namely the Weibull (Wei), log-normal (LN) and inverse Gaussian (IG). The scale parameter is taken to be equal to 1 in all cases, since the Gamma distribution is scale invariant. Note that the same distributions have been used previously by Vonta *et al.* (2012) in order to examine the performance of the test statistic [10.10] for the simple null hypothesis [10.6] as well as by Koutrouvelis and Karagrigoriou (2012) for the analysis of the inverse Gaussian distribution. For the partition of the data range, we use $m = 3$ intervals and consider four models for the resulting multinomial distribution, namely the equiprobable $(1/3, 1/3, 1/3)$, symmetric $(0.2, 0.6, 0.2)$, left skewed $(0.1, 0.5, 0.4)$ and right skewed $(0.4, 0.5, 0.1)$ model.

In order to investigate the power for various alternatives and the size of the proposed test, we generated $10,000$ samples from each distribution with sample sizes $n = 25, 50, 75, 100, 500$. The results for every sample size n are provided at the web page of the Laboratory of Statistics and Data Analysis (http://actuarweb.aegean.gr/labstada/publications.html). Here, we report only the results for $n = 25$ (Table 10.1) and the models

a) equiprobable $(1/3, 1/3, 1/3)$ and

b) left skewed $(0.1, 0.5, 0.4)$

as well as for $n = 50$ (Table 10.2), and the models

c) symmetric $(0.2, 0.6, 0.2)$ and

d) right skewed $(0.4, 0.5, 0.1)$.

In particular, in Table 10.1 we observe that for the equiprobable model, $X^2(\hat{\boldsymbol{\theta}}_{X^2})$ and $CR(\hat{\boldsymbol{\theta}}_{\text{CR}}^{\lambda})$ have the best performance among all competing tests for every alternative, with the first being slightly better. Also, we observe that $T_{\Phi_1}^{0.01}(\hat{\boldsymbol{\theta}}_{\Phi_2}^{0.50})$ and $G^2(\hat{\boldsymbol{\theta}})$ have a similar behavior. In terms of size (denoted by (S)), the best results are obtained for $T_{\Phi_1}^{0.01}(\hat{\boldsymbol{\theta}}_{\Phi_2}^{0.50})$, while $X^2(\hat{\boldsymbol{\theta}}_{X^2})$ and $CR(\hat{\boldsymbol{\theta}}_{\text{CR}}^{\lambda})$ appear to be liberal. Similar results (not shown here) are obtained for bigger sample sizes ($n \geq 50$), with the exception that against the Weibull alternative (Wei(0.77)) the best results are obtained for $T_{\Phi_1}^{0.50}(\hat{\boldsymbol{\theta}}_{\Phi_2}^{0.01})$ and $T_{\Phi_1}^{0.01}(\hat{\boldsymbol{\theta}}_{\Phi_2}^{0.50})$ with the first being slightly better.

For the left skewed model and the Weibull alternative, $T_{\Phi_1}^{0.01}(\hat{\boldsymbol{\theta}}_{\Phi_2}^{0.50})$ has the best performance in most cases. For the remaining alternatives, the best results are obtained for $X^2(\hat{\boldsymbol{\theta}}_{X^2})$, while $T_{\Phi_1}^{0.01}(\hat{\boldsymbol{\theta}}_{\Phi_2}^{0.50})$ and $G^2(\hat{\boldsymbol{\theta}})$ have a similar behavior with the first being slightly better. As the sample size n increases (results not shown here), $T_{\Phi_1}^{0.01}(\hat{\boldsymbol{\theta}}_{\Phi_2}^{0.50})$ performs better than $X^2(\hat{\boldsymbol{\theta}}_{X^2})$ and $CR(\hat{\boldsymbol{\theta}}_{\text{CR}}^{\lambda})$ for every alternative considered. It is interesting to note that for alternatives such as inverse Gaussian and log-normal, the best results when $n \geq 50$ are obtained for $T_{\Phi_1}^{0.50}(\hat{\boldsymbol{\theta}}_{\Phi_2}^{0.01})$. In terms of size, when $n = 25$ the best results are obtained for $X^2(\hat{\boldsymbol{\theta}}_{X^2})$, while for bigger sample sizes the best results are obtained for $G^2(\hat{\boldsymbol{\theta}})$ in most cases. The test statistic $T_{\Phi_1}^{0.01}(\hat{\boldsymbol{\theta}}_{\Phi_2}^{0.50})$ appears to be liberal for every sample size.

Table 10.2 provides the results for the symmetric and right skewed models with sample size $n = 50$. The results reveal that for the symmetric model in most cases

$T_{\Phi_1}^{0.50}(\hat{\theta}_{\Phi_2}^{0.01})$ has the best performance among all competing tests, not only in terms of power but also in terms of size. Test statistics $X^2(\hat{\theta}_{X^2})$ and $CR(\hat{\theta}_{\mathrm{CR}}^{\lambda})$ have an edge against the Weibull alternative (Wei(0.77)), but they appear to be liberal. As the sample size increases, although no test is clearly better than the others, $T_{\Phi_1}^{0.10}(\hat{\theta}_{\Phi_2}^{0.01})$ has a size quite close to the nominal level while $T_{\Phi_1}^{0.50}(\hat{\theta}_{\Phi_2}^{0.01})$ appears to have slightly higher power in most cases as compared to all tests examined.

Shape parameter	$T_{\Phi_1}^{0.10}(\hat{\theta}_{\Phi_2}^{0.01})$	$T_{\Phi_1}^{0.50}(\hat{\theta}_{\Phi_2}^{0.01})$	$T_{\Phi_1}^{0.01}(\hat{\theta}_{\Phi_2}^{0.50})$	$G^2(\hat{\theta})$	$X^2(\hat{\theta}_{X^2})$	$CR(\hat{\theta}_{\mathrm{CR}}^{\lambda})$
Equiprobable model (1/3, 1/3, 1/3)						
(S) Gamma(0.44)	4.74	4.30	4.93	4.83	6.12	5.63
Wei(0.77)	12.00	11.89	12.27	12.01	12.43	12.21
LN(0.729)	13.23	11.88	13.44	13.44	17.44	16.00
IG(1)	12.74	11.54	12.96	12.97	16.75	15.48
(S) Gamma(1)	4.88	4.38	4.94	4.93	6.21	5.72
Wei(1)	5.12	4.72	5.25	5.19	6.70	6.27
LN(0.551)	36.61	33.92	37.06	37.09	44.69	42.33
IG(2.25)	31.67	29.14	32.05	32.05	39.24	36.91
(S) Gamma(2)	4.71	4.36	4.84	4.78	6.14	5.63
Wei(1.259)	9.14	8.15	9.31	9.29	12.33	11.19
LN(0.423)	69.06	66.46	69.48	69.50	77.25	74.84
IG(4.5)	64.53	61.56	65.11	65.13	72.93	70.47
Left skewed model (0.1, 0.5, 0.4)						
(S) Gamma(0.44)	3.99	4.06	5.74	3.84	4.58	4.33
Wei(0.77)	10.55	10.53	16.17	10.27	11.56	11.14
LN(0.729)	25.44	24.50	26.03	25.61	27.68	26.99
IG(1)	31.59	30.54	32.24	31.75	34.40	33.41
(S) Gamma(1)	3.95	3.90	5.61	3.90	4.65	4.41
Wei(1)	4.01	3.85	5.84	3.97	4.75	4.49
LN(0.551)	64.73	63.18	65.51	64.99	67.63	66.79
IG(2.25)	63.68	62.08	64.48	63.93	66.54	65.74
(S) Gamma(2)	3.97	3.58	5.73	3.91	4.74	4.49
Wei(1.259)	7.76	7.41	8.18	7.79	8.80	8.35
LN(0.423)	94.44	93.98	94.67	94.53	95.38	95.16
IG(4.5)	93.45	92.86	93.65	93.48	94.48	94.07

Table 10.1. *Size (S) and powers (%) for the equiprobable and left skewed models,* $n = 25$

For the right skewed model, in most cases $T_{\Phi_1}^{0.01}(\hat{\theta}_{\Phi_2}^{0.50})$ performs better in terms of power than all other competing tests, but it appears to be liberal. The $X^2(\hat{\theta}_{X^2})$ test statistic appears to have a slight edge against the Weibull alternative (Wei(0.77)). Similar results are obtained for bigger sample sizes.

Shape parameter	$T_{\Phi_1}^{0.10}(\hat{\boldsymbol{\theta}}_{\Phi_2}^{0.01})$	$T_{\Phi_1}^{0.50}(\hat{\boldsymbol{\theta}}_{\Phi_2}^{0.01})$	$T_{\Phi_1}^{0.01}(\hat{\boldsymbol{\theta}}_{\Phi_2}^{0.50})$	$G^2(\hat{\boldsymbol{\theta}})$	$X^2(\hat{\boldsymbol{\theta}}_{X^2})$	$CR(\hat{\boldsymbol{\theta}}_{\text{CR}}^{\lambda})$
Symmetric model (0.2, 0.6, 0.2)						
(S) Gamma(0.44)	5.33	5.08	5.59	5.46	5.73	5.64
Wei(0.77)	28.66	25.01	30.41	29.37	31.84	31.03
LN(0.729)	46.77	47.20	46.72	46.71	46.59	46.59
IG(1)	42.82	43.28	42.74	42.72	42.42	42.43
(S) Gamma(1)	5.16	4.93	5.30	5.19	5.45	5.34
Wei(1)	5.15	4.81	5.34	5.22	5.55	5.41
LN(0.551)	97.12	97.20	97.11	97.11	97.08	97.06
IG(2.25)	94.34	94.54	94.34	94.34	94.32	94.32
(S) Gamma(2)	5.21	4.89	5.37	5.26	5.54	5.45
Wei(1.259)	29.13	29.48	29.05	29.04	28.89	28.89
LN(0.423)	100.00	100.00	100.00	100.00	100.00	100.00
IG(4.5)	99.98	99.98	99.98	99.98	99.98	99.98
Right skewed model (0.4, 0.5, 0.1)						
(S) Gamma(0.44)	4.61	3.66	5.34	4.95	5.19	5.10
Wei(0.77)	30.67	21.64	35.41	32.07	37.44	35.96
LN(0.729)	22.36	21.04	23.20	22.69	21.30	21.75
IG(1)	15.52	14.21	16.46	15.94	14.86	15.20
(S) Gamma(1)	4.95	4.22	5.72	5.23	5.37	5.34
Wei(1)	5.05	3.84	5.84	5.33	5.28	5.36
LN(0.551)	82.16	80.78	83.02	82.51	80.82	81.40
IG(2.25)	73.18	71.73	74.15	73.57	71.82	72.31
(S) Gamma(2)	5.33	4.36	6.20	5.58	5.63	5.66
Wei(1.259)	32.07	30.53	33.23	32.55	30.58	31.35
LN(0.423)	99.73	99.69	99.73	99.73	99.70	99.70
IG(4.5)	99.49	99.40	99.56	99.51	99.40	99.44

Table 10.2. *Size (S) and powers (%) for the symmetric and right skewed models, $n = 50$*

In conclusion, for every sample size and the right skewed model, $T_{\Phi_1}^{0.01}(\hat{\boldsymbol{\theta}}_{\Phi_2}^{0.50})$ appears to be the best choice among all tests considered, while for the symmetric model $T_{\Phi_1}^{0.50}(\hat{\boldsymbol{\theta}}_{\Phi_2}^{0.01})$ should be preferred. For these two models, $X^2(\hat{\boldsymbol{\theta}}_{X^2})$ appears to have a slight edge against the Weibull (Wei(0.77)) alternative. Although for small sample sizes and the left skewed model $X^2(\hat{\boldsymbol{\theta}}_{X^2})$ appears to have the best performance among all tests in most cases, when the sample size increases the best choice for every alternative is the $T_{\Phi_1}^{0.01}(\hat{\boldsymbol{\theta}}_{\Phi_2}^{0.50})$, while for alternatives such as the inverse Gaussian and the log-normal, the best choice is $T_{\Phi_1}^{0.50}(\hat{\boldsymbol{\theta}}_{\Phi_2}^{0.01})$. Finally, for the equiprobable model, $X^2(\hat{\boldsymbol{\theta}}_{X^2})$ has the best performance among all tests in most cases but as the sample size increases, $T_{\Phi_1}^{0.01}(\hat{\boldsymbol{\theta}}_{\Phi_2}^{0.50})$ has slightly better power against the Weibull (Wei(0.77)) alternative.

We conclude this section with a discussion about the number m of classes required in tests of fit. The problem of determining the optimum number of classes has a long history. Mann and Wald (1942), Cochran (1952), Dahiya and Gurland (1973) and Harrison (1985) proposed similar techniques for the case of the chi-squared test, which arrive at diverse conclusions with recommendations ranging from 3 to 24 classes. In almost all these cases, the optimum number of classes depends on various factors like the type of the alternative hypothesis, the "distance" between the alternative hypothesis and the data, the minimum power to be achieved, the significance level of the test and the sample size. One should always be aware of the danger associated with the use of too many classes in cases where the observations are spread too thinly over the data range. The power of the test together with the size are the key factors in deciding if the increase of the number of classes is useful. It should be noted that even in cases, where the number of classes is allowed to increase as the sample size increases, specific assumptions should be imposed in order to secure satisfactory asymptotic results. One such assumption is the well-known sparseness assumption according to which for $m = m(n)$, $\lim_{n\to\infty} n/m = c < \infty$ (see Holst 1972; Dale 1986; Read and Cressie 1988). Typically, for classes at most equal to 10, the results are quite satisfactory.

10.7. References

Akaike, H. (1973). Information theory and an extension of maximum likelihood principle. In *Proceedings of the 2nd International Symposium on Information Theory*, Petrov, B.N. and Csaki, F. (eds). Akademiai Kiado, Budapest, 267–281.

Ali, S.M., and Silvey, S.D. (1966). A general class of coefficients of divergence of one distribution from another. *Journal of the Royal Statistical Society, Series B (Methodological)*, 28(1), 131–142.

Alin, A., and Kurt, S. (2008). Ordinary and penalized minimum power-divergence estimators in two-way contingency tables. *Computational Statistics*, 23(3), 455–468.

Basu, A., Harris, I.R., Hjort, N.L., and Jones, M.C. (1998). Robust and efficient estimation by minimising a density power divergence. *Biometrika*, 85(3), 549–559.

Basu, A., Mandal, A., and Pardo, L. (2010). Hypothesis testing for two discrete populations based on the Hellinger distance. *Statistics & Probability Letters*, 80(3), 206–214.

Bekara, M., Knockaert, L., Seghouane, A.-K., and Fleury, G. (2006). A model selection approach to signal denoising using Kullback's symmetric divergence. *Signal Processing*, 86(7), 1400–1409.

Birch, M.W. (1964). A new proof of the Pearson-Fisher theorem. *The Annals of Mathematical Statistics*, 35(2), 817–824.

Bregman, L. (1967). The relaxation method of finding the common point of convex sets and its application to the solution of problems in convex programming. *USSR Computational Mathematics and Mathematical Physics*, 7(3), 200–217.

Broniatowski, M., Toma, A., and Vajda, I. (2012). Decomposable pseudodistances and applications in statistical estimation. *Journal of Statistical Planning and Inference*, 142(9), 2574–2585.

Cavanaugh, J.E. (2004). Criteria for linear model selection based on Kullback's symmetric divergence. *Australian & New Zealand Journal of Statistics*, 46(2), 257–274.

Cochran, W.G. (1952). The χ^2 test of goodness of fit. *The Annals of Mathematical Statistics*, 23(3), 315–345.

Cressie, N., and Read, T.R.C. (1984). Multinomial goodness-of-fit tests. *Journal of the Royal Statistical Society. Series B (Methodological)*, 46(3), 440–464.

Csiszár, I. (1963). Eine informationstheoretische ungleichung und ihre anwendung auf beweis der ergodizitaet von markoffschen ketten. *Magyer Tud. Akad. Mat. Kutato Int. Koezl.*, 8, 85–108.

Dahiya, R.C., and Gurland, J. (1973). How many classes in the Pearson chi-square test? *Journal of the American Statistical Association*, 68(343), 707–712.

Dale, J.R. (1986). Asymptotic normality of goodness-of-fit statistics for sparse product multinomials. *Journal of the Royal Statistical Society. Series B (Methodological)*, 48(1), 48–59.

Fischer, A. (2010). Quantization and clustering with Bregman divergences. *Journal of Multivariate Analysis*, 101(9), 2207–2221.

Harrison, R.H. (1985). Choosing the optimum number of classes in the chi-square test for arbitrary power levels. *Sankhy: The Indian Journal of Statistics, Series B*, 47(3), 319–324.

Holst, L. (1972). Asymptotic normality and efficiency for certain goodness-of-fit tests. *Biometrika*, 59(1), 137–145.

Jiménez-Gamero, M.D., and Batsidis, A. (2017). Minimum distance estimators for count data based on the probability generating function with applications. *Metrika*, 80(5), 503–545.

Jones, M.C., Hjort, N.L., Harris, I.R., and Basu, A. (2001). A comparison of related density-based minimum divergence estimators. *Biometrika*, 88(3), 865–873.

Kagan, A.M. (1963). On the theory of Fisher's amount of information. *Soviet Mathematics Doklady*, 4, 991–993.

Kateri, M. (2018). ϕ-divergence in contingency table analysis. *Entropy*, 20(324).

Koutrouvelis, I.A., and Karagrigoriou, A. (2012). Cumulant plots and goodness-of-fit tests for the inverse Gaussian distribution. *Journal of Statistical Computation and Simulation*, 82(3), 343–358.

Kullback, S., and Leibler, R.A. (1951). On information and sufficiency. *The Annals of Mathematical Statistics*, 22(1), 79–86.

Kupperman, M. (1957). Further application to information theory to multivariate analysis and statistical inference. PhD Thesis, George Washington University.

Kupperman, M. (1958). Probabilities of hypotheses and information-statistics in sampling from exponential-class populations. *The Annals of Mathematical Statistics*, 29(2), 571–575.

Liese, F., and Vajda, I. (1987). *Convex Statistical Distances*. Teubner-Texte sur Mathematik, Leibzig.

Mann, H.B., and Wald, A. (1942). On the choice of the number of class intervals in the application of the chi square test. *The Annals of Mathematical Statistics*, 13(3), 306–317.

Mantalos, P., Mattheou, K., and Karagrigoriou, A. (2010). Forecasting ARMA models: A comparative study of information criteria focusing on MDIC. *Journal of Statistical Computation and Simulation*, 80(1), 61–73.

Mattheou, K., and Karagrigoriou, A. (2010). A new family of divergence measures for tests of fit. *Australian & New Zealand Journal of Statistics*, 52(2), 187–200.

Mattheou, K., Lee, S., and Karagrigoriou, A. (2009). A model selection criterion based on the BHHJ measure of divergence. *Journal of Statistical Planning and Inference*, 139(2), 228–235.

Matusita, K. (1967). On the notion of affinity of several distributions and some of its applications. *Annals of the Institute of Statistical Mathematics*, 19, 181–192.

Menéndez, M., Morales, D., Pardo, L., and Vajda, I. (2001). Approximations to powers of φ-disparity goodness-of-fit tests. *Communications in Statistics–Theory and Methods*, 30(1), 105–134.

Meselidis, C., and Karagrigoriou, A. (2020). Statistical inference for multinomial populations based on the (Φ, α)-power divergence family. *Journal of Statistical Computation and Simulation*. Available at: https://doi.org/10.1080/00949655.2020.1747069.

Morales, D., Pardo, L., and Vajda, I. (1995). Asymptotic divergence of estimates of discrete distributions. *Journal of Statistical Planning and Inference*, 48(3), 347–369.

Nayak, T.K. (1983). Applications of entropy functions in measurement and analysis of diversity. PhD Thesis, University of Pittsburgh.

Nayak, T.K. (1985). On diversity measures based on entropy functions. *Communications in Statistics–Theory and Methods*, 14(1), 203–215.

Neath, A.A., Cavanaugh, J.E., and Weyhaupt, A.G. (2015). Model evaluation, discrepancy function estimation, and social choice theory. *Computational Statistics*, 30(1), 231–249.

Neyman, J., Pearson, E.S. (1928), On the use and interpretation of certain test criteria for purposes of statistical inference: Part I. *Biometrika*, 20A(1/2), 175–240.

Pearson, K. (1900). On the criterion that a given system of deviations from the probable in the case of a correlated system of variables is such that it can be reasonably supposed to have arisen from random sampling. *Philosopy Magazine*, 50, 157–172.

Qiao, Y., and Minematsu, N. (2010). A study on invariance of f-divergence and its application to speech recognition. *IEEE Transactions on Signal Processing*, 58(7), 3884–3890.

Rathie, P., and Kannappan, P. (1972). A directed-divergence function of type β. *Information and Control*, 20(1), 38–45.

Read, T.R., and Cressie, N.A. (1988). *Goodness-of-Fit Statistics for Discrete Multivariate Data*. Springer-Verlag, New York.

Rényi, A. (1961). On measures of entropy and information. In *Proceedings of the Fourth Berkeley Symposium on Mathematical Statistics and Probability, Volume 1: Contributions to the Theory of Statistics*. University of California Press, Berkeley, A, 547–561.

Shang, J., and Cavanaugh, J.E. (2008). Bootstrap variants of the Akaike information criterion for mixed model selection. *Computational Statistics & Data Analysis*, 52(4), 2004–2021.

Shiva, S.G.S., Ahmed, N.U., and Georganas, N.D. (1973). Order preserving measures of information. *Journal of Applied Probability*, 10(3), 666–670.

Simon, G. (1973). Additivity of information in exponential family probability laws. *Journal of the American Statistical Association*, 68(342), 478–482.

Stehlík, M., Střelec, L., and Thulin, M. (2014). On robust testing for normality in chemometrics. *Chemometrics and Intelligent Laboratory Systems*, 130, 98–108.

Stehlík, M., Economou, P., Papi, L., Aronov, J., Nicolis, O., Antoch, J., Cézová, E., and Kiseľák, J. (2018). Statistical testing of availability for mining technological systems with air quality constraints. *Applied Stochastic Models in Business and Industry*, 34(3), 278–292.

Toma, A. (2014). Model selection criteria using divergences. *Entropy*, 16(5), 2686–2698.

Toma, A., and Leoni-Aubin, S. (2010). Robust tests based on dual divergence estimators and saddlepoint approximations. *Journal of Multivariate Analysis*, 101(5), 1143–1155.

Toma, A., and Leoni-Aubin, S. (2014). Minimum pseudodistance estimators and applications to portfolio optimization. In *New Perspectives on Stochastic Modeling and Data Analysis*, Bozeman, V.J.R. and Skiadas, C. (eds). ISAST International, Athens, Greece, 79–86.

Toma, A., Karagrigoriou, A., and Trentou, P. (2020). Robust criteria for model selection using pseudodistances. *Entropy*, 22, 304.

Vonta, F., Mattheou, K., and Karagrigoriou, A. (2012). On properties of the (Φ, a)-power divergence family with applications in goodness of fit tests. *Methodology and Computing in Applied Probability*, 14(2), 335–356.

Zografos, K., Ferentinos, K., and Papaioannou, T. (1990). Divergence statistics: Sampling properties and multinomial goodness of fit and divergence tests. *Communications in Statistics–Theory and Methods*, 19(5), 1785–1802.

11

Goodness-of-Fit Tests Based on Divergence Measures for Frailty Models

In this chapter, we first review results from Vonta and Karagrigoriou (2014) where new goodness-of-fit tests based on φ-divergence measures were defined for censored data. These results are then extended to the case where there is heterogeneity in the data described by a random effect called frailty. The goodness-of-fit tests are extended in order to test under the null hypothesis the fit of a frailty model to right censored data. The performance of the tests is assessed through simulations.

11.1. Introduction

Measures of divergence between two probability distributions have a very long history initiated by the pioneering work of Pearson, Mahalanobis, Lévy and Kolmogorov. Among the most popular measures of divergence are the Kullback–Leibler measure of divergence (Kullback and Leibler 1951) and the Csiszár's φ-divergence family of measures (Csiszár 1964; Ali and Silvey 1966). A unified analysis has been provided by Cressie and Read (1984) who introduced the power divergence family of statistics that depends on a parameter a and is used for goodness-of-fit tests for multinomial distributions. The Cressie and Read family includes among others the well-known Pearson's χ^2 divergence measure and for multinomial models the loglikelihood ratio statistic. The BHHJ divergence measure was proposed by Basu *et al.* (1998) and generalized to the BHHJ family of measures by Mattheou *et al.* (2009). The BHHJ family depends on an index a that controls the trade-off between robustness and efficiency when the measure is used as an estimating criterion for robust parameter estimation.

Chapter written by Filia VONTA.

In statistics, the problem of determining the appropriate distribution or the appropriate model for a given data set is extremely important for reducing the possibility of wrong inference. Additional issues are raised in biomedicine and biostatistics. Indeed, the existence of censoring schemes in survival modeling and/or the existence of heterogeneity in the data makes the determination of the proper distribution or model an extremely challenging problem. An important aspect is how to check the validity of a specific model assumption. Some research has been done in this regard. For example, Gail and Ware (1979) studied grouped censored survival data by comparing it with a known survival distribution, while Akritas (1988) constructed a Pearson-type goodness-of-fit measure for one-sample data that allows for random censorship and Chen *et al.* (2004) proposed a test of fit based on the Cressie and Read power divergence measure for censored data. Vonta and Karagrigoriou (2014) focus on a class of goodness-of-fit tests based on Csiszár's family of measures and propose a general family of test statistics for treating the case of right-censored data with applications that can be extended from survival analysis to reliability theory.

In this study, we review the results of Vonta and Karagrigoriou (2014) and we extend these results to the case of frailty models and propose a class of test statistics that can be used to test whether the data follow a frailty model.

11.2. The proposed goodness-of-fit test

Let X be an absolutely continuous, non-negative random variable that describes the lifetime of an individual or an item. Let $f(x)$, $F(x)$ and $h(x)$ be the density function, the cumulative distribution function (cdf) and the hazard intensity function of X, respectively. If we would like to test the null hypothesis that the distribution function of X is equal to F_0, we could utilize divergence measures for this purpose.

For example, the Kullback–Leibler distance between F and F_0 (Kullback and Leibler 1951) is defined by

$$I_{F,F_0}^{KL} = \int_0^\infty f(x) \log\left(\frac{f(x)}{f_0(x)}\right) dx \qquad [11.1]$$

where log denotes the natural logarithm and f_0 is the density under the null hypothesis. Without loss of generality, we assume that the support of f and f_0 is $(0, +\infty)$.

A generalization of this distance is defined by (Csiszár 1964)

$$I_{F,F_0}^{\varphi} = \int_0^\infty f_0(x)\varphi\left(\frac{f(x)}{f_0(x)}\right) dx \qquad [11.2]$$

and is known as Csiszár's family of measures of divergence for some function φ.

Equivalently, for the discrete data case, let $\mathbf{x} = (x_1, \ldots, x_k)'$ be a random vector that follows a multinomial distribution with parameters N and $\mathbf{p}$, where $\mathbf{p} = (p_1, \ldots, p_k)'$ is a probability vector. Then $\sum_{i=1}^{k} x_i = N$ and $\sum_{i=1}^{k} p_i = 1$. Consider the null hypothesis $H_0 : \mathbf{p} \in \mathcal{P}_0$, where $\mathcal{P}_0$ is the set of possible values for $\mathbf{p}$ under the null hypothesis. Then, the appropriate φ-divergence measure based on Csiszár's class of functions is defined by

$$I_{F,F_0}^{\varphi} = \sum_{i=1}^{k} p_{i0}\varphi\left(\frac{\hat{p}_i}{p_{i0}}\right) \qquad [11.3]$$

where $\hat{\mathbf{p}} = \mathbf{x}/N = (\hat{p}_1, \ldots, \hat{p}_k)' = (x_1/N, \ldots, x_k/N)'$ is the MLE of $\mathbf{p}$ under the null hypothesis $H_0 : \mathbf{p} = (p_{10}, \ldots, p_{k0})'$. The test statistic associated with the above measure is given by

$$T_{F,F_0}^{\varphi} = \frac{2N}{\varphi''(1)} I_{F,F_0}^{\varphi}. \qquad [11.4]$$

We assume from now on that the function φ belongs to a class of functions Φ with properties:

1) $\varphi(x)$ is continuous, differentiable and convex for $x \geq 0$;

2) $\varphi(1) = 0$;

3) $\varphi'(1) = 0$.

From the above assumptions, we deduce that $\varphi(x) \geq 0$ for all $x > 0$ and $\varphi'(x) > 0$ for $x > 1$ and $\varphi'(x) < 0$ for $x < 1$.

When the function φ is defined as $\varphi(u) = u \log u$ or $\varphi(u) = u \log u + 1 - u$, measure [11.3] reduces to the discrete version of the Kullback–Leibler measure of divergence, which is defined as

$$I_{F,F_0}^{KL} = \sum_{i=1}^{k} \hat{p}_i \log\left(\frac{\hat{p}_i}{p_{i0}}\right). \qquad [11.5]$$

If $\varphi(u) = \frac{1}{2}(1-u)^2$, Csiszár's measure [11.3] yields the Pearson's chi-square divergence, also known as Kagan's divergence (Kagan 1963). If $\varphi(u) = (1-\sqrt{u})^2$, we obtain the Matusita's divergence (Matusita 1967), and if $\varphi(u) = -\log(u) + u - 1$ or $\varphi(u) = -\log(u)$, we obtain the Kullback–Leibler divergence, also known as the minimum discrimination information. Finally, if

$$\varphi(u) \doteq \varphi_1(u) = \left(u^{a+1} - u - a(u-1)\right)/(a(a+1)) \qquad [11.6]$$

with $a \neq 0, -1$ we obtain the Cressie and Read power divergence test statistic (Cressie and Read 1984) given by

$$T_{F,F_0}^{CR}(a) = \frac{2}{a(a+1)} \sum_{i=1}^{k} x_i \left(\left(\frac{\hat{p}_i}{p_{i0}}\right)^a - 1\right), \; a \neq 0, -1. \qquad [11.7]$$

Other functions that we consider are

$$\varphi(u) \doteq \varphi_2(u) = u^{1+a} - (1+\frac{1}{a})u^a + \frac{1}{a}, \quad a \neq 0 \qquad [11.8]$$

and

$$\varphi(u) \doteq \varphi_3(u) = 1 - (1+\frac{1}{a})u + \frac{u^{1+a}}{a}, \quad a \neq 0. \qquad [11.9]$$

The function φ_2 is related to a recently proposed measure of divergence (see Basu *et al.* 1998, BHHJ power divergence) while both $\varphi_2(\cdot)$ and $\varphi_3(\cdot)$ are special cases of the BHHJ family of divergence measures proposed by Mattheou *et al.* (2009), the discrete version of which is given by

$$I_{F,F_0}^{BHHJ} = \sum_{i=1}^{k} p_{i0}^{1+\alpha}\varphi\left(\frac{\hat{p}_i}{p_{i0}}\right), \ \alpha \geq 0 \qquad [11.10]$$

and the continuous, for $\alpha \geq 0$, by

$$I_X^{\alpha}(f_0, f) = E_{f_0}\left(f_0^{\alpha}(X)\varphi\left(\frac{f(X)}{f_0(X)}\right)\right) = \int f_0^{1+\alpha}(z)\varphi\left(\frac{f(z)}{f_0(z)}\right) dz. \qquad [11.11]$$

Appropriately chosen functions $\varphi(\cdot)$ give rise to known measures, some of which were mentioned above, while for $\alpha = 0$ the BHHJ family reduces to the Csiszár's family (discrete version) given by [11.3]. The test statistic associated with [11.10] is given by

$$T_{F,F_0}^{\varphi} = \frac{2N}{\varphi''(1)} I_{F,F_0}^{BHHJ}. \qquad [11.12]$$

REMARK 11.1.– Let $\Phi^\star \subset \Phi$ the class of functions $\varphi(\cdot)$ satisfying properties (1) and (2) above. It is not difficult to see that for every function $\varphi \in \Phi^\star$ which is differentiable at 1, one can always define the function $\psi(u) = \varphi(u) - \varphi'(1)(u-1)$ such that $\psi(\cdot) \in \Phi$, $\psi'(1) = 0$ and $I_{F,F_0}^{\psi} = I_{F,F_0}^{\varphi}$. Observe that for the function $\varphi(u) = u \log u \in \Phi^\star$, the discrepancy measure [11.4] is reduced to the measure [11.5]. In this case, the function $\psi \in \Phi$ for which $I^\psi = I^\varphi$ is given by $\psi(u) = u \log u + 1 - u$. As a result, from now on and without loss of generality we focus solely on functions that belong to the Csiszár family, that is, the class of functions Φ that satisfy all three conditions (1)—(3). Finally, we observe that for the function $\varphi(u) = -\log(u) + u - 1$, the discrepancy measure [11.4] is equal to the measure $I_{F_0,F}^{\varphi}$.

We turn now to the formulation of our problem for the censored case. Let us assume that the subjects are observed during a time interval $[0, \tau]$, which is partitioned into k subintervals of the form $E_i = (\tau_{i-1}, \tau_i]$, $i = 1, \ldots, k$ with $0 = \tau_0 < \tau_1 < \ldots < \tau_k = \tau$ in a fashion similar to the one used for the standard

Kaplan–Meier estimation method. The sole difference is that we assume that the censoring events instead of the deaths, occur at τ_i, $i = 1, \ldots, k$. Let n_i be the number of subjects at risk at the beginning of the ith interval and d_i the number of failures during the ith interval. Let $h_i^c = 1 - h_i$ and h_i be the survival and hazard rate for the ith interval. For each i, let $\mathcal{F}_i = \sigma\{d_1, \ldots, d_{i-1}, n_1, \ldots, n_i\}$ be the σ-field generated by $d_1, \ldots, d_{i-1}, n_1, \ldots, n_i$. Then, conditionally on $\mathcal{F}_i$, the number of failures d_i follows a binomial distribution:

$$P(d_i = j|\mathcal{F}_i) = \binom{n_i}{j} h_i^j (1 - h_i)^{n_i - j}, \; j = 0, \ldots, n_i. \qquad [11.13]$$

It is important to point out that any type of data can be allocated in subintervals by dividing the range of data into k categories or groups. Various techniques can be used for constructing subintervals such as those based on equidistant or equiprobable intervals or those based on data-driven methods. In that sense, data related to biomedicine, engineering, economics and finance that usually come from continuous distributions can be transformed into grouped data and goodness-of-fit based on the above (discrete) measures can be applied. The proper modeling of such data plays a key role in determining the optimum goodness-of-fit test. In that sense, the search for the ideal value of k, namely the optimal number of categories, is vital so that the resulting test of fit for various null hypotheses given, will result in achieving the nominal size as well as the maximum power for a wide range of possible alternatives.

In our case, suppose that there are N subjects with failure and censoring time denoted by X_i and C_i, $i = 1, \ldots, N$, respectively, with F the distribution function of X_i. The data consist of pairs (T_i, δ_i), where $T_i = \min(X_i, C_i)$ and $\delta_i = I_{[T_i \leq C_i]}$, $i = 1, \ldots, N$. Under the assumption that censoring occurs only at τ_i, [11.13] holds with

$$h_i = (F(\tau_i) - F(\tau_{i-1}))/(1 - F(\tau_{i-1})), \qquad i = 1, \ldots, k. \qquad [11.14]$$

We now propose for the simple null hypothesis

$$H_0 : F = F_0 \text{ or equivalently } H_0 : \boldsymbol{h} = \boldsymbol{h}_0 = (h_{10}, \ldots, h_{k0}),'$$

a general class of test statistics for censored data based on Csiszár's family of measures defined as

$$D^{\varphi}(\mathbf{d},\mathbf{n},\mathbf{h}) = \frac{2}{\varphi''(1)} \sum_{i=1}^{k} n_i \left\{ h_i \varphi\left(\frac{d_i/n_i}{h_i}\right) + h_i^c \varphi\left(\frac{d_i^c/n_i}{h_i^c}\right) \right\} \qquad [11.15]$$

where $\mathbf{d} = (d_1, \ldots, d_k)'$, $\mathbf{n} = (n_1, \ldots, n_k)'$, $\mathbf{h} = (h_1, \ldots, h_k)'$, $d_i^c = n_i - d_i$ and $h_i^c = 1 - h_i$. Observe that the proposed test statistic is a summation of k Csiszár's test statistics as in [11.4].

Although a simple null hypothesis suggested above appears frequently in practice, it is quite common to test the composite null hypothesis that the unknown distribution belongs to a parametric family $\{F_{\boldsymbol{\theta}}\}_{\boldsymbol{\theta}\in\Theta}$, where Θ is an open subset in R^m. In this case, we can again consider a partition of the original sample space with k disjoint intervals. In such a case, the above general class of test statistics takes the form

$$D^{\varphi}(\mathbf{d},\mathbf{n},\mathbf{h}(\boldsymbol{\theta})) = \frac{2}{\varphi''(1)}\sum_{i=1}^{k} n_i \left\{ h_i(\boldsymbol{\theta})\varphi\left(\frac{d_i/n_i}{h_i(\boldsymbol{\theta})}\right) + h_i^c(\boldsymbol{\theta})\varphi\left(\frac{d_i^c/n_i}{h_i^c(\boldsymbol{\theta})}\right)\right\} \qquad [11.16]$$

Let $\mathcal{H} = \{\boldsymbol{h}(\boldsymbol{\theta}) = (h_1(\boldsymbol{\theta}), \dots, h_k(\boldsymbol{\theta})) \ : \ \boldsymbol{\theta} \in \Theta\}$ and assume that $h_i(\boldsymbol{\theta})$ are twice continuously differentiable in θ. Let finally $\boldsymbol{\theta}_0$ denote the true parameter and $\mathbf{h}_0 = \mathbf{h}(\boldsymbol{\theta}_0)$.

Observe that in the composite case, the probabilities of failure in the various intervals depend on the unknown m–dimensional parameter $\boldsymbol{\theta}$ so that a consistent estimator $\hat{\boldsymbol{\theta}}$ of $\boldsymbol{\theta}$ is required. In regard to the estimating method applied for obtaining such an estimator, the traditional maximum likelihood estimator (MLE), under the null distribution, can be evaluated and implemented. Note that one may alternatively consider the wider class of φ–divergence estimators. More specifically, for the partition $\{E_i\}_{i=1,\dots,k}$ of the original sample space, the minimum φ–divergence estimator of $\boldsymbol{\theta}$ under the null hypothesis

$$H_0 : F = F_0(\boldsymbol{\theta}) \text{ or equivalently } H_0 : \boldsymbol{h} = \boldsymbol{h}_0(\boldsymbol{\theta}) = (h_{10}(\boldsymbol{\theta}), \dots, h_{k0}(\boldsymbol{\theta}))'$$

is any $\hat{\boldsymbol{\theta}}_{\varphi} \in \Theta$ satisfying

$$\hat{\boldsymbol{\theta}}_{\varphi} = \arg\min_{\boldsymbol{\theta}\in\Theta}\sum_{i=1}^{k} n_i \left\{ h_{i0}(\boldsymbol{\theta})\varphi\left(\frac{d_i/n_i}{h_{i0}(\boldsymbol{\theta})}\right) + h_{i0}^c(\boldsymbol{\theta})\varphi\left(\frac{d_i^c/n_i}{h_{i0}^c(\boldsymbol{\theta})}\right)\right\}. \qquad [11.17]$$

Obviously, the resulting estimator depends on the φ-function chosen. Observe that for φ having the special form given in [11.9] and for $a \to 0$ the resulting estimator is the usual maximum likelihood estimator for the grouped data. Note also that for φ as in [11.8], the measure [11.16] is related to a censored data version of the BHHJ measure of divergence. Finally, it should be pointed out that the function φ used for the estimation of the unknown parameter in [11.17] is not necessarily the same as the function φ used for the test statistic [11.16].

11.3. Main results

In this section, we provide the asymptotic distribution of the proposed test statistic for both under a simple and under a composite null hypothesis. We also establish the asymptotic distribution for a specific type of alternatives known as local contiguous alternatives. In what follows, we use for simplicity the notation θ and θ_0 instead of $\boldsymbol{\theta}$ and $\boldsymbol{\theta}_0$ and $\hat{\theta}_{\varphi}$ for the estimator of θ with $\varphi \in \Phi$ not necessarily being the same as the function φ in [11.16].

Simple null hypothesis

THEOREM 11.1.– Under assumption [11.13] and if $n_i \to \infty$ for $i = 1, \ldots, k$, the proposed test statistic $D^{\varphi}(\mathbf{d},\mathbf{n},\mathbf{h})$ defined in [11.15], under the null hypothesis H_0 : $\mathbf{h} = \mathbf{h}_0 = (h_{10}, \ldots, h_{k0})'$, converges in distribution to a χ^2_k random variable.

Composite null hypothesis

THEOREM 11.2.– Suppose $n_i/N \to \gamma_i \in (0,1), i = 1, \ldots, k$. Assume that the $m \times m$ matrix

$$V = \sum_{i=1}^{k} \frac{\gamma_i \nabla_\theta h_i(\theta) \nabla_\theta h_i(\theta)^T}{h_i(\theta) h_i^c(\theta)}\bigg|_{\theta=\theta_0}$$

is strictly positive definite. Then, under assumption [11.13] in a neighborhood of θ_0, $\hat{\theta}_\varphi$ given in [11.17] is uniquely defined and

$$\sqrt{N}(\hat{\theta}_\varphi - \theta_0) \xrightarrow{\mathcal{L}} \mathcal{N}(0, V^{-1}).$$

The following proposition is a natural consequence of theorem 11.2 and Slutsky's theorem.

PROPOSITION 11.1.– Under the assumptions of theorem 11.2, we have

$$\sqrt{N}\sqrt{\hat{V}_N}(\hat{\theta}_\varphi - \theta_0) \xrightarrow{\mathcal{L}} \mathcal{N}(0, I_m) \qquad [11.18]$$

where

$$\hat{V}_N = \sum_{i=1}^{k} \frac{n_i \nabla_\theta h_i(\theta) \nabla_\theta h_i(\theta)^T}{N h_i(\theta) h_i^c(\theta)}\bigg|_{\theta=\hat{\theta}_\varphi}.$$

THEOREM 11.3.– Under the assumptions of theorem 11.2, the asymptotic distribution of the test statistic [11.16] under the composite null hypothesis is given by

$$D^{\varphi}(\mathbf{d},\mathbf{n},\mathbf{h}(\hat{\theta}_\varphi)) \xrightarrow{\mathcal{L}} \chi^2_{k-m}. \qquad [11.19]$$

Contiguous alternatives

Consider now the hypothesis

$$H_0 : h_i = h_{i0} \ vs. \ H_1 : h_i = h_{ib}, \ i = 1, \ldots, k.$$

Suppose that the null hypothesis indicates that $h_i = h_{i0}$, $i = 1, 2, \ldots, k$ when in fact it is $h_i = h_{ib}$, $\forall i$. As it is well known if h_{i0} and h_{ib} are fixed, then as the sample size tends to infinity the power of the test tends to 1. In order to examine the situation when the power is not close to 1, we must make it continually difficult for the test as the sample size increases. This can be done by allowing the alternative hypothesis steadily closer to the null hypothesis. As a result, we define a sequence of alternative hypotheses as follows:

$$H_{1,N} : h_i = h_{iN} = h_{i0} + \beta_i/\sqrt{n_i}, \ \forall i \qquad [11.20]$$

where β_i are constants, which is known as Pitman transition alternative or Pitman (local) alternative or local contiguous alternative Akritas (1988) to the null hypothesis $H_0 : h_i = h_{i0}$. In vector notation, the null hypothesis and the local contiguous alternative hypotheses take the form

$$H_0 : \boldsymbol{h} = \boldsymbol{h}_0 \ vs. \ H_{1,N} : \boldsymbol{h} = \boldsymbol{h}_N = \boldsymbol{h}_0 + \boldsymbol{\beta},$$

where $\boldsymbol{h}_N = (h_{1N}, h_{2N}, \ldots, h_{kN})'$ and $\boldsymbol{\beta} = (\beta_1/\sqrt{n_1}, \ldots, \beta_k/\sqrt{n_k})'$ is a fixed vector such that $\sum_{i=1}^{k} h_{i0} + \beta_i/\sqrt{n_i} \leq k$. Observe that as n_i tends to infinity $\forall i$, the local contiguous alternative converges to the null hypothesis at the rate $O(N^{-1/2})$.

In order to derive the asymptotic distribution of the test statistic [11.16] under the local contiguous alternatives $H_{1,N}$, we first obtain the asymptotic distribution of $\hat{p}_i \equiv d_i/n_i$.

THEOREM 11.4.– Under the local contiguous alternative hypotheses [11.20] and assumption [11.13], we have

$$(\hat{\boldsymbol{P}} - \boldsymbol{H}_0) \xrightarrow{\mathcal{L}} \mathcal{N}(\boldsymbol{\beta_0}, \Sigma)$$

where $\boldsymbol{\beta}_0 = (\beta_1, \ldots, \beta_k)'$, $\hat{\boldsymbol{P}} = (n_1^{1/2}\hat{p}_1, \ldots, n_k^{1/2}\hat{p}_k)'$ and $\boldsymbol{H}_0 = (n_1^{1/2}h_{10}, \ldots, n_k^{1/2}h_{k0})'$. The variance–covariance matrix is given by $\Sigma = [\sigma_{ij}]_{m \times m}$,

$$\sigma_{ij} = \begin{cases} h_{i0}h_{i0}^c, & i = j \\ 0 & , i \neq j \end{cases}.$$

LEMMA 11.1.– Suppose that $X \sim \mathcal{N}(\xi, Q)$ where $Q_{r_0 \times r_0}$ is a projection matrix of rank $r \leq r_0$ and $Q\xi = \xi$. Then, $X'X \sim \mathcal{X}^2_{r,\xi'\xi}$.

THEOREM 11.5.– Under assumption [11.13] and if $n_i/N \rightarrow \gamma_i \in (0,1)$ for $i = 1, \ldots, k$, the proposed test statistic $D^{\varphi}(\mathbf{d,n,h}_0)$, under the contiguous alternative hypotheses given in [11.20], converges asymptotically in distribution to a non-central $\chi_k^2(\xi)$ random variable with non-centrality parameter $\xi = \sum_{i=1}^{k} \beta_i^2/(h_{i0}h_{i0}^c)$.

REMARK 11.2.– The proofs of the above series of theorems and lemmas depend on assumption [11.13] and the use of Taylor expansions of the function φ, and φ' when necessary, around 1 in order for the test statistic $D^{\varphi}(\mathbf{d,n,h})$ or $D^{\varphi}(\mathbf{d,n,h}(\theta))$ to be written in a convenient way so that standard asymptotic theory techniques could be applied to them.

11.4. Frailty models

One method of analyzing the survival times of a group of individuals is by their hazard intensity function, which represents the rate of instantaneous failure of an individual. A widely used model in survival analysis is the Cox proportional hazards model (Cox 1972), for which the hazard intensity function is given by

$$h(t|\mathbf{X}) = h_0(t)e^{\mathbf{X}^\top \beta}$$

where $\beta \in R^p$ is the parameter of interest, $\mathbf{X}$ is a p-dimensional vector of covariates or explanatory variables and $h_0(t)$ is the unknown baseline hazard intensity function. Consider a Cox proportional hazards model with no covariates, that is,

$$h(t|\mathbf{X} = \mathbf{0}) = h_0(t)$$

denoted in the sequel $h(t)$ for convenience of notation. Equivalently, the marginal survival function for failure times is given as

$$S(t) = e^{-H(t)} \quad [11.21]$$

where $H(t)$ is the baseline cumulative hazard function.

A general class of models arises naturally from the Cox model when, in order to explain population heterogeneity, one introduces an unobservable frailty parameter η into the model through the hazard intensity function as follows:

$$h(t|\eta) = \eta h(t), \qquad \eta > 0.$$

Then,

$$S(t|\eta) = e^{-\eta H(t)}$$

and thus Vonta (1996)

$$S(t) = \int_0^\infty e^{-yH(t)} dF_\eta(y) = e^{-G(H(t))} \quad [11.22]$$

where $F_\eta(.)$ is the cdf of the positive random effect η and the function G is defined as

$$G(w) = -log\Big(\int_0^\infty e^{-yw} dF_\eta(y)\Big).$$

The concept of frailty was first introduced by Vaupel *et al.* (1979). The well-known Clayton–Cuzick model (Clayton and Cuzick 1985) is obtained when η is taken to be distributed with a Gamma distribution with mean 1 and variance b. The function G takes in this case the form $G(w) = \log(1 + bw)^{1/b}$. The case $G(w) = w$ reduces the model to the Cox proportional hazards model. More examples can be found in Hougaard (1986) and the references therein. The distribution function of η and therefore equivalently the function G are assumed from now on to be completely known. The function G is assumed to be a strictly increasing, concave function with $G(0) = 0$ and $G(\infty) = \infty$.

Following the formulation of the previous section, in [11.14] let us define the distribution function F or, equivalently the survival function S through the frailty model defined in [11.22]. Then,

$$h_i = (F(\tau_i) - F(\tau_{i-1}))/(1 - F(\tau_{i-1})) =$$

$$\frac{S(\tau_{i-1}) - S(\tau_i)}{S(\tau_{i-1})} = \frac{e^{-G(H(\tau_{i-1}))} - e^{-G(H(\tau_i))}}{e^{-G(H(\tau_{i-1}))}} \qquad [11.23]$$

for $i = 1, \ldots, k$.

We concentrate from now on, on the parametric frailty model, that is, the case where the cumulative hazard function (or equivalently the hazard function) is of known functional form except of a finite dimensional parameter $\boldsymbol{\theta} \in R^m$. More specifically, the null hypothesis that the data follow a parametric frailty model is defined similarly to the previous section as

$$H_0 : \boldsymbol{h} = \boldsymbol{h}(\boldsymbol{\theta}) = (h_1(\boldsymbol{\theta}), \ldots, h_k(\boldsymbol{\theta})) =$$

$$\left(\frac{e^{-G(H(\tau_0,\boldsymbol{\theta}))} - e^{-G(H(\tau_1,\boldsymbol{\theta}))}}{e^{-G(H(\tau_0,\boldsymbol{\theta}))}}, \ldots, \frac{e^{-G(H(\tau_{k-1},\boldsymbol{\theta}))} - e^{-G(H(\tau_k,\boldsymbol{\theta}))}}{e^{-G(H(\tau_{k-1},\boldsymbol{\theta}))}}\right). \qquad [11.24]$$

The test statistic that we propose is the same one used for the composite hypothesis in the previous section given in [11.16], namely,

$$D^{\varphi}(\mathbf{d},\mathbf{n},\mathbf{h}(\boldsymbol{\theta})) = \frac{2}{\varphi''(1)} \sum_{i=1}^{k} n_i \left\{ h_i(\boldsymbol{\theta}) \varphi\left(\frac{d_i/n_i}{h_i(\boldsymbol{\theta})}\right) + h_i^c(\boldsymbol{\theta}) \varphi\left(\frac{d_i^c/n_i}{h_i^c(\boldsymbol{\theta})}\right) \right\} \qquad [11.25]$$

with the hazard function defined as in [11.24].

11.5. Simulations

For the simulations, we considered a parametric frailty model for which the cumulative hazard function was assumed to be the one of an exponential model, namely, $H(t) = \theta t$ with $\theta = 1$. We considered two well-known frailty distributions, the Gamma$(1/c, 1/c)$ distribution and the inverse Gaussian IG(b, b) distribution. The corresponding G functions take the form

$$G(x, c) = (1/c)log(1 + cx) \qquad [11.26]$$

and

$$G(x, b) = \sqrt{4b(b + x)} - 2b \qquad [11.27]$$

with $c, b > 0$. The mean of the frailty parameter is 1 for both distributions and the variance is c for the Gamma model and $1/(2b)$ for the inverse Gaussian model. We assumed both small and large variances of the frailty parameter.

We considered five different φ functions, namely, the Kullback–Leibler, the Pearson, the Cressie and Read with the parameter α = 2/3 that is considered optimal for this function, the BHHJ function φ_2 with the parameter α = 10/9 (which is optimal according to Tsanousa (2010) and denoted from now on as φ_{21} and finally the BHHJ function φ_2 for α = 0.8, which we denote by φ_{22}. We considered in each case 20,000 simulations with various sample sizes n = 20, 50, 100, 200 in order to explore the case of small or large sample sizes. We considered as censoring proportions (denoted as censor) 10%, 30% and 50% in order to examine how the size of the censoring proportion is affecting the results. The cutoff points for the intervals E_i, $i = 1, \ldots, k$ are selected in such a way in order to have intervals of equal size based on the interval $[0, \tau]$, where τ was the maximum observation in the sample. The optimal number of the subintervals k seems to be dependent on the sample size n and we have used for $n = 20$, $k = 3$ or 5, while for $n = 50$, k was taken 3, 5 or 7. For $n = 100$ or 200, we used $k = 5, 7$ or 10. In our simulations, we have used a wide range of subintervals with k ranging from 3 to more than 10 in case of large sample sizes. Based on our results, we recommend that the value of k should not be very big in order to have enough observations in each subinterval for good estimation and for simplicity. We have found the use of $k = 3$ or 4 for $n = 20$, $k = 4$ or 5 for $n = 50$ and $k = 5$ or 7 for $n = 100$ or $n = 200$ to be optimal in that respect.

For the case of the frailty models, we discovered that the asymptotic distribution given in theorem 11.3 in the cases of large frailty variance has not been very satisfactory in order to find critical values for the proposed tests. This matter needs further investigation. We are presenting in Table 11.1 the 90th, 95th and 99th percentiles of the empirical distribution of the test statistic for the inverse Gaussian distribution with $b = 2$, which corresponds to the case of small variance, for various values n and k and for censoring proportion 50%. Similar tables such as Table 11.1 can be obtained for various values of n and k and any censoring proportion. We presented here the censoring proportion 50% because it is quite high and it is therefore a more challenging case for the proposed tests. In Table 11.2, we present for comparative purposes the corresponding percentiles of the asymptotic distribution.

n	k	$censor$	φ	$90th$	$95th$	$99th$
20	3	50	CR	4.66598	5.786858	8.723995
20	3	50	φ_{22}	4.942743	6.095835	9.075358
20	3	50	φ_{21}	4.483589	5.67788	8.685081
20	3	50	PR	4.60642	5.724714	8.67288
20	3	50	KB	5.629743	6.972144	9.875265
20	5	50	CR	7.660473	9.18156	12.73291
20	5	50	φ_{22}	8.072183	9.713788	13.35699
20	5	50	φ_{21}	7.448709	9.14944	12.71321
20	5	50	PR	7.555404	9.16733	12.77813
20	5	50	KB	9.14464	10.85669	14.81828
50	3	50	CR	4.657846	5.980671	8.725693
50	3	50	φ_{22}	4.840794	6.262405	9.217035
50	3	50	φ_{21}	4.415151	5.616867	8.453647
50	3	50	PR	4.514183	5.753471	8.51079
50	3	50	KB	5.366287	7.056272	10.55444
50	5	50	CR	7.725531	9.321537	12.97159
50	5	50	φ_{22}	8.080895	9.754848	13.55486
50	5	50	φ_{21}	7.43864	9.088949	13.00625
50	5	50	PR	7.556042	9.128345	13.01081
50	5	50	KB	9.04746	10.87835	15.05373
50	7	50	CR	10.26856	12.12561	16.33264
50	7	50	φ_{22}	10.80308	12.73131	17.13855
50	7	50	φ_{21}	10.09035	12.1512	16.95942
50	7	50	PR	10.19673	12.08631	16.54584
50	7	50	KB	12.08524	14.17972	18.49094
100	5	50	CR	8.138804	9.824922	13.26228
100	5	50	φ_{22}	8.474665	10.23182	13.77963
100	5	50	φ_{21}	7.708753	9.351616	13.0387
100	5	50	PR	7.864626	9.548602	13.10474
100	5	50	KB	9.49978	11.47385	15.48319
100	7	50	CR	10.72277	12.51354	16.32408
100	7	50	φ_{22}	11.17318	12.99713	16.91543
100	7	50	φ_{21}	10.32048	12.24789	16.56357
100	7	50	PR	10.48063	12.33386	16.39418
100	7	50	KB	12.42471	14.52161	18.86532
100	10	50	CR	14.34369	16.51612	21.49615
100	10	50	φ_{22}	15.02623	17.23647	22.39702
100	10	50	φ_{21}	14.25156	16.76366	22.41504
100	10	50	PR	14.23023	16.64651	21.9652
100	10	50	KB	16.69011	19.08121	24.36199

Table 11.1. *Empirical percentiles for the inverse Gaussian frailty with b = 2*

k	$90th$	$95th$	$99th$
3	4.60517	5.991465	9.21034
5	7.77944	9.487729	13.2767
7	10.64464	12.59159	16.81189
10	14.68366	16.91898	21.66599

Table 11.2. *Theoretical percentiles from the χ^2_{k-1} distribution*

11.5.1. *Linear models for the estimation of critical values*

From the previous tables, we observe that the empirical percentiles depend on the sample size n, the number k of intervals chosen and the censoring proportion. We continue by defining multiple linear models in order to find percentiles that do not depend on these parameters for a specific φ function, percentile value and frailty distribution. The dependent variable in these models consists of the empirical percentiles found for various values of n, k and censoring proportion (see Table 11.1). The independent variables are the parameters n, k and certain functions of n and k and the censoring proportion $censor$.

The general linear model considered is given by

$$y = \beta_0 + \beta_1 n + \beta_2 k + \beta_3 \sqrt{n} + \beta_4 \sqrt{k} + \beta_5 censor + \beta_6 \log(n) + \beta_7 \log(k) + \epsilon$$

Our goal is to discover the "best" model based on the AIC criterion from the point of view of not being very complicated and at the same time adequate to describe all cases. We have worked with both the inverse Gaussian and gamma frailty models and all five φ functions. Based on simulations we discovered those models that seem to work well for both frailty distributions, all five φ functions and for the three percentiles 90th, 95th and 99th. From all the models considered, we decided to continue with the model with independent variables

$$n, k, censor, \log(k)$$

because it is simple, resulted in big values of adjusted R^2 and minimized the AIC criterion in many cases. We will call this model from now on Model (I). We present in Tables 11.3 and 11.4 the estimated coefficients of Model (I) in order to find universal 95th percentiles of the test statistic for the case of the inverse Gaussian frailty distribution with parameter value $b = 0.125$ (small variance) and $b = 2$ (large variance), respectively, and for the five divergence measures.

We provide here as well the values of the coefficients of determination R^2 for each one of the models presented in Table 11.3, which are 0.983, 0.986, 0.975, 0.980 and 0.985, respectively. Note that the variable $\log(k)$ is not significant for φ_{21} and PR and this resulted in lower coefficients of determination from the other three models but not too low to prevent us from considering these models for the sake of uniformity between all cases. The values of the coefficients of determination R^2 for each one of the models presented in Table 11.4 are 0.982, 0.984, 0.984, 0.984 and 0.982, respectively. Although again we notice that the variable $\log(k)$ is not significant for φ_{21} and borderline for PR, but with much smaller p-values than in the previous table, we see that we have comparable coefficients of determination in all five models.

$coefficients$	φ	$intercept$	n	k	$censor$	$\log(k)$
	CR	-1.8228	-0.0055	1.2385	8.6892	2.2613
	φ_{22}	-1.8444	-0.0066	1.2935	8.2821	2.4598
	φ_{21}	-2.5671	-0.0077	1.9004	10.5067	1.4706
	PR	-2.2556	-0.0069	1.6162	9.8038	1.8099
	KB	-1.2525	-0.0045	1.1093	5.7119	2.9501
$p-values$	CR	0.05665	0.00173	1.28e-06	3.46e-15	0.07036
	φ_{22}	0.04072	0.00014	2.01e-07	2.29e-15	0.03760
	φ_{21}	0.09020	0.00555	2.40e-06	2.93e-12	0.44957
	PR	0.06674	0.00231	1.05e-06	9.48e-14	0.25281
	KB	0.13663	0.00357	1.18e-06	5.54e-12	0.01024

Table 11.3. *Coefficients and their significance (p-values) for Model (I) for the 95th percentile. Inverse Gaussian with b = 0.125*

$coefficients$	φ	$intercept$	n	k	$censor$	$\log(k)$
	CR	-2.6496	0.0104	0.8918	6.9603	2.3679
	φ_{22}	-2.5912	0.0103	0.9299	6.5161	2.6164
	φ_{21}	-2.7264	0.0053	1.0912	7.3142	1.7363
	PR	-2.6984	0.0075	0.9915	7.1484	2.0457
	KB	-2.3535	0.0138	0.9979	5.3660	3.2098
$p-values$	CR	0.00650	3.73e-07	0.00011	5.23e-13	0.05474
	φ_{22}	0.00636	2.97e-07	4.71e-05	1.38e-12	0.03099
	φ_{21}	0.00287	0.00115	2.16e-06	2.42e-14	0.12366
	PR	0.00331	2.20e-05	1.04e-05	5.07e-14	0.07387
	KB	0.03019	2.65e-08	0.000144	3.43e-09	0.02460

Table 11.4. *Coefficients and their significance (p-values) for Model (I) for the 95th percentile. Inverse Gaussian with b = 2*

11.5.2. *Size of the test*

In this section, we explore the behavior of the hypothesis test defined in [11.24] and [11.25], in terms of the size of the test. In order to derive the critical values of the test, we used the estimated coefficients of Model (I) that were obtained separately for the five φ functions, the two frailty distributions and percentile values.

n	k	Censor (%)	CR	φ_{22}	φ_{21}	Pearson	Kullback
20	3	10	0.07685	0.0684	0.07325	0.0782	0.06095
20	5	10	0.0552	0.0541	0.04995	0.05245	0.0541
50	3	10	0.0546	0.05025	0.06625	0.06345	0.0379
50	5	10	0.054	0.05165	0.05535	0.0544	0.0465
50	7	10	0.04975	0.04925	0.04965	0.0502	0.04715
100	5	10	0.04015	0.0401	0.04605	0.04345	0.0423
100	7	10	0.0423	0.04225	0.04605	0.04405	0.0401
100	10	10	0.04025	0.04055	0.0424	0.0421	0.03985
200	5	10	0.03915	0.04145	0.04155	0.041	0.04395
200	7	10	0.03725	0.0389	0.0387	0.03785	0.04125
200	10	10	0.03695	0.0389	0.0384	0.0373	0.0423
20	3	30	0.03585	0.0368	0.03255	0.03555	0.04345
20	5	30	0.03235	0.03375	0.0298	0.0311	0.03765
50	3	30	0.02605	0.025	0.03125	0.0284	0.02265
50	5	30	0.02945	0.0302	0.03205	0.0316	0.0307
50	7	30	0.0305	0.0322	0.03195	0.0315	0.0342
100	5	30	0.0203	0.02225	0.0241	0.0219	0.0265
100	7	30	0.02375	0.0251	0.0281	0.0266	0.02745
100	10	30	0.027	0.02895	0.03025	0.02875	0.0289
200	5	30	0.01855	0.02185	0.0181	0.0179	0.0298
200	7	30	0.01945	0.02275	0.02135	0.0199	0.02895
200	10	30	0.0222	0.0253	0.02735	0.02475	0.0315
20	3	50	0.0194	0.02155	0.0168	0.0189	0.0284
20	5	50	0.01815	0.02055	0.01965	0.0195	0.02635
50	3	50	0.0137	0.0144	0.01475	0.0142	0.0144
50	5	50	0.01785	0.01875	0.01925	0.01885	0.02045
50	7	50	0.0206	0.0223	0.02155	0.02155	0.0246
100	5	50	0.0118	0.0129	0.01485	0.01385	0.01585
100	7	50	0.0147	0.0163	0.0177	0.0161	0.0196
100	10	50	0.01765	0.02	0.02045	0.01965	0.0215
200	5	50	0.00925	0.0117	0.00915	0.00865	0.0194
200	7	50	0.01	0.01215	0.0124	0.0102	0.01925
200	10	50	0.0144	0.0162	0.01855	0.0161	0.0219

Table 11.5. *Size of the test: inverse Gaussian distribution with b = 2*

If, for example, we are interested in obtaining the critical value for the 5% level of significance, the inverse Gaussian frailty distribution with $b = 0.125$ and the Cressie–Read measure of divergence, we employ from Table 11.3 and Model (I) the model

$$\hat{y} = -1.8228 - 0.0055n + 1.2385k + 8.6892censor + 2.2613\log(k) \quad [11.28]$$

where $\hat{y}$ is the estimated critical value. Then by entering the desired values for n, k and $censor$ in model [11.28] we estimate through the regression model the critical value of the test statistic for $a = 5\%$. In order to obtain the size of the test, we run 20,000 simulations in each case. In Table 11.5, we present the size of the test for various cases for the inverse Gaussian distribution with $b = 2$.

From Table 11.5, we note that the sizes are good as long as the censoring proportion is small but they deteriorate as the censoring proportion increases, giving smaller than expected size regardless of the sample size or the number of subintervals chosen. No major differences exist among the five divergence measures φ considered. Similar results hold for the significance level 1% but are not shown here. More work needs to be done in terms of obtaining better size of the tests proposed as well as good power.

11.6. References

Akritas, M.G. (1988). Pearson-type goodness-of-fit tests: The univariate case. *Journal of the American Statistical Association*, 83(401), 222–230.

Ali, S.M. and Silvey, S.D. (1966). A general class of coefficients of divergence of one distribution from another. *Journal of the Royal Statistical Society: Series B (Methodological)*, 28(1), 131–142.

Basu, A., Harris, I.R., Hjort, N.L., and Jones, M. (1998). Robust and efficient estimation by minimising a density power divergence. *Biometrika*, 85(3), 549–559.

Chen, H.-S., Lai, K., and Ying, Z. (2004). Goodness-of-fit tests and minimum power divergence estimators for survival data. *Statistica Sinica*, 14, 231–248.

Clayton, D. and Cuzick, J. (1985). Multivariate generalizations of the proportional hazards model. *Journal of the Royal Statistical Society: Series A (General)*, 148(2), 82–108.

Cox, D.R. (1972). Regression models and life-tables. *Journal of the Royal Statistical Society: Series B*, 34(2), 187–202.

Cressie, N. and Read, T.R. (1984). Multinomial goodness-of-fit tests. *Journal of the Royal Statistical Society: Series B (Methodological)*, 46(3), 440–464.

Csiszár, I. (1964). Eine informationstheoretische ungleichung und ihre anwendung auf beweis der ergodizitaet von markoffschen ketten. *Magyer Tud. Akad. Mat. Kutato Int. Koezl.*, 8, 85–108.

Gail, M. and Ware, J. (1979). Comparing observed life table data with a known survival curve in the presence of random censorship. *Biometrics*, 385–391.

Hougaard, P. (1986). Survival models for heterogeneous populations derived from stable distributions. *Biometrika*, 73(2), 387–396.

Kagan, A. (1963). On the theory of Fisher's amount of information. In *Doklady Academii Nauk SSSR*, 51, 277–278.

Kullback, S. and Leibler, R.A. (1951). On information and sufficiency. *The Annals of Mathematical Statistics*, 22(1), 79–86.

Mattheou, K., Lee, S., and Karagrigoriou, A. (2009). A model selection criterion based on the BHHJ measure of divergence. *Journal of Statistical Planning and Inference*, 139(2), 228–235.

Matusita, K. (1967). On the notion of affinity of several distributions and some of its applications. *Annals of the Institute of Statistical Mathematics*, 19(1), 181.

Tsanousa, A. (2010). Statistical inference via ϕ-measures of divergence. Master's Thesis. National Technical University of Athens (in Greek).

Vaupel, J.W., Manton, K.G., and Stallard, E. (1979). The impact of heterogeneity in individual frailty on the dynamics of mortality. *Demography*, 16(3), 439–454.

Vonta, F. (1996). Efficient estimation in a non-proportional hazards model in survival analysis. *Scandinavian Journal of Statistics*, 23(1), 49–61.

Vonta, I. and Karagrigoriou, A. (2014). Goodness-of-fit tests via ϕ-measures of divergence for censored data. *Journal of Statistical Computation and Simulation*, 84(5), 946–963.

List of Authors

Ouerdia ARKOUN
LMRS - UMR 6085 CNRS
University of Rouen-Normandy
and
BIRL - Sup'Biotech
Villejuif
France

Romain AZAÏS
The RDP Institute
University of Lyon, ENS Lyon,
UCB Lyon 1
CNRS, INRAE, INRIA
France

Irina BĂNCESCU
"Costin C. Kiriţescu" National Institute
for Economic Research
Bucharest
Romania

Vlad Stefan BARBU
LMRS - UMR 6085 CNRS
University of Rouen-Normandy
France

Salim BOUZEBDA
LMAC
Sorbonne University Association
University of Technology of Compiègne
France

Michel BRONIATOWSKI
LPSM-UMR 8001 CNRS
Sorbonne University
Paris
France

Jean-Yves BRUA
LMRS - UMR 6085 CNRS
University of Rouen-Normandy
France

Peggy CÉNAC
IMB - UMR 5584 CNRS
University of Burgundy
Dijon
France

Brigitte CHAUVIN
LMV - UMR 8100 CNRS
University of Paris-Saclay
Versailles
France

Guglielmo D'AMICO
Department of Pharmacy
Universitá G. d'Annunzio Chieti-Pescara
Italy

Riccardo DE BLASIS
Department of Management
Universitá Politecnica delle Marche
Ancona
Italy

Brenda Ivette GARCIA-MAYA
LMAC
Sorbonne University Association
University of Technology of Compiègne
France

Valérie GIRARDIN
LMNO - UMR 6139 CNRS
University of Caen-Normandy
France

Alex KARAGRIGORIOU
Laboratory of Statistics and Data Analysis
Department of Statistics and Actuarial-Financial Mathematics
University of the Aegean
Samos
Greece

Bruno LEGGIO
The RDP Institute
University of Lyon, ENS Lyon
UCB Lyon 1
CNRS, INRAE, INRIA
France

Nikolaos LIMNIOS
LMAC
Sorbonne University Association
University of Technology of Compiègne
France

Christos MESELIDIS
Laboratory of Statistics and Data Analysis
Department of Statistics and Actuarial-Financial Mathematics
University of the Aegean
Samos
Greece

Frédéric PACCAUT
LAMFA - UMR 7352 CNRS
University of Picardy Jules Verne
Amiens
France

Serguei PERGAMENCHTCHIKOV
LMRS - UMR 6085 CNRS
University of Rouen-Normandy
France
and
International Laboratory of Statistics of Stochastic Processes and Quantitative Finance
National Research Tomsk State University
Russia

Nicolas POUYANNE
LMV - UMR 8100 CNRS
University of Paris-Saclay
Versailles
France

Vasile PREDA
"Costin C. Kiriţescu" National Institute for Economic Research
and
"Gheorghe Mihoc-Caius Iacob" Institute of Mathematical Statistics and Applied Mathematics
Bucharest
Romania

Rachid SENOUSSI
BioSP
INRAE
Avignon
France

Nicolas VERGNE
LMRS - UMR 6085 CNRS
University of Rouen Normandy
France

Filia VONTA
Department of Mathematics
National Technical University of Athens
Greece

Index

I, L

M

N, O, P

Q, R

S

T

V, W

Other titles from

iSTE

in

Mathematics and Statistics

2020

KOROLIOUK Dmitri
Dynamics of Statistical Experiments

MANOU-ABI Solym Mawaki, DABO-NIANG Sophie, SALONE Jean-Jacques
Mathematical Modeling of Random and Deterministic Phenomena

2019

BANNA Oksana, MISHURA Yuliya, RALCHENKO Kostiantyn, SHKLYAR Sergiy
Fractional Brownian Motion: Approximations and Projections

GANA Kamel, BROC Guillaume
Structural Equation Modeling with lavaan

KUKUSH Alexander
Gaussian Measures in Hilbert Space: Construction and Properties

LUZ Maksym, MOKLYACHUK Mikhail
Estimation of Stochastic Processes with Stationary Increments and Cointegrated Sequences

MICHELITSCH Thomas, PÉREZ RIASCOS Alejandro, COLLET Bernard, NOWAKOWSKI Andrzej, NICOLLEAU Franck
Fractional Dynamics on Networks and Lattices

VOTSI Irene, LIMNIOS Nikolaos, PAPADIMITRIOU Eleftheria, TSAKLIDIS George
Earthquake Statistical Analysis through Multi-state Modeling
(Statistical Methods for Earthquakes Set – Volume 2)

2018

AZAÏS Romain, BOUGUET Florian
Statistical Inference for Piecewise-deterministic Markov Processes

IBRAHIMI Mohammed
Mergers & Acquisitions: Theory, Strategy, Finance

PARROCHIA Daniel
Mathematics and Philosophy

2017

CARONI Chysseis
First Hitting Time Regression Models: Lifetime Data Analysis Based on Underlying Stochastic Processes
(Mathematical Models and Methods in Reliability Set – Volume 4)

CELANT Giorgio, BRONIATOWSKI Michel
Interpolation and Extrapolation Optimal Designs 2: Finite Dimensional General Models

CONSOLE Rodolfo, MURRU Maura, FALCONE Giuseppe
Earthquake Occurrence: Short- and Long-term Models and their Validation
(Statistical Methods for Earthquakes Set – Volume 1)

D'AMICO Guglielmo, DI BIASE Giuseppe, JANSSEN Jacques, MANCA Raimondo
Semi-Markov Migration Models for Credit Risk
(Stochastic Models for Insurance Set – Volume 1)

GONZÁLEZ VELASCO Miguel, del PUERTO GARCÍA Inés, YANEV George P.
Controlled Branching Processes
(Branching Processes, Branching Random Walks and Branching Particle Fields Set – Volume 2)

HARLAMOV Boris
Stochastic Analysis of Risk and Management
(Stochastic Models in Survival Analysis and Reliability Set – Volume 2)

KERSTING Götz, VATUTIN Vladimir
Discrete Time Branching Processes in Random Environment
(Branching Processes, Branching Random Walks and Branching Particle Fields Set – Volume 1)

MISHURA YULIYA, SHEVCHENKO Georgiy
Theory and Statistical Applications of Stochastic Processes

NIKULIN Mikhail, CHIMITOVA Ekaterina
Chi-squared Goodness-of-fit Tests for Censored Data
(Stochastic Models in Survival Analysis and Reliability Set – Volume 3)

SIMON Jacques
Banach, Fréchet, Hilbert and Neumann Spaces
(Analysis for PDEs Set – Volume 1)

2016

CELANT Giorgio, BRONIATOWSKI Michel
Interpolation and Extrapolation Optimal Designs 1: Polynomial Regression and Approximation Theory

CHIASSERINI Carla Fabiana, GRIBAUDO Marco, MANINI Daniele
Analytical Modeling of Wireless Communication Systems
(Stochastic Models in Computer Science and Telecommunication Networks Set – Volume 1)

GOUDON Thierry
Mathematics for Modeling and Scientific Computing

KAHLE Waltraud, MERCIER Sophie, PAROISSIN Christian
Degradation Processes in Reliability
(Mathematial Models and Methods in Reliability Set – Volume 3)

KERN Michel
Numerical Methods for Inverse Problems

RYKOV Vladimir
Reliability of Engineering Systems and Technological Risks
(Stochastic Models in Survival Analysis and Reliability Set – Volume 1)

2015

DE SAPORTA Benoîte, DUFOUR François, ZHANG Huilong
Numerical Methods for Simulation and Optimization of Piecewise Deterministic Markov Processes

DEVOLDER Pierre, JANSSEN Jacques, MANCA Raimondo
Basic Stochastic Processes

LE GAT Yves
Recurrent Event Modeling Based on the Yule Process
(Mathematical Models and Methods in Reliability Set – Volume 2)

2014

COOKE Roger M., NIEBOER Daan, MISIEWICZ Jolanta
Fat-tailed Distributions: Data, Diagnostics and Dependence
(Mathematical Models and Methods in Reliability Set – Volume 1)

MACKEVIČIUS Vigirdas
Integral and Measure: From Rather Simple to Rather Complex

PASCHOS Vangelis Th
Combinatorial Optimization – 3-volume series – 2nd edition
Concepts of Combinatorial Optimization / Concepts and Fundamentals – volume 1
Paradigms of Combinatorial Optimization – volume 2
Applications of Combinatorial Optimization – volume 3

2013

COUALLIER Vincent, GERVILLE-RÉACHE Léo, HUBER Catherine, LIMNIOS Nikolaos, MESBAH Mounir
Statistical Models and Methods for Reliability and Survival Analysis

JANSSEN Jacques, MANCA Oronzio, MANCA Raimondo
Applied Diffusion Processes from Engineering to Finance

SERICOLA Bruno
Markov Chains: Theory, Algorithms and Applications

2012

BOSQ Denis
Mathematical Statistics and Stochastic Processes

CHRISTENSEN Karl Bang, KREINER Svend, MESBAH Mounir
Rasch Models in Health

DEVOLDER Pierre, JANSSEN Jacques, MANCA Raimondo
Stochastic Methods for Pension Funds

2011

MACKEVIČIUS Vigirdas
Introduction to Stochastic Analysis: Integrals and Differential Equations

MAHJOUB Ridha

Recent Progress in Combinatorial Optimization – ISCO2010

RAYNAUD Hervé, ARROW Kenneth
Managerial Logic

2010

BAGDONAVIČIUS Vilijandas, KRUOPIS Julius, NIKULIN Mikhail
Nonparametric Tests for Censored Data

BAGDONAVIČIUS Vilijandas, KRUOPIS Julius, NIKULIN Mikhail
Nonparametric Tests for Complete Data

IOSIFESCU Marius *et al.*
Introduction to Stochastic Models

VASSILIOU PCG
Discrete-time Asset Pricing Models in Applied Stochastic Finance

2008

ANISIMOV Vladimir
Switching Processes in Queuing Models

FICHE Georges, HÉBUTERNE Gérard
Mathematics for Engineers

HUBER Catherine, LIMNIOS Nikolaos *et al.*
Mathematical Methods in Survival Analysis, Reliability and Quality of Life

JANSSEN Jacques, MANCA Raimondo, VOLPE Ernesto
Mathematical Finance

2007

HARLAMOV Boris
Continuous Semi-Markov Processes

2006

CLERC Maurice
Particle Swarm Optimization